The Geography of Transport Systems

University of Liverpool

Withdrawn from stock

Mobility is fundamental to economic and social activities, including commuting, manufacturing and supplying energy. Each movement has an origin, a potential set of intermediate locations, a destination and a nature which is linked with geographical attributes. Transport systems composed of infrastructures, modes and terminals are so embedded in the socio-economic life of individuals, institutions and corporations that they are often invisible to the consumer. This is paradoxical as the perceived invisibility of transportation is derived from its efficiency. Understanding how mobility is linked with geography is the main purpose of this textbook.

This second edition of *The Geography of Transport Systems* maintains the overall structure of its predecessor, with chapters dealing with specific conceptual dimensions and methodologies, but the contents have been revised and updated. This edition also offers new topics and approaches that have emerged as critical issues in contemporary transport systems, including security, energy, supply chain management and GIS-T. Relevant case studies have also been included to underline real world issues related to transport geography.

Some key points of the second edition:

- Updated and revised conceptual and methodological material to reflect the most current issues in transport geography.
- A case study for each chapter addresses a real world transportation geography issue.
- Reorganization of the text to improve readability and continuity.
- Updated and improved figures and maps.
- Continuously updated and revised supporting website.

Mainly aimed at an undergraduate audience, *The Geography of Transport Systems* provides a comprehensive and accessible introduction to the field with a broad overview of its concepts, methods and areas of application. It is highly illustrated and a companion website has also been enhanced for the book. It contains PowerPoint slides, exercises, databases and GIS datasets and can be accessed at: http://people.hofstra.edu/geotrans

Jean-Paul Rodrigue is an Associate Professor in the Department of Global Studies and Geography at Hofstra University, New York.
Claude Comtois is Professor of Geography at the University of Montreal, Canada, and is affiliated with the Research Centre on Enterprise Networks, Logistics and Transportation of the same institution.
Brian Slack is Distinguished Professor Emeritus at Concordia University, Montreal, Canada.

The Geography of Transport Systems

Second edition

Jean-Paul Rodrigue, Claude Comtois
and Brian Slack

Routledge
Taylor & Francis Group

LONDON AND NEW YORK

First edition published 2006
by Routledge

Second edition published 2009 by Routledge
2 Park Square, Milton Park, Abingdon, Oxon, OX14 4RN

Simultaneously published in the USA and Canada
by Routledge
711 Third Avenue, New York, NY 10017

Routledge is an imprint of the Taylor & Francis Group, an informa business

© 2006, 2009 Jean-Paul Rodrigue, Claude Comtois and Brian Slack

Typeset in 8.75/10pt Times by Graphicraft Limited, Hong Kong

British Library Cataloguing in Publication Data
A catalogue record for this book is available from the British Library

Library of Congress Cataloging-in-Publication Data
Rodrigue, Jean-Paul, 1967–
The geography of transport systems/Jean-Paul Rodrigue, Claude Comtois, and
Brian Slack. – 2nd ed.

  p. cm.
  Includes index.
  1. Transportation geography. I. Comtois, Claude, 1954– II. Slack, Brian, 1939– III. Title.
HE323.R63 2009
388.01–dc22
2008040189

ISBN 10: 0-415-48323-9 (hbk)
ISBN 10: 0-415-48324-7 (pbk)
ISBN 10: 0-203-88415-9 (ebk)

ISBN 13: 978-0-415-48323-0 (hbk)
ISBN 13: 978-0-415-48324-7 (pbk)
ISBN 13: 978-0-203-88415-7 (ebk)

To Gordana, Suzanne and Mabel

Contents

Plates

Figures

Tables

Preface

The Geography of Transport Systems is now into its second edition. Substantial efforts have been made to build on the first edition's success by improving the content and its structure. Like the first edition, we have elected for a synthetic writing style, instead of a narrative, where the goal is to provide a structured framework to the reader. Great care has been made to avoid factual information, particularly in the usage of graphs related to statistical data, so that the textbook can retain its relevance in spite of continuous and often unforeseeable changes in the transport industry. A large quantity of statistical information is available on the companion website, which is constantly updated.

Transportation is concerned with mobility, particularly how this mobility is taking place in the context of a wide variety of conditions. Mobility is a geographical endeavor since it trades space for a cost. Technological and economic forces have changed this balance many times in the past, but in recent decades a growing amount of space has been made accessible at a similar cost. It is thus not surprising to realize that at the same time that technology permitted improvements in transport speed, capacity and efficiency, individuals and corporations have been able to take advantage of this improved mobility. A driving force of the global economy resides in the capacity of transport systems to ship large quantities of freight and to accommodate vast numbers of passengers. The world has become interconnected at several scales. This new geographical dimension transcends a more traditional perspective of transportation mainly focused on the city or the nation. At the beginning of the twenty-first century, the geography of transportation is thus fundamentally being redefined by global, regional and local issues.

Presenting these issues to students or the public remains a challenging task. This book has specifically been designed with this in mind. Its origins are rather unusual since it began in 1997 as an online initiative to provide material about transport geography and was simply titled "Transport Geography on the Web". The material was considerably revised and expanded over the years, often thanks to comments and queries we received, as the site gained a wider audience. It has already endured the decade-long test of being exposed to the scrutiny of a global audience including practitioners, policy makers, educators and, most importantly, students. For many years and as these words were written, the site ranked first in Google under the topic of transport geography, and other key terms such as intermodal transportation or transport modes, implying its popularity as a trusted source of information. Its contents are appearing in a growing number of transport-related curriculums, underlining the relevance of the material covered and that a demand was being fulfilled. The step of moving to a textbook was a natural one especially after receiving many requests in this direction.

Like the first edition, the textbook is articulated along two core approaches to transport geography, one conceptual and the other methodological. The conceptual parts present what we think are some of the most relevant issues explaining contemporary transport geography. In addition to the more conventional topics related to transport modes and terminals, as well as urban transportation, the book also substantially focuses on emerging issues such as globalization, supply chain management, energy and the environment. Many of these issues have been superficially covered, if at all, in the past, but their importance cannot be underestimated in a transport geography that involves an increasingly integrated world.

The methodological parts address how transportation information is used to assist transport operators allocate their resources (investments, vehicles) or to influence public policy. This includes a wide array of methods ranging from qualitative to quantitative. Since transport is a field of application, the use of methodologies is particularly relevant as they relate to real world issues. The merging between methodologies and information technologies has led to many new opportunities, notably with the emergence of transportation geographic information systems (GIS-T). It has become a very active field of investigation and application.

In an effort to illustrate the applied dimension of transport geography, the second edition also contains a number of case studies selected for their relevance and their potential appeal to a wide international audience. The first one deals with a perspective about how transportation geography can be taught and how it can fit into a variety of academic curriculums. Others, each in their own way, illustrate how transportation and geography merge into the consideration of real world problems.

It is our hope that the reader will have a better understanding of the nature, function, importance and challenges of contemporary transportation systems. The online companion site will insure that this book will not be a static endeavor and will be revised and updated as changes take place in this fascinating field which is transport geography.

New York, August 2008

① Transportation and geography

Movements of people, goods and information have always been fundamental components of human societies. Contemporary economic processes have been accompanied by a significant increase in mobility and higher levels of accessibility. Although this trend can be traced back to the industrial revolution, it significantly accelerated in the second half of the twentieth century as trade was liberalized, economic blocs emerged and the comparative advantages of global labor and resources were used more efficiently. However, these conditions are interdependent with the capacity to manage, support and expand movements of passengers and freight as well as their underlying information flows. Societies have become increasingly dependent on their transport systems to support a wide variety of activities ranging, among others, from commuting, supplying energy needs, to distributing parts between factories. Developing transport systems has been a continuous challenge to satisfy mobility needs, to support economic development and to participate in the global economy.

Concept 1 – What is transport geography?

The purpose of transportation

The ideal transport mode would be instantaneous, free, have an unlimited capacity and always be available. It would render space obsolete. This is obviously not the case. Space is a constraint for the construction of transport networks. Transportation appears to be an economic activity different from the others. It trades space with time and thus money.

(translated from Merlin, 1992)

As the above quotation underlines, the unique purpose of transportation is to overcome space, which is shaped by a variety of human and physical constraints such as distance, time, administrative divisions and topography. Jointly, they confer a friction to any movement, commonly known as the friction of distance. However, these constraints and the friction they create can only be partially circumscribed. The extent to which this is done has a cost that varies greatly according to factors such as the distance involved and the nature of what is being transported. There would be no transportation without geography and there would be no geography without transportation. The goal of transportation is thus to transform the geographical attributes of freight, people or information, from an origin to a destination, conferring them an added value in the process. The convenience at which this can be done – transportability – varies considerably.

Transportability. Refers to the ease of movement of passengers, freight or information. It is related to transport costs as well as to the attributes of what is being transported (fragility, perishability, price). Political factors can also influence transportability such as laws, regulations, borders and tariffs. When transportability is high, activities are less constrained by distance.

The specific purpose of transportation is to fulfill a demand for mobility, since transportation can only exists if it moves people, freight and information around. Otherwise it has no purpose. This is because transportation is dominantly the outcome of a derived demand (Figure 1.1).

What takes place in one sector has impacts on another; demand for a good or service in one sector is derived from another. For instance, a consumer buying a good in a store will likely trigger the replacement of this product, which will generate demands for activities such as manufacturing, resource extraction and, of course, transport. What is different about transport is that it cannot exist alone and a movement cannot be stored. An unsold product can remain on the shelf of a store until a customer buys it (often with discount incentives), but an unsold seat on a flight or unused cargo capacity in the same flight remain unsold and cannot be brought back as additional capacity later. In this case an opportunity has been missed since the amount of transport being offered has exceeded the demand for it. The derived demand of transportation is often very difficult to reconcile with an equivalent supply and actually transport companies would prefer to have some additional capacity to accommodate unforeseen demand (often at much higher prices). There are two major types of derived transport demand:

Direct derived demand. Refers to movements that are directly the outcome of economic activities, without which they would not take place. For instance, work-related activities commonly involve commuting between the place of residence and the workplace. There is a supply of work in one location (residence) and a demand of labor in another (workplace), transportation (commuting) being directly derived from this relationship. For freight transportation, all the components of a supply chain

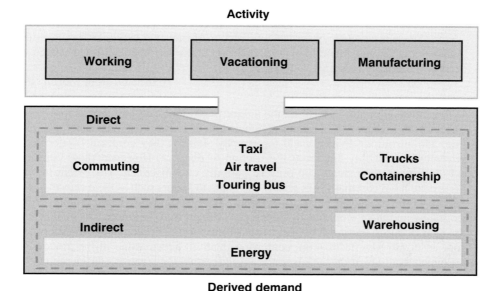

Figure 1.1 Transportation as a derived demand

require movements of raw materials, parts and finished products on modes such as trucks, rail or containerships. Thus, transportation is directly the outcome of the functions of production and consumption.

Indirect derived demand. Considers movements created by the requirements of other movements. The most obvious example is energy where fuel consumption from transportation activities must be supplied by an energy production system requiring movements from zones of extraction, to refineries and storage facilities and, finally, to places of consumption. Warehousing can also be labeled as an indirect derived demand since it is a "non movement" of a freight element. Warehousing exists because it is virtually impossible to move commodities instantly from where they are produced to where they are consumed.

Distance, a core attribute of transportation, can be represented in a variety of ways, ranging from a simple Euclidean distance – a straight line between two locations – to what can be called logistical distance; a complete set of tasks required to be done so that distance can be overcome (Figure 1.2).

- **Euclidean distance**. A simple function of a straight line between two locations where distance is expressed in geographical units such as kilometers. Commonly used to provide an approximation of distance, but almost never has a practical use.
- **Transport distance**. A more complex representation where a set of activities related to circulation, such as loading, unloading and transhipment, are considered. Additional elements such as costs and time are also part of the transport distance. On Figure 1.2, the transport distance between locations A and B includes pickup, travel by mode 1, transhipment, travel by mode 2 and finally, delivery. The same applies to the circulation of people, although the involved activities will be different. For instance, someone using air travel between two locations will require going to an airport, may transit through an intermediate airport and will finally need to reach his/her destination from the terminal airport. Transport distance is jointly expressed in geographical units, in cost and in time.
- **Logistical distance**. A complex representation that encompasses all the tasks required so that a movement between two locations can take place. Logistical distance

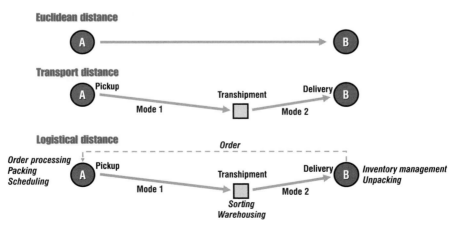

Figure 1.2 Different representations of distance

thus includes flows, but also a set of activities necessary for the management of these flows. For freight movements, among the most significant tasks are order processing, packing, sorting and inventory management. Geographical distance units are less relevant in its assessment, but the factors of costs and time are very significant. Time not only involves the delay related to management and circulation, but also how it is used to service the transport demand, namely the scheduling of pickups and deliveries. On Figure 1.2, the logistical distance between locations A and B includes an order from B, which is processed, packed and scheduled to be picked up. At the intermediate transhipment location, sorting and warehousing are performed, and finally, at the destination the delivery will be unpacked and used. For the transportation of passengers, logistical distance also concerns a specific array of tasks. Taking again an air travel example, a ticket would first need to be purchased, commonly several weeks in advance. Other common time and cost tasks concern packing, checking in, security checks, boarding and disembarking, picking up luggage and, finally, unpacking. Thus, a three-hour flight can in reality be a movement planned several weeks in advance and its full realization can take twice as much time if all the related logistical activities are considered.

Any movement must thus consider its geographical setting which in turn is linked to spatial flows and their patterns. Urbanization, multinational corporations, the globalization of trade and the international division of labor are all forces shaping and taking advantage of transportation at different, but often related, scales.

Consequently, the fundamental purpose of transport is geographic in nature, because it facilitates movements between different locations. Transport thus plays a role in the structure and organization of space and territories, which may vary according to the level of development. In the nineteenth century, the purpose of the emerging modern forms of transportation, mainly railways and maritime shipping, was to expand coverage, and create and consolidate national markets. In the twentieth century, the objective shifted to selecting itineraries, prioritizing transport modes, increasing the capacity of existing networks and responding to mobility needs and this at a scale that was increasingly global. In the twenty-first century, transportation must cope with a globally oriented economic system in a timely and cost effective way, but also with several local problems such as congestion and capacity constraints.

The importance of transportation

Transport represents one of the most important human activities worldwide. It is an indispensable component of the economy and plays a major role in spatial relations between locations. Transport creates valuable links between regions and economic activities, between people and the rest of the world. Transport is a multidimensional activity whose importance is:

- **Historical**. Transport modes have played several different historical roles in the rise of civilizations (Egypt, Rome and China), in the development of societies (creation of social structures) and also in national defense (Roman Empire, American road network).
- **Social**. Transport modes facilitate access to healthcare, welfare and cultural or artistic events, thus performing a social service. They shape social interactions by favoring or inhibiting the mobility of people. Transportation thus supports and may even shape social structures.

- **Political**. Governments play a critical role in transport as sources of investment and as regulators. The political role of transportation is undeniable as governments often subsidize the mobility of their populations (highways, public transit, etc.). While most transport demand relates to economic imperatives, many communication corridors have been constructed for political reasons such as national accessibility or job creation. Transport thus has an impact on nation building and national unity, but it is also a political tool.
- **Economic**. The evolution of transport has always been linked to economic development. It is an industry in its own right (car manufacturing, air transport companies, etc.). The transport sector is also an economic factor in the production of goods and services. It contributes to the value-added of economic activities, facilitates economies of scale, influences land (real estate) value and the geographic specialization of regions. Transport is a factor shaping economic activities, but is also shaped by them.
- **Environmental**. Despite the manifest advantages of transport, its environmental consequences are also significant. They include air and water quality, noise level and public health. All decisions relating to transport need to be evaluated taking into account the corresponding environmental costs. Transport is a dominant factor in contemporary environmental issues.

Substantial empirical evidence indicates that the importance of transportation is growing. The following contemporary trends can be identified regarding this issue:

- **Growth of the demand**. The last 50 years have seen a considerable growth of the transport demand related to individual (passengers) as well as freight mobility. This growth is jointly the result of larger quantities of passengers and freight being moved, but also the longer distances over which they are carried. Recent trends underline an ongoing process of mobility growth, which has led to the multiplication of the number of journeys involving a wide variety of modes that service transport demands.
- **Reduction of costs**. Even if several transportation modes are very expensive to own and operate (ships and planes for instance), costs per unit transported have dropped significantly over recent decades. This has made it possible to overcome larger distances and further exploit the comparative advantages of space. As a result, despite the lower costs, the share of transport activities in the economy has remained relatively constant in time.
- **Expansion of infrastructures**. The above two trends have obviously extended the requirements for transport infrastructures both quantitatively and qualitatively. Roads, harbors, airports, telecommunication facilities and pipelines have expanded considerably to service new areas and add capacity to existing networks. Transportation infrastructures are thus a major component of land use, notably in developed countries.

Facing these contemporary trends, an important part of the spatial differentiation of the economy is related to where resources (raw materials, capital, people, information, etc.) are located and how well they can be distributed. Transport routes are established to distribute resources between places where they are abundant and places where they are scarce, but only if the costs are lower than the benefits.

Consequently, transportation has an important role to play in the conditions that affect global, national and regional economic entities. It is a strategic infrastructure that is so embedded in the socio-economic life of individuals, institutions and corporations that it is often invisible to the consumer, but always part of all economic and social functions. This is paradoxical, since the perceived invisibility of transportation is derived from its efficiency. If transport is disrupted or ceases to operate, the consequences can be dramatic. The paradox gives rise to several fallacies about transportation; two major ones shown on Figure 1.3 include:

- **Access is not accessibility**. Many transport systems have universal access since no specific user can have a competitive advantage over others; access is the same for everyone. For instance, a public highway system can in theory be accessed by anyone, be it by a major trucking company with a large fleet, its competitors or by an individual driving an automobile. Thus, access is uniform wherever one is located in regard to the transport system as long as there is a possibility to enter or to exit. On the other hand, accessibility varies according to one's location within the transport system. Access is thus uniform while accessibility is not; the latter is a relative concept. On the transport network shown on Figure 1.3, locations a, b and c all have access to the system. However, location b appears to be more accessible than the other two due to its central location in relation to the network.
- **Distance is not time**. Distance often tends to be interchanged with time when measuring the performance of transport systems, which is a conceptual error. While distance remains constant, time can vary due to improvements in transport technology (positive effect), because of congestion (negative effect) or regulations such as speed limits. A simple and common way to express this relationship is speed: the unit of distance traveled per unit of time. Driving one kilometer through Manhattan is not the same as driving one kilometer through an Interstate in Iowa even if in both cases the same unit of distance has been traveled. Distance is thus a uniform attribute of the geography, while time is relative. On the transport network shown on Figure 1.3, while distance is a uniform attribute, each segment has a travel time expressed as speed which, due to congestion, varies differently from distance.

Transportation in geography

Transportation interests geographers for two main reasons. First, transport infrastructures, terminals, equipment and networks occupy an important place in space and constitute

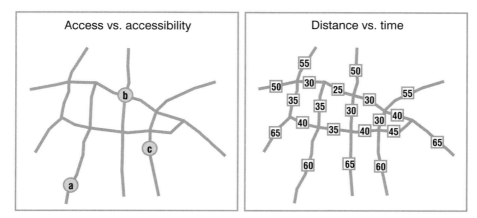

Figure 1.3 Two common fallacies in transport geography

the basis of a complex spatial system. Second, since geography seeks to explain spatial relationships, transport networks are of specific interest because they are the main support of these interactions.

> **Transport geography** is a subdiscipline of geography concerned about movements of freight, people and information. It seeks to link spatial constraints and attributes with the origin, the destination, the extent, the nature and the purpose of movements.

Transport geography, as a discipline, emerged from economic geography in the second half of the twentieth century. Traditionally, transportation has been an important factor behind the economic representations of the geographic space, namely in terms of the location of economic activities and the monetary costs of distance. The growing mobility of passengers and freight justified the emergence of transport geography as a specialized field of investigation. In the 1960s, transport costs were recognized as key factors in location theories and transport geography began to rely increasingly on quantitative methods, particularly over network and spatial interactions analysis. However, from the 1970s globalization challenged the centrality of transportation in many geographical and regional development investigations. As a result, transportation became under-represented in economic geography in the 1970s and 1980s, even if mobility of people and freight and low transport costs were considered as important factors behind the globalization of trade and production.

Since the 1990s, transport geography has received renewed attention, especially because the issues of mobility, production and distribution are interrelated in a complex geographical setting. It is now recognized that transportation is a system that considers the complex relationships between its core elements (Figure 1.4). Transport geography must be systematic as one element of the transport system is linked with numerous others. An approach to transportation thus involves several fields where some are at the core of transport geography while others are more peripheral. However, three central concepts to transport systems can be identified:

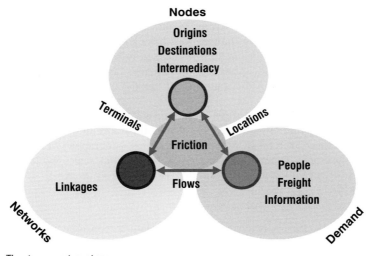

Figure 1.4 The transport system

- **Transportation nodes**. Transportation primarily links locations, often characterized as nodes. They serve as access points to a distribution system or as transhipment/ intermediary locations within a transport network. This function is mainly serviced by transport terminals where flows originate, end or are being transhipped from one mode to the other. Transport geography must consider its places of convergence and transhipment.
- **Transportation networks**. Consider the spatial structure and organization of transport infrastructures and terminals. Transport geography must include in its investigation the infrastructures supporting and shaping movements.
- **Transportation demand**. Considers the demand for transport services as well as the modes used to support movements. Once this demand is realized, it becomes an interaction which flows through a transport network. Transport geography must evaluate the factors affecting its derived demand function.

The analysis of these concepts relies on methodologies often developed by other disciplines such as economics, mathematics, planning and demography. Each provides a different dimension to transport geography. For instance, the spatial structure of transportation networks can be analyzed with graph theory, which was initially developed for mathematics. Further, many models developed for the analysis of movements, such as the gravity model, were borrowed from physical sciences. Multidisciplinarity is consequently an important attribute of transport geography, as in geography in general.

The role of transport geography is to understand the spatial relations that are produced by transport systems. A better understanding of spatial relations is essential to assist private and public actors involved in transportation mitigate transport problems, such as capacity, transfer, reliability and integration of transport systems. There are three basic geographical considerations relevant to transport geography:

- **Location**. All activities have a location with its own characteristics conferring a potential supply and/or a demand for resources, products, services or labor. A location will determine the nature, the origin, the destination, the distance and even the possibility of a movement to be realized. For instance, a city provides employment in various sectors of activity in addition to consuming resources.
- **Complementarity**. Locations must require exchanging goods, people or information. This implies that some locations have a surplus while others have a deficit. The only way an equilibrium can be reached is by movements between locations having surpluses and locations having demands. For instance, a complementarity is created between a store (surplus of goods) and its customers (demand of goods).
- **Scale**. Movements generated by complementarity are occurring at different scales, pending the nature of the activity. Scale illustrates how transportation systems are established over local, regional and global geographies. For instance, home-to-work journeys generally have a local or regional scale, while the distribution network of a multinational corporation is most likely to cover several regions of the world.

Consequently, transport systems, by their nature, consume land and support the relationships between locations.

Concept 2 – Transportation and space

Physical constraints

Transport geography is concerned with movements that take place over space. The physical features of this space impose major constraints on transportation systems, in terms of what mode can be used, the extent of the service, its costs, capacity and reliability. Three basic spatial constraints of the terrestrial space can be identified:

- **Topography**. Features such as mountains and valleys have strongly influenced the structure of networks, the cost and feasibility of transportation projects. The main land transport infrastructures are built usually where there are the least physical impediments, such as on plains, along valleys or through mountain passes. Water transport is influenced by water depths and the location of obstacles such as reefs. Coastlines exert an influence on the location of port infrastructure. Aircraft require airfields of considerable size for takeoff and landing. Topography can impose a natural convergence of routes that will create a certain degree of centrality and may assist a location in becoming a trade center as a collector and distributor of goods. Topography can complicate, postpone or prevent the activities of the transport industry. Physical constraints fundamentally act as absolute and relative barriers to movements. Land transportation networks are notably influenced by the topography, as highways and railways tend to be impeded by grades higher than 3 percent and 1 percent respectively. Under such circumstances, land transportation tends to be of higher density in areas of limited topography.
- **Hydrology**. The properties, distribution and circulation of water play an important role in the transport industry (see Figure 1.5). Maritime transport is influenced greatly by the availability of navigable channels through rivers, lakes and shallow seas. Several rivers such as the Mississippi, the St. Lawrence, the Rhine, the Mekong or the Yangtze are important navigable routeways into the heart of continents and historically have been the focus of human activities that have taken advantage of the transport opportunities. Port sites are also highly influenced by the physical attributes of the site where natural features (bays, sand bars and fjords) protect port installations. Since it is at these installations that traffic is transhipped, the location of ports is a dominant element in the structure of maritime networks. Where barriers exist, such as narrows, rapids or land breaks, water transport can only overcome these obstacles with heavy investments in canals or dredging. Conversely, waterways serve

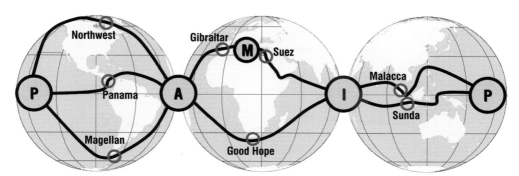

Figure 1.5 The geographical space of maritime transportation

as barriers to land transportation necessitating the construction of bridges, tunnels and detours, etc.

- **Climate**. Its major components include temperature, wind and precipitation. Their impacts on transportation modes and infrastructure range from negligible to severe. Freight and passenger movement can be seriously curtailed by hazardous conditions such as snow, heavy rainfall, ice or fog. Jet streams are also a major physical component that international air carriers must take into consideration. For an aircraft, the speed of wind can affect travel costs. When the wind is pushing the airplane towards its destination, it can reduce flight time up to several hours for intercontinental flights. Climate is also an influence over transportation networks by influencing construction and maintenance costs.

The geographical space of maritime transportation is primary defined by its absolute barriers. Some 71 percent of the Earth's surface is covered by water, dominantly oceanic masses, but the profile of continental masses seriously constrains maritime access to different parts of the world. The maritime system can be summarized by 4 points (oceans) representing centers of gravity between elements of the world-system (Atlantic (A), Pacific (P), Indian (I) and Mediterranean (M)) and 8 links (passages) representing intermediate locations between the oceanic masses (Figure 1.5). They are all strategic locations within the maritime space. Physical constraints fundamentally act as absolute and relative barriers to movements (Figure 1.6):

- **Absolute barriers** are geographical features that entirely prevent a movement. They must either be bypassed or be overcome by specific infrastructures. For instance, a river is considered as an absolute barrier for land transportation and can only be overcome if a tunnel or a bridge is constructed. A body of water forms a similar absolute barrier and could be overcome if ports are built and a maritime service (ferry, cargo ships, etc.) is established. Conversely, land acts as an absolute barrier for maritime transportation; discontinuities (barriers) can be overcome with costly infrastructures such as navigation channels and canals.

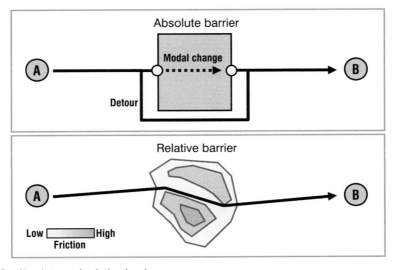

Figure 1.6 Absolute and relative barriers

- **Relative barriers** are geographical features that force a degree of friction on a movement. In turn, this friction is likely to influence the path (route) selected to link two locations (A and B on Figure 1.6). Topography is a classic example of a relative barrier that influences land transportation routes along paths having the least possible friction (e.g. plains and valleys). For maritime transportation, relative barriers, such as straits, channels or ice, generally slow down circulation.

Since the Earth is a sphere, the shortest path between two points is calculated by the great circle distance, which corresponds to an arc linking two points on a sphere. The great circle distance is useful to establish the shortest path to use when traveling at the intercontinental air and maritime level (Figure 1.7). Because of the distortions caused by projections of the globe on a flat sheet of paper, a straight line on a map is not necessarily the shortest distance. Ships and aircraft usually follow the great circle geometry to minimize distance and save time and money to customers. For instance, Figure 1.7 shows the shortest path between New York and Moscow (about 7,540 km). This path corresponds to an air transportation corridor. Air travel over the North Atlantic between North America and Europe follows a similar path.

Overcoming the physical environment

Rapid scientific and technological developments have and continue to overcome the physical environment. Before the Middle Ages, road location was adapted to topography. Since then, efforts have been made to pave roads, bridge rivers and cut paths over mountain passes. Engineering measures using the arches and vaults of the Byzantine and Gothic church constructions in the twelfth century permitted bridge building across wide streams or deep river valleys. Road building has been at the core of technological efforts to overcome the environment. Roads have always been the support for local and even long distance travel. From the efforts to mechanize individual transport to the development of integrated highways, road building has transformed the environment. Land transportation was further facilitated with the development of technical solutions for preventing temporary interruptions in road transport provision through routeways protection. In the last 20 years, the development of road transport and the growth in just-in-time and door-to-door services have increased engineering demands for constructing multilevel and high-speed highways.

Innovations in maritime transport can be found around the world. The earliest developments came in the transformation of waterways for transportation purposes through the development of canal locks. Adverse natural gradients in inland waterways can be overcome through the use of locks. Further improvements in navigation came with the cutting of artificial waterways. Some of the earliest examples can be found in the Dutch canals, the Martesana Canals of Lombardy, the Canal de Briare in France or the Imperial Canal of China. Further improvements in navigation technology and the nature of ships increased the speed, range and capacity of ocean transport. However, the increasing size of ships has resulted in canals such as the Panama and the Suez being unable to accommodate the largest modern and efficient world's maritime carriers. Several canal authorities have thus embarked on expansion programs that have severe environmental consequences. Increasing attention has also been paid to creating new passages between semi-enclosed seas. In Canada and Russia, the growing competition between the sea and land corridors is not only reducing tariffs and encouraging international trade but prompting the governments to reassess traditional ocean connections. Passages through the Arctic Ocean are being investigated with a view to creating new international connections (Figure 1.8).

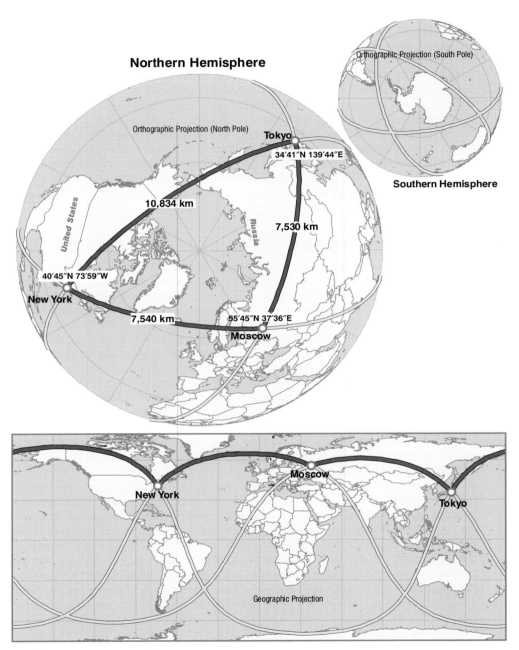

Figure 1.7 The great circle distance between New York, Moscow and Tokyo

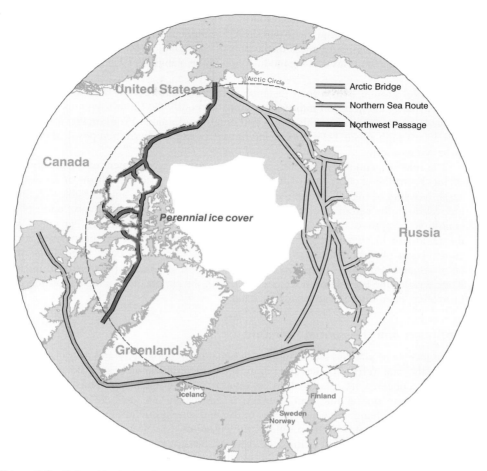

Figure 1.8 Polar shipping routes

Global climate change hinting at a warming of global temperatures is offering new opportunities for international transportation networks, notably with a trend of receding ice around the North Pole. If this trend in the reduction of the perennial ice cover continues the Arctic could be used more reliably for navigation, at least during summer months. The Northwest Passage crossing Canada's Arctic Ocean by 2020 could become usable on a regular basis, thereby lessening maritime shipping distances substantially. In 2007 the Northwest Passage was open during the summer months for the first time in recorded history, although it remains to be seen how stable this opening is. A maritime journey between London and Tokyo would take 8,500 miles using the Northwest Passage, but 15,000 miles using the Panama Canal. The Northern Sea route (or Northeast Passage) passing through the Russian Arctic could also become a possibility. It would reduce a maritime journey between London and Tokyo from 13,000 miles using the Suez Canal to 8,000 miles, cutting transit time by 10–15 days. In addition, an Arctic Bridge could also be used, linking the Russian port of Murmansk or the Norwegian port of Narvik to the Canadian port of Churchill.

Artificial islands are also created with a view to permit port installations in deep waters. In China, it became clear that dredging the Yangtze River Delta was insufficient to insure the competitiveness of the port of Shanghai, so the development of a new port site in Hangzhou Bay and the modification of the Yangshan islands landscape became indispensable.

As level ground over a long distance is important for increasing the efficiency of railway routes, the transport industry has come to modify the Earth's features by building bridges and tunneling, by embanking and creating drainage. From medieval Germany to France's high-speed TGV, increasing motive power has permitted physical obstacles to be overcome.

The role of technology has been determinant in the development of the air transport sector. From the experiments of the Montgolfier brothers to the advent of jet aircraft, aerial crossing of rugged terrain over considerable distance became possible. Technical innovation in the aeronautic industry has permitted planes to avoid adverse atmospheric conditions, improve speed, increase stage length and raise carrying capacity. With the rapid rise in air passenger and freight transport, emphasis has been given to the construction of airport terminals and runways. As airports occupy large areas, their environmental imprint is important. The construction of Chep Lap Kok airport in Hong Kong entailed the leveling of mountainous land for the airport site. Kansai airport servicing Osaka has been built on an artificial island.

Transportation and the spatial structure

The concepts of site and situation are fundamental to geography and to transportation. While the site refers to the geographical characteristics of a specific location, its situation concerns its relationships in regard to other locations. Thus, all locations are relative to one another but situation is not a constant attribute as transportation developments change levels of accessibility, and thus the relations between locations. The development of a location reflects the cumulative relationships between transport infrastructure, economic activities and the built-environment. The following factors are particularly important in shaping the spatial structure:

- **Costs**. The spatial distribution of activities is related to factors of distance, namely its friction. Locational decisions are taken in an attempt to minimize costs, often related to transportation.
- **Accessibility**. All locations have a level of accessibility, but some are more accessible than others. Thus, because of transportation, some locations are perceived as more valuable than others.
- **Agglomeration**. There is a tendency for activities to agglomerate to take advantage of the value of specific locations. The more valuable a location, the more likely agglomeration will take place. The organization of activities is essentially hierarchical, resulting from the relationships between agglomeration and accessibility at the local, regional and global levels.

Many contemporary transportation networks are inherited from the past, notably transport infrastructures. Even if over the last 200 years new technologies have revolutionized transportation in terms of speed, capacity and efficiency, the spatial structure of many networks has not much changed. This inertia in the spatial structure of some transportation networks can be explained by two major factors:

- **Physical attributes**. Natural conditions can be modified and adapted to suit human uses, but they are a very difficult constraint to escape, notably for land transportation. It is thus not surprising to find that most networks follow the easiest (least cost) paths, which generally follow valleys and plains. Considerations that affected road construction a few hundred years ago are still in force today, although they are sometimes easier to circumscribe.

- **Historical considerations**. New infrastructures generally reinforce historical patterns of exchange, notably at the regional level. For instance, the current highway network of France has mainly followed the patterns set by the national roads network built early in the twentieth century. This network was established over the Royal roads network, itself mainly following roads built by the Romans. At the urban level, the pattern of streets is often inherited from an older pattern, which itself may have been influenced by the pre-existing rural structure (lot pattern and rural roads).

While physical and historical considerations are at play, the introduction of new transport technology or the addition of new transport infrastructure may lead to a transformation of existing networks. Recent developments in transport systems such as container shipping, jumbo aircrafts and the extensive application of information technology to transport management have created a new transport environment and a new spatial structure. These transport technologies and innovations have intensified global interactions and modified the relative location of places. In this highly dynamic context, two processes are taking place at the same time:

- **Specialization**. Linked geographical entities are able to specialize in the production of commodities for which they have an advantage, and trading for what they do not produce. As a result, efficient transportation systems are generally linked with higher levels of regional specialization. The globalization of production clearly underlines this process as specialization occurs as long as the incurred saving in production costs are higher than the incurred additional transport costs.
- **Segregation**. Linked geographical entities may see the reinforcement of one at the expense of others, notably through economies of scale. This outcome often contradicts regional development policies aiming at providing uniform accessibility levels within a region.

The continuous evolution of transportation technology may not necessarily have expected effects on the spatial structure, as two forces are at play: concentration and dispersion. A common myth tends to relate transportation solely as a force of dispersion, favoring the spread of activities in space. This is not always the case. In numerous instances, transportation is a force of concentration, notably for business activities. Since transport infrastructures are generally expensive to build, they are established first to service the most important locations. Even if it was a strong factor of dispersion, the automobile has also favored the concentration of several activities at specific places and in large volumes. Shopping centers are a relevant example of this process where central locations emerge in a dispersed setting.

Space/time relationships

One of the most basic relationships of transportation involves how much space can be overcome within a given amount of time. The faster the mode, the larger is the distance that can be overcome within the same amount of time. Transportation, notably improvements in transport systems, changes the relationship between time and space. When this relationship involves easier, faster and cheaper access between places, this result is defined as a space/time convergence because the amount of space that can be overcome for a similar amount of time increases significantly. Significant regional and continental gains were achieved during the eighteenth and nineteenth centuries with the establishment of national and continental railway systems as well as with the growth of maritime shipping, a process which continued into the twentieth century with air and road transport systems. The outcome has been significant differences in space/time relationships, mainly between developed and developing countries, reflecting differences in the efficiency of transport systems.

At the international level, globalization processes have been supported by improvements in transport technology. The result of more than 200 years of technological improvements has been a space/time convergence of global proportions in addition to the regional and continental processes previously mentioned.

Circumnavigation is a good proxy for space/time convergence (Figure 1.9). Prior to the steamship, circumnavigating the globe would take about one sailing year, a journey greatly delayed by rounding the Cape of Good Hope and the Strait of Magellan. The late nineteenth and early twentieth centuries provided a series of innovations that would greatly improve circumnavigation, notably the construction of the Suez (1869) and Panama (1914) canals as well as steam propulsion. Circumnavigation was reduced to about 100 days at the beginning of the twentieth century and to 60 days by 1925 with fast liner services. The introduction of the jet plane in the second half of the twentieth century reduced circumnavigation to about 24 hours. This enabled the extended exploitation of the advantages of the global market, notably in terms of resources and labor. Significant reductions in transport and communication costs occurred concomitantly. There is thus a relationship between the space/time convergence and the integration of a region in global trade. Five major factors are of particular relevance in this process:

● **Speed**. The most straightforward factor relates to the increasing speed of many transport modes, a condition that particularly prevailed in the first half of the twentieth century. More recently, speed has played a less significant role as many modes are not going much faster. For instance, an automobile has a similar operating speed today as it had 60 years ago and a commercial jet plane operates at a similar speed to one 30 years ago.

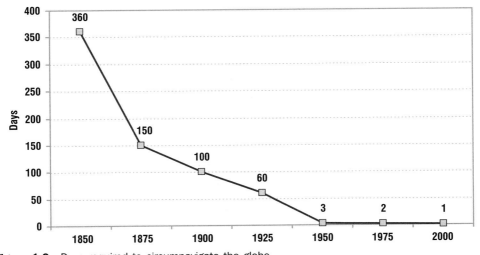

Figure 1.9 Days required to circumnavigate the globe

- **Economies of scale**. Being able to transport larger amounts of freight and passengers at lower costs has improved considerably the capacity and efficiency of transport systems.
- **Expansion of transport infrastructures**. Transport infrastructures have expanded considerably to service areas that were not previously serviced or were insufficiently serviced. A paradox of this feature is that although the expansion of transport infrastructures may have enabled distribution systems to expand, it also increased the average distance over which passengers and freight are being carried.
- **Efficiency of transport terminals**. Terminals, such as ports and airports, have shown a growing capacity to handle large quantities in a timely manner. Thus, even if the speed of many transport modes has not increased, more efficient transport terminals may have helped reduce transport time.
- **Substitution of transportation by telecommunications**. This has enabled several economic activities to bypass spatial constraints in a very significant manner. Electronic mail is an example where the transmission of information does not have a physical form (outside electrons or photons) once the supporting infrastructure is established. There is obviously a limit to this substitution, but several corporations are trying to use the advantages of telecommuting as much as they can because of the important savings involved.

However, space/time convergence can also be inverted under specific circumstances, which means that a process of space/time divergence takes place. For instance, congestion is increasing in many metropolitan areas, implying additional delays for activities such as commuting. Traffic in congested urban areas is moving at the same speed that it did 100 years ago on horse carriages. Air transportation, despite having dramatically contributed to the space/time convergence is also experiencing growing delays. Flight times are getting longer between many destinations, mainly because of takeoff, landing and gate access delays. Airlines are simply posting longer flight times to factor in congestion. An express mail package flown from Washington to Boston in about an hour (excluding delays at takeoff and landing due to airport congestion) can have an extra one hour delay as it is carried from Logan Airport to downtown Boston, a distance of only two miles. More stringent security measures at airports have also imposed additional delays, which tends to penalize more short distance flights. The "Last Mile" can be the longest in many transport segments.

Concept 3 – The geography of transportation networks

Transport networks

Transportation systems are commonly represented using networks as an analogy for their structure and flows.

The term **network** refers to the framework of routes within a system of locations, identified as nodes. A route is a single link between two nodes that are part of a larger network that can refer to tangible routes such as roads and rails, or less tangible routes such as air and sea corridors.

The territorial structure of any region corresponds to a network of all its economic interactions. The implementation of networks, however, is rarely premeditated but the consequence of continuous improvements as opportunities arise, investments are made and as conditions change. The setting of networks is the outcome of various strategies, such as providing access and mobility to a region, reinforcing a specific corridor or technological developments making a specific mode (and its network) more advantageous over others. A transport network denotes either a permanent track (e.g. roads, rail and canals) or a scheduled service (e.g. airline, transit, train). It can be extended to cover various types of links between points along which movements can take place.

In transport geography, it is common to identify several types of transport structures that are linked with transportation networks with key elements such as nodes, links, flows, hubs or corridors. Network structure ranges from centripetal to centrifugal in terms of the accessibility they provide to locations. A centripetal network favors a limited number of locations while a centrifugal network tends not to convey any specific locational advantages. The recent decades have seen the emergence of transport hubs, a strongly centripetal form, as a privileged network structure for many types of transport services, notably for air transportation (Figure 1.10). Although hub-and-spoke networks often result in improved network efficiency, they have drawbacks linked with their vulnerability to disruptions and delays at hubs, an outcome of the lack of direct connections. Evidence underlines that the emergence of hub-and-spoke networks is a transitional form of network development rationalizing limited volumes through a limited number of routes. When traffic becomes sufficient, direct point-to-point services tend to be established as they better reflect the preference of users.

Hubs, as a network structure, allow a greater flexibility within the transport system, through a concentration of flows. For instance, on Figure 1.10, a point-to-point network involves 16 independent connections, each to be serviced by vehicles and infrastructures. By using a hub-and-spoke structure, only eight connections are required. The main advantages of hubs are:

- Economies of scale on **connections** by offering a high frequency of services. For instance, instead of one service per day between any two pairs in a point-to-point network, four services per day could be possible.
- Economies of scale at the **hubs**, enabling the potential development of an efficient distribution system since the hubs handle larger quantities of traffic.
- Economies of scope in the use of **shared transhipment facilities**. This can take several dimensions such as lower costs for the users as well as higher quality infrastructures.

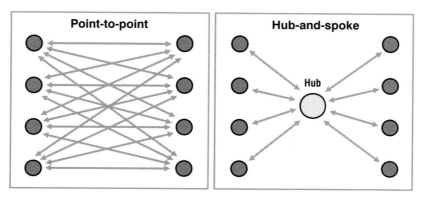

Figure 1.10   Point-to-point and hub-and-spoke networks

Many transportation services have adapted to include a hub-and-spoke structure. The most common examples involve air passenger and freight services which have developed such a structure at the global, national and regional levels, such as those used by UPS, FedEx and DHL. However, potential disadvantages may also occur such as additional transhipment as less point-to-point services are offered, which for some connections may involve delays and potential congestion as the hub becomes the major point of transhipment.

Inequalities between locations can often be measured by the quantity of links between nodes and the related revenues generated by traffic flows. Many locations within a network have higher accessibility, which is often related to better opportunities. However, economic integration processes tend to change inequalities between regions, mainly through a reorientation of the structure and flows within transportation networks at the transnational level (Figure 1.11).

Prior to economic integration processes (such as a free trade agreement) networks tended to service their respective national economies with flows representing this structure. With economic integration, the structure of transportation networks is modified with new transnational linkages. Flows are also modified. In some cases, there could be a relative decline of national flows and a comparative growth of transnational flows.

The efficiency of a network can be measured through graph theory and network analysis. These methods rest on the principle that the efficiency of a network depends partially on the lay-out of nodes and links. Obviously some network structures have a higher degree of accessibility than others, but careful consideration must be given to the basic relationship between the revenue and costs of specific transport networks. Rates thus tend to be influenced by the structure of transportation networks since the hub-and-spoke structure, particularly, had a notable impact on transport costs, namely through economies of scale.

The topology and typology of networks

Transportation networks, like many networks, are generally embodied as a set of locations and a set of links representing connections between those locations. The arrangement

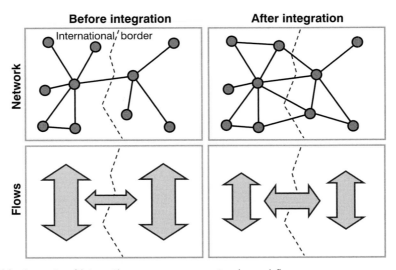

Figure 1.11   Impacts of integration processes on networks and flows

and connectivity of a network is known as its topology. Each transport network has consequently a specific topology. The most fundamental elements of such a structure are the network geometry and the level of connectivity. Transport networks can be classified in specific categories depending on a set of topological attributes that describe them. It is thus possible to establish a basic typology of transport networks that relates to its geographical setting as well as its modal and structural characteristics.

There are many criteria that can be used to classify transportation networks (Figure 1.12). Its level of abstraction can be considered with concrete network representations closely matching the reality (such as a road map) while conversely an abstract network would only be a symbolization of the nodes and flows (such as the network of an airline). Since transportation networks have a geographical setting, they can be defined according to their location relative to main elements of a territory (such as the Rhine delta). Networks also have an orientation and an extent that approximates their geographical coverage or their market area. The number of nodes and edges is relevant to express the complexity and structure of transportation networks with a branch of mathematics, graph theory, developed to infer structural properties from these numbers. Since networks are the support of movements they can be considered from a modal perspective, their edges being an abstraction of routes (roads, rail links, maritime routes) and their nodes an abstraction of terminals (ports, railyards). Specific modes can further be classified in terms of types of road (highway, road, street, etc.) and level of control (speed limit, vehicle restrictions, etc.). Flows on a network have a volume and a direction, enabling links to be ranked by their importance and the general direction of flows (e.g. centripetal or centrifugal) to be evaluated. Each segment and network has a physical capacity related to the volume it can support under normal conditions. The load (or volume to capacity) is the relation between the existing volume and the capacity. The closer it is to a full load (a ratio of 1), the more congested it is. The structure of some networks imposes a hierarchy reflecting the importance of each of its nodes and a pattern reflecting their spatial arrangement. Finally, networks have a dynamic where both their nodes and links can change due to new circumstances.

Further, there are three types of spaces on which transport networks are set and where each represents a specific mode of territorial occupation:

- **Clearly defined and delimited**. The space occupied by the transport network is strictly reserved for its exclusive usage and can be identified on a map. Ownership can also be clearly established. Major examples include road, canal and railway networks.
- **Vaguely defined and delimited**. The space of these networks may be shared with other modes and is not the object of any particular ownership, only of rights of passage. Examples include air and maritime transportation networks.
- **Without definition**. Space has no tangible meaning, except for the distance it imposes. Little control and ownership are possible, but agreements must be reached for common usage. Examples are radio, television and cellular networks, which rely on specific wave frequencies granted by governing agencies.

Networks provide a level of transport service which is related to its costs. An optimal network would be a network servicing all possible locations but such a service would have high capital and operational costs. Transport infrastructures are established over discontinuous networks since many were not built at the same time, by the same entity or with the same technology. Therefore, operational networks rarely service all parts of the territory directly. Some compromise must often be found among a set of alternatives considering a variety of route combinations and level of service.

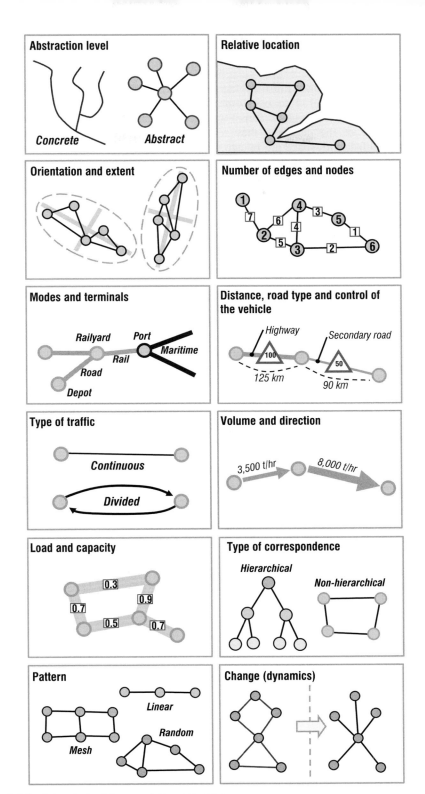

Figure 1.12 A typology of transportation networks

Networks and space

Transportation networks illustrate the territorial organization of economic activities and the efforts incurred to overcome distance. These efforts can be measured in absolute (distance) or relative (time) terms and are proportional to the efficiency and the structure of the networks they represent. The relationships transportation networks establish with space are related to their continuity, their topographic space and the spatial cohesion they establish. The territory is a topological space having two or three dimensions, depending on the transport mode considered (roads are roughly set over a two-dimensional space while air transport is set over a three-dimensional space). However, flows and infrastructures are linear; having one dimension since they conceptually link two points. The establishment of a network is thus a logical outcome for a one-dimensional feature to service a territory by forming a lattice of nodes and links. In order to have such a spatial continuity in a transport network, three conditions are necessary (Figure 1.13):

- **Ubiquity**. The possibility to reach any location from any other location on the network, thus providing a general access. Access can be a simple matter of vehicle ownership or bidding on the market to purchase a thoroughfare from one location to another.
- **Fractionalization**. The possibility for a traveler or a unit of freight to be transported without depending on a group. It becomes a balance between the price advantages of economies of scale and the convenience of a dedicated service.
- **Instantaneity**. The possibility to undertake transportation at the desired or most convenient moment. There is a direct relationship between fractionalization and instantaneity since the more fractionalized a transport system is, the more likely time convenience can be accommodated.

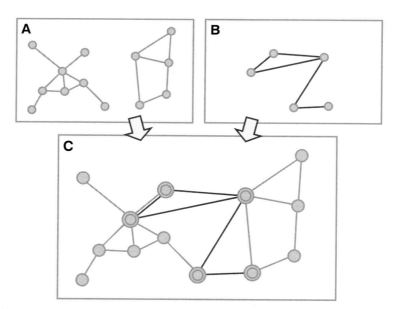

Figure 1.13 Networks and spatial continuity

These three conditions are never perfectly met as some transport modes fulfill them better than others. For instance, the automobile is the most flexible and ubiquitous mode for passenger transportation, but has important constraints such as low capacity and high levels of space and energy consumption. In comparison, public transit is more limited in the spatial coverage of its service, implies batch movements (bus loads, train loads, etc.) and follow specific schedules (limited instantaneity), but is more cost and energy efficient. Freight transportation also varies in its spatial continuity, ranging from massive loads of raw materials (oil and ores) that can be handled only in a limited number of ports to highly flexible parcel movements. Containerization has been a remarkable attempt to address the issue of ubiquity (the system permits intermodal movements), fractionalization (each container is a load unit) and instantaneity (units can be loaded by trucks at any time of the day and containerships make frequent port calls). On Figure 1.13 networks A and B are servicing the same territory, but both have a level of discontinuity (especially A). If a transfer between those two networks is possible then their combination (network C) increases the level of spatial continuity. This implies intermodal nodes.

An important cause of discontinuity is linked to the spatial distribution of economic activities, notably industrial and urban, which tend to agglomerate. Congestion may also alter those conditions. Road congestion in a metropolitan area may impair ubiquity as some locations may be very difficult to reach since their accessibility is reduced. Fractionalization may also be reduced under such circumstances as people would consider public transit and carpooling and would thus move as batches. Further, as commuters cope with increasing congestion, several trips may be delayed or canceled altogether reducing instantaneity.

Transportation networks have always been a tool for spatial cohesion and occupation. The Roman and Chinese Empires relied on transportation networks to control their respective territories, mainly to collect taxes and move commodities and military forces. During the colonial era, maritime networks became a significant tool of trade, exploitation and political control, which was later on expanded by the development of modern transportation networks within colonies. In the nineteenth century, transportation networks also became a tool of nation building and political control. For instance, the extension of railways in the American hinterland had the purpose to organize the territory, extend settlements and distribute resources to new markets. In the twentieth century, road and highways systems (such as the Interstate system in the United States and the autobahn in Germany) were built to reinforce this purpose. In the later part of the twentieth century, air transportation networks played a significant role in weaving the global economy. For the early twenty-first century, telecommunication networks have become means of spatial cohesion and interactions abiding well to the requirements of global supply chains.

Network expansion

As transport networks expand, existing transport infrastructures are being upgraded to cope with spatial changes. Airports and ports are being transformed, expanded or relocated. In the air transport sector, emphasis is been given to integrate airports within fully fledged multimodal transport systems, networking air with rail and road transport. In maritime transport, networks are also being modified with increasing attention being paid to:

- Exploiting sea leg routes across the Arctic Ocean (Figure 1.8).
- Expanding the Panama and Suez Canals.

- Increasing traffic on inland maritime waterways.
- Creating new inland passages between semi-enclosed or enclosed seas.

The growing competition between the sea and land corridors are not only reducing tariffs and encouraging international trade but prompting many governments to reassess their land-based connections and seek shorter transit routes.

Existing land routes are also being extended. Passages through extreme rigorous terrain are being investigated with a view to create fully fledged land-based continental connections, notably through railways. These land network expansions are driven by economic globalization and inter-regional cooperation and eventually become multimodal transcontinental corridors for rail, road, pipelines and trunk telecommunications routes. But the impact of increasing world trade on land network expansion, notably over railways, is scale specific. The expansion of railways has permitted inter- and intra-continental connections in the form of:

- **Landbridges**: a land movement across a continent linking origin and destination overseas.
- **Minibridges**: to cover movement linking two ends of a continent.
- **Microbridges**: covering traffic from/to a port to inland destination or origin.

Over the last twenty years new rail routes in North America, Eurasia, Latin America and Africa have been developed or are being considered. There is scope for shippers to increase their trade through these new routes, particularly if rising insurance premiums, charter rates and shipping risks prompt them to opt for a land route instead of the sea route through the Suez or Panama Canals. These developments linked to the integration of regional economies to the world market are part of a rationalization and specialization process of rail traffic presently occurring around the world. But the success of these rail network expansions depends on the speed of movement and the unitization of general cargo by containerization. Railways servicing ports tend more and more to concentrate on the movement of container traffic. This strategy followed by some rail transport authorities allows on the one hand an increase in the delivery of goods, and on the other hand the establishment of door-to-door services through a better distribution of goods among different transport modes.

New arterial links are constructing and reshaping new trade channels, underpinning outward cargo movements and the distribution of goods. As some coastal gateways are now emerging as critical logistics service centers that rationalize distribution systems to fit new trading patterns, the land network development and cross-border crossings throughout the world have far-reaching geopolitical implications.

Method 1 – Definition and properties of graph theory

Basic graph definition

A graph is a symbolic representation of a network and of its connectivity. It implies an abstraction of the reality so it can be simplified as a set of linked nodes.

Graph theory is a branch of mathematics concerned about how networks can be encoded and their properties measured.

The goal of a graph is representing the structure, not the appearance, of a network. The conversion of a real network into a planar graph is a straightforward process which follows some basic rules: (1) The most important rule is that every terminal and intersection point becomes a node. (2) Each connected node is then linked by a straight segment.

The outcome of this abstraction, as portrayed on Figure 1.14, is the actual structure of the network. The real network, depending on its complexity, may be confusing in terms of revealing its connectivity (what is linked with what). A graph representation reveals the connectivity of a network in the best possible way. Other rules can also be applied, depending on the circumstances. (3) A node that is not a terminal or an intersection point can be added to the graph if along that segment an attribute is changing. For instance, it would be recommended to represent as a node the shift from two lanes to four lanes along a continuous road segment, even if that shift does not occur at an intersection or terminal point. (4) A "dummy node" can be added for aesthetical purposes, especially when it is required that the graph representation remains comparable to the real network. (5) Although the relative location of each node can remain similar to their real world counterpart (as in Figure 1.14), this is not required.

In transport geography most networks have an obvious spatial foundation, namely road, transit and rail networks, which tend to be defined more by their links than by their nodes. This it is not necessarily the case for all transportation networks. For instance, maritime and air networks tend to be defined more by their nodes than by their links since links are often not clearly defined. A telecommunication system can also be represented as a network, while its spatial expression can have limited importance and would actually be difficult to represent. Mobile telephone networks or the Internet, possibly the most complex graphs to be considered, are relevant cases of networks having a structure that can be difficult to symbolize. However, cellular phones and antennas can be represented as nodes while the links could be individual phone calls. Servers, the core of the Internet, can also be represented as nodes within a graph while the physical infrastructure between them, namely fiber optic cables, can act as links. Consequently, all transport networks can be represented by graph theory in one way or another.

The following elements are fundamental in understanding graph theory:

- **Graph**. A graph G is a set of vertexes (nodes) v connected by edges (links) e. Thus $G = (v, e)$.
- **Vertex (node)**. A node v is a terminal point or an intersection point of a graph. It is the abstraction of a location such as a city, an administrative division, a road intersection or a transport terminal (stations, terminuses, harbors and airports).

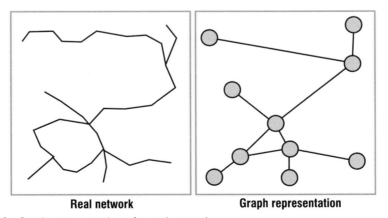

Real network   **Graph representation**

Figure 1.14 Graph representation of a real network

- **Edge (link)**. An edge e is a link between two nodes. The link (i, j) is of initial extremity i and of terminal extremity j. A link is the abstraction of a transport infrastructure supporting movements between nodes. It has a direction that is commonly represented as an arrow. When an arrow is not used, it is assumed the link is bi-directional.

The graph on Figure 1.15 has the following definition: $G = (v, e)$; $v = (1,2,3,4,5)$; $e = (1,2), (1,3), (2,2), (2,5), (4,2), (4,3), (4,5)$.

Sub-graph. A subset of a graph G where p is the number of sub-graphs. For instance $G' = (v', e')$ can be a distinct sub-graph of G. Unless the global transport system is considered in its whole, every transport network is in theory a sub-graph of another. For instance, the road transportation network of a city is a sub-graph of a regional transportation network, which is itself a sub-graph of a national transportation network.

Buckle. A link that makes a node correspond to itself is a buckle.

Planar graph. A graph where all the intersections of two edges are a vertex. Since this graph is located within a plane, its topology is two-dimensional.

Non-planar graph. A graph where there are no vertices at the intersection of at least two edges. This implies a third dimension in the topology of the graph since there is the possibility of having a movement "passing over" another movement such as for air transport. A non-planar graph has potentially many more links than a planar graph.

Links and their structures

A transportation network enables flows of people, freight or information, which occur along its links. Graph theory must thus offer the possibility of representing movements as linkages, which can be considered over several aspects:

- **Connection**. A set of two nodes, as every node is linked to another. Considers if a movement between two nodes is possible, whatever its direction. Knowing the connections within a graph makes it possible to find whether a node can be reached from another node.

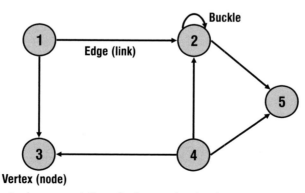

Figure 1.15 Basic graph representation of a transport network

- **Path**. A sequence of links that are traveled in the same direction. For a path to exist between two nodes, it must be possible to travel an uninterrupted sequence of links. Finding all the possible paths in a graph is a fundamental attribute in measuring accessibility and traffic flows.

On graph A of Figure 1.16 there are five links [(1,2), (2,1), (2,3), (4,3), (4,4)] and three connections [(1–2), (2–3), (3–4)]. On graph B, there is a path between 1 and 3, but on graph C there is no path between 1 and 3.

> **Chain**. A sequence of links having a connection in common with each other. Direction does not matter.
> **Length of a link, connection or path**. Refers to the label associated with a link, a connection or a path. This label can be distance, the amount of traffic, the capacity or any attribute of that link. The length of a path is the number of links (or connections) in this path.
> **Cycle**. Refers to a chain where the initial and terminal node is the same and which does not use the same link more than once.
> **Circuit**. A path where the initial and terminal node corresponds. It is a cycle where all the links are traveled in the same direction. Circuits are very important in transportation because several distribution systems use circuits to cover as much territory as possible in one direction (delivery route).

On the graph of Figure 1.17, 2-3-6-5-2 is a cycle but not a circuit. 1-2-4-1 is a cycle and a circuit.

Basic structural properties

The organization of nodes and links in a graph convey a structure that can be labeled. The basic structural properties of a graph are:

- **Symmetry and asymmetry**. A graph is symmetrical if each pair of nodes linked in one direction is also linked in the other. By convention, a line without an arrow represents a link where it is possible to move in both directions. However, both

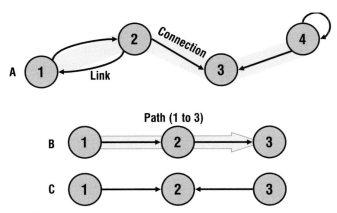

Figure 1.16 Connections and paths

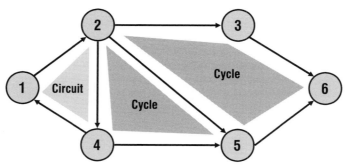

Figure 1.17 Cycles and circuits

directions have to be defined in the graph. Most transport systems are symmetrical but asymmetry can often occur as is the case for maritime (pendulum) and air services. Asymmetry is rare on road transportation networks, unless one-way streets are considered.

- **Completeness**. A graph is complete if two nodes are linked in at least one direction. A complete graph has no sub-graph.
- **Connectivity**. A complete graph is described as connected if for all its distinct pairs of nodes there is a linking chain. Direction does not have importance for a graph to be connected, but may be a factor for the level of connectivity. If $p > 1$ the graph is not connected because it has more than one sub-graph. There are various levels of connectivity, depending on the degree to which each pair of nodes is connected.
- **Complementary**. Two sub-graphs are complementary if their union results in a complete graph. Multimodal transportation networks are complementary as each sub-graph benefits from the connectivity of other sub-graphs.
- **Root**. A node r where every other node is the extremity of a path coming from r is a root. Direction has an importance. A root is generally the starting point of a distribution system, such as a factory or a warehouse.
- **Trees**. A connected graph without a cycle is a tree. A tree has the same number of links than nodes plus one ($e = v - 1$). If a link is removed, the graph ceases to be connected. If a new link between two nodes is provided, a cycle is created. A branch of root r is a tree where no links connect any node more than once.
- **Articulation node**. In a connected graph, a node is an articulation node if the sub-graph obtained by removing this node is no longer connected. It therefore contains more than one sub-graph ($p > 1$). An articulation node is generally a port or an airport, or an important hub of a transportation network, which serves as a bottleneck.
- **Isthmus**. In a connected graph, an isthmus is a link that, when removed, creates two sub-graphs with at least one connection.

Method 2 – Measures and indices of graph theory

Measures

Several measure and indices can be used to analyze network efficiency. Many of them were initially developed by Kansky (1963) and can be used for:

- Expressing the relationship between values and the network structures they represent.
- Comparing different transportation networks at a specific point in time.
- Comparing the evolution of a transport network at different points in time.

As well as the number of nodes and edges, three basic measures are used to define the structural attributes of a graph: the diameter, the number of cycles and the order of a node.

Diameter (d). The length of the shortest path between the most distanced nodes of a graph is the diameter. d measures the extent of a graph and the topological length between two nodes.

The diameter can provide a measurement of the development of a network in time. The greater the diameter, the less linked a network tends to be. In the case of a complex graph, the diameter can be found with a topological distance matrix (Shimbel distance), which computes for each node pair its minimal topological distance. Graphs in which the extent remains constant, but with a higher connectivity, have lower diameter values.

Number of Cycles (u). The maximum number of independent cycles in a graph. This number (u) is estimated through the number of nodes (v), links (e) and sub-graphs (p): $u = e - v + p$.

Trees and simple networks have a value of 0 since they have no cycles. The more complex a network, the higher the value of u, so it can be used as an indicator of the level of development and complexity of a transport system.

Order (degree) of a node (o). The number of attached links in a graph. This is a simple but effective measure of nodal importance. The higher its value, the more a node is important in a graph as many links converge to it. Hub nodes have a high order, while terminal points have an order that can be as low as 1. A perfect hub would have its order equal to the summation of all the orders of the other nodes in the graph and a perfect spoke would have an order of 1.

Indexes

Indexes are more complex methods to represent the structural properties of a graph since they involve the comparison of a measure over another.

Detour index. A measure of the efficiency of a transport network in terms of how well it overcomes distance or the friction of space. The closer the detour index gets to 1, the more the network is spatially efficient. Networks having a detour index of 1 are rarely, if ever, seen and most networks would fit on an asymptotic curve getting close to 1, but never reaching it.

$DI = DT/DD$

For instance, the straight distance (DD) between two nodes may be 40 km but the transport distance (DT; real distance) is 50 km. The detour index is thus 0.8 (40/50). The complexity of the topography is often a good indicator of the level of detour.

Network density. Measures the territorial occupation of a transport network in terms of km of links (*L*) per square kilometers of surface (*S*). The higher it is, the more a network is developed.

Pi index. The relationship between the total length of the graph *L(G)* and the distance along its diameter *D(d)*. It is labeled as Pi because of its similarity with the real Pi (3.14), which expresses the ratio between the circumference and the diameter of a circle. A high index shows a developed network. It is a measure of distance per units of diameter and an indicator of the shape of a network.

Eta index. Average length per link. Adding new nodes will cause a decrease of eta as the average length per link declines.

$$\eta = \frac{L(G)}{e}$$

Theta index. Measures the function of a node, that is the average amount of traffic per intersection. The higher theta is, the greater the load of the network.

$$\theta = \frac{Q(G)}{v}$$

Beta index. Measures the level of connectivity in a graph and is expressed by the relationship between the number of links (*e*) over the number of nodes (*v*). Trees and simple networks have a beta value of less than 1. A connected network with one cycle has a value of 1. More complex networks have a value greater than 1. In a network with a fixed number of nodes, the higher the number of links, the higher the number of paths possible in the network. Complex networks have a high value of beta.

The four graphs of Figure 1.18 have a growing level of connectivity. Graph A and B are not fully connected and their beta value is lower than 1. Graph C is connected and has a beta value of 1. Graph D is even more connected with a beta value of 1.25.

Alpha index. A measure of connectivity which evaluates the number of cycles in a graph in comparison with the maximum number of cycles. The higher the alpha index, the more a network is connected. Trees and simple networks will have a value

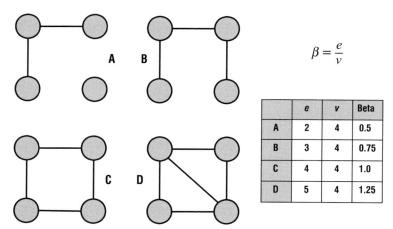

$$\beta = \frac{e}{v}$$

	e	*v*	Beta
A	2	4	0.5
B	3	4	0.75
C	4	4	1.0
D	5	4	1.25

Figure 1.18 Beta index

of 0. A value of 1 indicates a completely connected network. The alpha index measures the level of connectivity independently of the number of nodes. It is very rare for a network to have an alpha value of 1, because this would imply very serious redundancies.

The graphs of Figure 1.19 have a growing level of connectivity. While graph A has no cycles, graph D has the maximum possible number of cycles for a planar graph.

Gamma index. A measure of connectivity that considers the relationship between the number of observed links and the number of possible links. The value of gamma is between 0 and 1 where a value of 1 indicates a completely connected network and would be extremely unlikely in reality. Gamma is an efficient value to measure the progression of a network in time.

The graphs of Figure 1.20 have a growing level of connectivity with graph D having the maximum number of links (10) and a gamma index of 1.0.

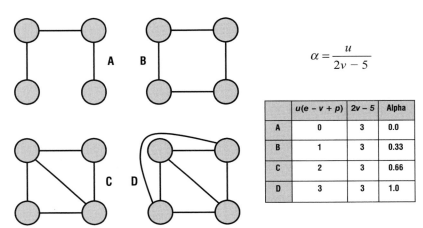

$$\alpha = \frac{u}{2v - 5}$$

	u(e – v + p)	2v – 5	Alpha
A	0	3	0.0
B	1	3	0.33
C	2	3	0.66
D	3	3	1.0

Figure 1.19 Alpha index

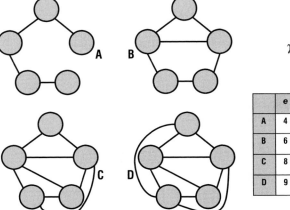

$$\gamma = \frac{e}{3(v-2)}$$

	e	3(v−2)	Gamma
A	4	9	0.44
B	6	9	0.66
C	8	9	0.88
D	9	9	1.0

Figure 1.20 Gamma index

Method 3 – Geographic information systems for transportation (GIS-T)[1]

Introduction

In a broad sense a geographic information system (GIS) is an information system specializing in the input, management, analysis and reporting of geographical (spatially related) information. Among the wide range of potential applications GIS can be used for, transportation issues have received a lot of attention. A specific branch of GIS applied to transportation issues, commonly labeled as GIS-T, has emerged.

> **Geographic information systems for transportation (GIS-T)** refers to the principles and applications of applying geographic information technologies to transportation problems (Miller and Shaw, 2001).

The four major components of a GIS – encoding, management, analysis and reporting – have specific considerations for transportation (Figure 1.21):

- **Encoding**. Deals with issues concerning the representation of a transport system and its spatial components. To be of use in a GIS, a transport network must be correctly encoded, implying a functional topology composed of nodes and links. Other elements relevant to transportation, namely qualitative and quantitative data, must also be encoded and associated with their respective spatial elements. For instance, an encoded road segment can have data related to its width, number of lanes, direction, peak hour traffic, etc.
- **Management**. The encoded information often is stored in a database and can be organized along spatial (by region, country, census units, etc.), thematic (for highway, transit, railway, terminals, etc.) or temporal (by year, month, week, etc.)

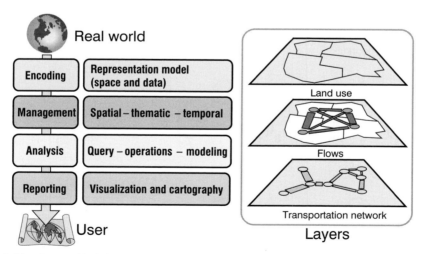

Figure 1.21 Geographic information systems and transportation

1 Dr. Shih-Lung Shaw (University of Tennessee) is the main contributor of this section.

considerations. It is important to design a GIS database that organizes a large amount of heterogeneous data in an integrated and seamless environment such that the data can be easily accessed to support various transportation application needs.

- **Analysis**. Considers the wide array of tools and methodologies available for transport issues. They can range from a simple query over an element of a transport system (what is the peak hour traffic of a road segment?) to a complex model investigating the relationships between its elements (if a new road segment was added, what would be the impacts on traffic and future land use developments?).
- **Reporting**. A GIS would not be complete without all its visualization and data reporting capabilities for both spatial and non-spatial data. This component is particularly important as it offers interactive tools to convey complex information in a map format. A GIS-T thus becomes a useful tool to inform people who otherwise may not be able to visualize the hidden patterns and relationships embedded in the datasets (potential relationships among traffic accidents, highway geometry, pavement condition and terrain).

Information in a GIS is often stored and represented as layers, which are a set of geographical features linked with their attributes. On Figure 1.21 a transport system is represented as three layers related to land use, flows (spatial interactions) and the network. Each has its own features and related data.

GIS-T research can be approached from two different, but complementary, directions. While some GIS-T research focuses on issues of how GIS can be further developed and enhanced in order to meet the needs of transportation applications, other GIS-T research investigates the questions of how GIS can be used to facilitate and improve transportation studies. In general, topics related to GIS-T studies can be grouped into three categories:

- **Data representations**. How can various components of transport systems be represented in a GIS-T?
- **Analysis and modeling**. How can transport methodologies be used in a GIS-T?
- **Applications**. What types of applications are particularly suitable for GIS-T?

GIS-T data representations

Data representation is a core research topic of GIS. Before a GIS can be used to tackle real world problems, data must be properly represented in a digital computing environment. One unique characteristic of GIS is the capability of integrating spatial and non-spatial data in order to support both display and analysis needs. There have been various data models developed for GIS. The two basic approaches are object-based data models and field-based data models.

- An object-based data model treats geographic space as populated by **discrete and identifiable objects**. Features are often represented as points, lines and/or polygons.
- A field-based data model treats geographic space as populated by **real world features** that vary continuously over space. Features can be represented as regular tessellations (e.g. a raster grid) or irregular tessellations (e.g. triangulated irregular network – TIN).

Representing the "real world" in a data model has been a challenge for GIS since their inception in the 1960s. A GIS data model enables a computer to represent real geographical elements as graphical elements. As shown on Figure 1.22, two representational models are possible: raster (grid-based) and vector (line-based):

- **Raster**. Based on a cellular organization that divides space into a series of units. Each unit is generally similar in size to another. Grid cells are the most common raster representation. Features are divided into cellular arrays and a coordinate (X,Y) is assigned to each cell, as well as a value. This allows for registration with a geographic reference system. A raster representation also relies on tessellation: geometric shapes that can completely cover an area. Although many shapes are possible (triangles and hexagons), the square is the most commonly used. The problem of resolution is common to raster representations. For a small grid, the resolution is coarse but the required storage space is limited. For a large grid the resolution is fine, but at the expense of a much larger storage space. On Figure 1.22, the real world (shown as an aerial photograph) is simplified as a grid where each cell color relates to an entity such as road, highway and river.
- **Vector**. The concept assumes that space is continuous, rather than discrete, which gives an infinite (in theory) set of coordinates. A vector representation is composed of three main elements: points, lines and polygons. Points are spatial objects with no area but can have attached attributes since they are a single set of coordinates (X and Y) in a coordinate space. Lines are spatial objects made up of connected points (nodes) that have no width. Polygons are closed areas that can be made up of a circuit of line segments. On Figure 1.22, the real world is represented by a series of lines (roads and highway) and one polygon (the river). A real world entity could be represented by different types of vector features depending on the map scale used in an application (e.g. a road can be represented as a line at a smaller scale or as a polygon at a larger scale).

GIS-T studies have employed both object-based and field-based data models to represent the relevant geographic data. Some transportation problems tend to fit better with one type of GIS data model than the other. For example, **network analysis** based on graph theory typically represents a network as a set of nodes interconnected with

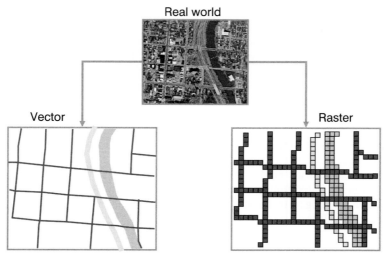

Figure 1.22 GIS data models

a set of links. The object-based GIS data model therefore is a better candidate for such transportation applications. Other types of transportation data exist which require extensions to the general GIS data models. One well-known example is **linear referencing data** (e.g. highway mileposts). Transportation agencies often measure locations of features or events along transportation network links (e.g. a traffic accident occurred at the 52.3 milepost on a specific highway). Such a one-dimensional linear referencing system (i.e. linear measurements along a highway segment with respect to a pre-specified starting point of the highway segment) cannot be properly handled by the two-dimensional Cartesian coordinate system used in most GIS data models. Consequently, the dynamic segmentation data model was developed to address the specific need of the GIS-T community. Origin–destination (O–D) flow data are another type of data that are frequently used in transportation studies. Such data have been traditionally represented in matrix forms (i.e. as a two-dimensional array in a digital computer) for analysis. Unfortunately, the relational data model widely adopted in most commercial GIS software does not provide adequate support for handling matrix data. Some GIS-T software vendors therefore have developed additional file formats and functions for users to work with matrix data in a GIS environment. The above examples illustrate how the conventional GIS approaches can be further extended and enhanced to meet the needs of transportation applications.

In recent years, developments of enterprise and multidimensional GIS-T data models also received increasing attention. Successful GIS deployments at the enterprise level (e.g. within a state department of transportation) demand additional considerations to embrace the diversity of application and data requirements. An enterprise GIS-T data model is designed to allow each application group to meet the established needs while enabling the enterprise to integrate and share data. The needs of integrating 1–D, 2–D, 3–D, and temporal data in support of various transportation applications also have called for the implementation of multidimensional (including spatio-temporal) data representations.

In short, one critical component of GIS-T is how transportation-related data in a GIS environment can be best represented in order to **facilitate and integrate the needs of various transportation applications**. Existing GIS data models provide a good foundation of supporting many GIS-T applications. However, due to some unique characteristics of transportation data and application needs, many challenges still exist to develop better GIS data models that will improve rather than limit what we can do with different types of transportation studies.

GIS-T analysis and modeling

GIS-T applications have benefited from many of the standard GIS functions (query, geocoding, buffer, overlay, etc.) to support data management, analysis and visualization needs. Like many other fields, transportation has developed its own **unique analysis methods and models**. Examples include shortest path and routing algorithms (e.g. traveling salesman problem, vehicle routing problem), spatial interaction models (e.g. gravity model), network flow problems (e.g. minimum cost flow problem, maximum flow problem, network flow equilibrium models), facility location problems (e.g. p-median problem, set covering problem, maximal covering problem, p-centers problem), travel demand models (e.g. the four-step trip generation, trip distribution, modal split and traffic assignment models) and land use – transportation interaction models.

While the basic transportation analysis procedures (e.g. shortest path finding) can be found in most commercial GIS software, other transportation analysis procedures and models (e.g. facility location problems) are available only selectively in some commercial software packages. Fortunately, the component GIS design approach adopted by GIS software companies provides a better environment for experienced GIS-T users to develop their own custom analysis procedures and models.

It is essential for GIS-T practitioners and researchers to have a thorough understanding of transportation analysis methods and models. For GIS-T practitioners, such knowledge can help them evaluate different GIS software products and choose the one that best meets their needs. It also can help them select appropriate analysis functions available in a GIS package and properly interpret the analysis results. GIS-T researchers, on the other hand, can apply their knowledge to help improve the design and analysis capabilities of GIS-T.

GIS-T applications

GIS-T is one of the leading GIS application fields. Many GIS-T applications have been implemented at various transportation agencies and private firms. They cover much of the broad scope of transportation and logistics, such as infrastructure planning and management, transportation safety analysis, travel demand analysis, traffic monitoring and control, public transit planning and operations, environmental impacts assessment, intelligent transportation systems (ITS), routing and scheduling, vehicle tracking and dispatching, fleet management, site selection and service area analysis, and supply chain management. Each of these applications tends to have its specific data and analysis requirements. For example, representing a street network as centerlines may be sufficient for transportation planning and vehicle routing applications. A traffic engineering application, on the other hand, may require a detailed representation of individual traffic lanes. Turn movements at intersections also could be critical to a traffic engineering study, but not to a region-wide travel demand study. These different application needs are directly relevant to the GIS-T data representation and the GIS-T analysis and modeling issues discussed above. When a need arises to represent transportation networks of a study area at different scales, what would be an appropriate GIS-T design that could support the analysis and modeling needs of various applications? In this case, it may be preferable to have a GIS-T data model that allows multiple geometric representations of the same transportation network. Research on enterprise and multidimensional GIS-T data models, as we have mentioned, aims at addressing these important issues of better data representations in support of various transportation applications.

With the rapid growth of the Internet and wireless communications in recent years, a growing number of Internet-based and wireless GIS-T applications can be found. Websites such as Google Maps and Mapquest are frequently used by people to get driving directions. Global positioning systems (GPS) for navigation also are available as a built-in device in a vehicle or as a portable device. Coupled with wireless communications, these devices can offer real-time traffic information and provide helpful location-based services (LBS) (e.g. finding the closest ATM location and reporting the estimated travel time to reach the ATM location based on the current traffic conditions). Another trend observed in recent years is the growing number of GIS-T applications in the private sector, particularly for logistics applications. Since many businesses involve operations at geographically dispersed locations (e.g. supplier sites, distribution centers/warehouses, retail stores and customer sites), GIS-T can be useful tools for a variety of logistics applications. Again, many of these logistics applications are based on GIS-T analysis and modeling procedures such as those for routing and facility location problems.

GIS-T is interdisciplinary in nature and has many possible applications. Transportation geographers, who have appropriate backgrounds in both geography and transportation, are well positioned to pursue GIS-T studies.

CASE STUDY Teaching transport geography

Transport geography education

Transportation geography is not a science, but dominantly a **field of application**. Transport geography has been part of the curriculum of many geography programs, providing a significant contribution, both conceptually and methodologically, to the discipline and to transportation studies in general. Indeed, many spatial theories rely on the concepts of distance, mobility and accessibility, which are perspectives on which transport geography offers a solid background. More recently, transportation networks and spatial organization issues have been notable contributions by the discipline to understand contemporary economic and social processes. The core issue in transport geography education, like any discipline, is related to **relevancy** and **coherence**. How relevant are the concepts and methods that are being taught and how coherently are they explained? Substantiating this question requires a critical overview of concepts, theories and methodologies in transport geography and how they fulfill the curriculum requirements and also societal needs.

The development of modern transport geography curriculum began in the 1960s, with Ned Taaffe and Howard Gauthier among the most significant contributors in the United States. This curriculum development, virtually from scratch, led to the seminal textbook, *Geography of Transportation* (1973), which brought rigor in the description and optimization of transport systems. Since, transport geography education has evolved with the priorities and concerns of the public and private sectors, with the focus shifting increasingly to **global issues**, but still acknowledging that they are deeply rooted in the local. The importance of supply chain management and logistics is a reality of contemporary world economics. This leads to provision of education in technical expertise related to information technology, inventory management and transport management. The objective is to provide knowledge and skills across business areas and industrial sectors within a supply chain context. This situation is conducive to programs in transport management shifting towards logistics management. More importantly, transport geographers must be able to anticipate rather than follow policy needs.

Where transportation geography is taught also has a significant impact on the curriculum since the **geographical setting changes the modal focus** of teaching transport geography. Under such circumstances, transport geography education must reflect the realities of the regional transport system as it is optimally the market in which students will find potential employment. In the case of Hong Kong, rail transportation has little importance and most of the focus is on public transit and international transportation issues. In the case of several European countries, the perspective tends to be more policy-oriented. European governments tend to have a more direct involvement over their transport systems through public and semi-public agencies. Transport geographers can thus be involved in the decision-making process in the public and private sectors through policy evaluation and formulation. The European emphasis is fairly different from the more privately owned system in North America where deregulation has been a dominant paradigm for the last 20 years.

Curriculum approaches

The last half-century has seen the creation of a solid curriculum in transport geo-graphy, as a subdiscipline of human and economic geography. Expanding the transport geography curriculum into the twenty-first century will require a continuation of this tradition with **new conceptual and methodological initiatives** since new problems will arise and new perspectives will be developed. To tackle these issues, a new generation of transportation geographers will obviously have to be trained with a particular emphasis on analytical and methodological perspectives. Economic integration and sustainability issues are receiving growing attention in the transport geography curriculum.

The quantitative revolution in the 1970s has led to a variety of methodological and technical dimensions in geographical education. Many of these have led to the mathematical abstraction and quantification of transport geography, but may also have substituted for relevance. Transport geography is a specialized part of spatial analysis and is focused on the importance of integrating analytical tools in the curriculum. Many of the tools and methods traditionally used in transport geography, such as spatial interaction, accessibility and network modeling, are now part of many GIS packages and readily available to investigate real world problems. GIS-T has become a funda-mental part of transport geography curriculums.

A look at geography curriculums also reveals that there are two transport geo-graphies, one **general** and the other **urban**. The latter comes from the growing influence of Urban Geography in geography curriculums, and growing traffic problems in cities. However, urban freight is largely absent from the perspective. In contrast, general transport geography has a more balanced orientation between passengers and freight, although modally there are differences, particularly if the approach switches to an extended geographical scale.

The development of concepts, theories and methods is a collective undertaking which involves seeking a **consensus** about what is relevant, as well as what is no longer relevant. From this large pool of knowledge, the transport geographer brings **coherence** in a curriculum by making choices concerning what should be introduced in accordance with the requirements of a curriculum. This mainly involves three challenges:

- **Theoretical and conceptual**. The core challenge of transport geography education is that it must underline how relevant are its theoretical and conceptual foundations in explaining contemporary events and processes. Prospects over this issue are very positive as empirical evidence underlines the growing mobility of people, freight and information at all geographical scales. This is a good indication of the relevancy of transport geography and it is thus important to ensure that it clearly gets conveyed to undergraduate audiences.
- **Methodological**. Another important aspect of transport geography education obviously relies on how information is analyzed, which includes a wide array of methods ranging from qualitative policy analysis to quantitative operations research. Methodologies previously tended to be taught from a more technical perspective and often in a very abstract manner. As methodologies are merging with information technologies there are opportunities to go beyond abstraction.
- **Technical and applied**. This involves using technology to replicate techniques and their procedures, but also using technology for educational purposes. It is quite clear that GIS-T remains a promising educational tool in transport geography, especially when it is used to demonstrate methodologies within simulations. Surprisingly, and despite the emphasis educational technologies have received, their level of integra-tion to transport geography education, even in its simplest form, remains fairly low.

Another challenge resides within the institutional structure, as transport geography remains what departments and programs commit to the discipline in terms of human and physical resources. This challenge is however linked to educational issues as a successful curriculum, however modest, promotes its own continuation and growth. The question remains how transport geographers, through their contribution to geographical education, will make sure the discipline receives a role proportional to its relevancy.

Transport geography in the classroom

Students being introduced to a discipline are particularly responsive to new concepts, ideas and fields of application brought into the classroom which will challenge and even change their vision of the world. This does not exclude a strong emphasis, with demonstrations and case studies, that concepts and methodologies are mutually imbedded in any scientific investigation. Then, it is for the student to make the strategic decision to pursue this investigation in upper level classes and at the graduate level. This decision may rarely take place if relevancy and coherence are not efficiently provided. While relevance is the responsibility of the whole scientific community, coherence is assumed by **individual transport geographers** within the classroom. As always, pertinent material cannot compensate for a lack of pedagogy.

The balance between concepts, methods and applications is obviously the prerogative of the educator to comply with a program's stated objectives, but the following approaches can be suggested depending on the general types of transport geography classes:

- **General introductory courses**. These transportation classes are generally offered to the undergraduate student population at large and tend to have no prerequisites. Considering the wide variety of students' backgrounds, they tend to be challenging classes but offer the possibility to attract students into a transportation or a geography program. Offering such courses should thus be seriously considered to place transportation issues within an academic community. In this context the curriculum should almost strictly focus on concepts by explaining the importance of transportation from the local to the global. A particular emphasis should be placed on presenting the relationships between transportation and geography, discussing their history, presenting major modes and terminals, as well as international and urban transportation systems. Methodological issues should not be discussed in detail, but simply in terms of how they are relevant to the discipline.
- **Specific introductory courses**. Regular transport geography classes as part of a geography curriculum. They are commonly offered at a more advanced level (e.g. second or third year) and thus assume that the students are already familiar with core geographical concepts linked to accessibility and spatial interactions. The goal is to expand these concepts through transportation issues with a balance between concepts and methodologies. If students have already received training in GIS, it is possible to provide some GIS-T methodologies and exercises, but methods can still be solved "by hand" or using spreadsheets. This book's website has specifically been designed for such a purpose and offers ample material to address a wide range of transport geography issues.
- **Topical intermediate courses**. Specialized classes often focusing on methods and fields of application. Many have prerequisites linked with quantitative methods and GIS. They thus offer an opportunity to teach a selected group of students already familiar with transport a series of customized concepts and methods. In many programs this is essentially a GIS-T class, but there are also opportunities to focus on topics

such as supply chain management, urban transportation, transport policy or transportation and land use.

- **Advanced courses**. Tend to be seminars and offered to small groups of students, commonly at the graduate level. Many students come from different programs, such as economics, engineering or political science, and take such seminars to expand their knowledge and help them address their specific research topic. Paradoxically, these courses tend to be less methodologically oriented and focus more on policy and management issues. A good approach involves the analysis of advanced research papers selected to cover the students' expressed interests. Students can be encouraged to develop projects in their fields of interest which will lead to a variety of approaches ranging from advanced methodologies (with GIS-T) to content analysis that can be presented and debated at the seminar.

Bibliography

Banister, D. (2002) *Transport Planning*, 2nd edn, London: Spon Press.

Black, W. (2003) *Transportation: A Geographical Analysis*, New York: Guilford.

Butler, J.A. and K.J. Dueker (2001) "Implementing the Enterprise GIS in Transportation Database Design", *Urban and Regional Information Systems Association (URISA) Journal*, 13(1), 17–28.

Haggett, P. (2001) *Geography: A Modern Synthesis*, 4th edn, New York: Prentice Hall.

Harrington, R. (1999) "Transport: Then, Now, and Tomorrow", *Royal Society of Arts Journal*, 146(5488). http://www.york.ac.uk/inst/irs/irshome/papers/carmen.htm.

Hoover, E.M. (1948) *The Location of Economic Activity*, New York: McGraw-Hill.

Hugill, P.J. (1992) *World Trade since 1431*, Baltimore, MD: Johns Hopkins University Press.

Kansky, K. (1963) *Structure of Transportation Networks: Relationships between Network Geography and Regional Characteristics*, University of Chicago, Department of Geography, Research Papers 84.

Keeling, D.J. (2008) "Transportation Geography – New Regional Mobilities", *Progress in Human Geography*, 32(2), 275–83.

Knowles, R., J. Shaw and I. Docherty (eds) (2008) *Transport Geographies: Mobilities, Flows and Spaces*, Malden, MA: Blackwell.

Leinbach, T. (1976) "Networks and Flows", *Progress in Human Geography*, 8, 179–207.

Lo, C.P. and A.K.W. Yeung (2002) *Concepts and Techniques of Geographic Information Systems*, Upper Saddle River, NJ: Prentice Hall.

Merlin, P. (1992) *Géographie des Transports, Que sais-je?*, Paris: Presses Universitaires de France.

Miller, H.J. and S.L. Shaw (2001) *Geographic Information Systems for Transportation: Principles and Applications*, New York: Oxford University Press.

Rimmer, P. (1985) "Transport Geography", *Progress in Human Geography*, 10, 271–7.

Rodrigue, J.-P. (2003) "Teaching Transport Geography: Conference Report and Viewpoint", *Journal of Transport Geography*, 11(1), 73–5.

Shaw, S.L. (2002) "Book Review: *Geographic Information Systems in Transportation Research*", *Journal of Regional Science*, 42(2), 418–21.

Taaffe, E.J., H.L. Gauthier and M.E. O'Kelly (1996) *Geography of Transportation*, 2nd edn, Upper Saddle River, NJ: Prentice Hall.

Thill, J.C. (ed.) (2000) *Geographic Information Systems in Transportation Research*, Oxford: Elsevier Science.

Tolley, R. and B. Turton (1995) *Transport Systems, Policy and Planning: A Geographical Approach*, Harlow, Essex: Longman.

Yu, H. (2006) "Spatio-Temporal GIS Design for Exploring Interactions of Human Activities", *Cartography and Geographic Information Science*, 33(1), 3–19.

Yu, H. and S.-L. Shaw (2008) "Exploring Potential Human Activities in Physical and Virtual Spaces: A Spatio-Temporal GIS Approach", *International Journal of Geographic Information Science*, 22(4), 409–30.

② Transportation and the spatial structure

An historical perspective on the evolution of transport systems underlines the consequences of technical innovations and how improvements in transportation were interdependent with contemporary economic and social changes. The current transport systems are thus the outcome of a long evolution marked by a period of rapid changes when a new transport technology was adopted. Such radical shifts are however not common, with rail, the internal combustion engine and the jet engine being the most salient examples. Future transportation systems will likely be shaped by the same forces than in the past but it remains to be seen which technologies will prevail. Transportation systems are composed of a complex set of relationships between the demand, the locations they service and the networks that support movements. Such conditions are closely related to the development of transportation networks, both in capacity and in spatial extent. This chapter consequently investigates the relationships between transportation and its related spatial structure.

Concept 1 – Historical geography of transportation

Efficiently distributing freight and moving people has always been an important factor for maintaining the cohesion of economic systems from empires to modern nation states. With technological and economic developments, the means to achieve such a goal have evolved considerably. The historical evolution of transportation is very complex and is related to the spatial evolution of economic systems. It is possible to summarize this evolution, from the pre-industrial era to transportation in the early twenty-first century, in five major stages, each linked with specific technological innovations in the transport sector.

Transportation in the pre-industrial era (pre-1800s)

Before the major technical transformations brought forward by the industrial revolution at the end of the eighteenth century, no forms of motorized transportation existed. Transport technology was mainly limited to harnessing animal labor for land transport and wind for maritime transport. The transported quantities were very limited and so was the speed at which people and freight were moving. The average overland speed by horse was between 8 to 15 kilometers per hour and maritime speeds were barely above these figures. Waterways were the most efficient transport systems available and cities next to rivers were able to trade over longer distances and maintain political, economic and cultural cohesion over a larger territory. It is not surprising to find that the first civilizations emerged along river systems for agricultural but also for trading purposes (Tigris–Euphrates, Nile, Indus, Ganges, Huang He).

Because the efficiency of the land transport system of this era was poor, the overwhelming majority of trade was local in scope. From the perspective of regional economic organization, the provision of cities in perishable agricultural commodities was limited to a radius of about 50 kilometers, at most. The size of cities also remained constant in time. Since people can walk about 5 km per hour and are not willing to spend more than one hour per day walking, the daily space of interaction would be constrained by a 2.5-km radius, or about 20 square kilometers. Thus, most rural areas centered around a village and cities rarely exceeded a 5-km diameter. The largest cities prior to the industrial revolution, such as Rome, Beijing, Constantinople and Venice, never surpassed an area of 20 square kilometers. International trade did exist, but traded commodities were high-value (luxury) goods such as spices, silk, wine and perfume, notably along the Silk Road (Figure 2.1).

The Silk Road was the most enduring trade route in human history, being used for about 1,500 years. Its name is taken from the prized Chinese textile that flowed from Asia to the Middle East and Europe, although many other commodities were traded along the route. The Silk Road consisted of a succession of trails followed by caravans through Central Asia, about 6,400 km in length. Travel was favored by the presence of steppes, although several arid zones had to be bypassed such as the Gobi and Takla Makan Deserts. Economies of scale, harsh conditions and security considerations required the organization of trade into caravans, which slowly trekked from one stage (town and/or oasis) to the next.

Although it is suspected that significant trade occurred for about 1,000 years beforehand, the Silk Road opened around 139 BC, once China was unified under the Han Dynasty. It started at Changan (Xian) and ended at Antioch or Constantinople (Istanbul), passing by commercial cities such as Samarkand and Kashgar. It was very rare that caravans traveled for the whole distance since the trade system functioned as a chain. Merchants with their caravans were shipping goods back and forth from one trade center to the other. Major commodities traded included silk (of course), gold, jade, tea and spices. Since the transport capacity was limited, over long distances and often unsafe, luxury goods were the only commodities that could be traded. The Silk Road also served as a vector for the diffusion of ideas and religions (initially Buddhism and then Islam), enabling civilizations from Europe, the Middle East and Asia to interact.

The initial use of the sea route linking the Mediterranean basin and India took place during the Roman era. Between the first and sixth centuries, ships were sailing between the Red Sea and India, aided by summer monsoon winds. Goods were transhipped at the town of Berenike along the Red Sea and moved by camels inland to the Nile. From that point, river boats moved the goods to Alexandria, from which trade could be undertaken with the Roman Empire. From the ninth century, maritime routes controlled by the Arab traders emerged and gradually undermined the importance of the Silk Road. Since ships were much less constraining than caravans in terms of capacity, larger quantities of goods could be traded. The main maritime route started at Canton (Guangzhou), passed through Southeast Asia, the Indian Ocean, the Red Sea and then reached Alexandria. A significant feeder went to the Spice Islands (Moluccas) in today's Indonesia. The diffusion of Islam was also favored through trade as many rules of ethics and commerce are embedded in the religion.

The Silk Road reached its peak during the Mongolian Empire (thirteenth century) when China and Central Asia were controlled by Mongol Khans, who were strong proponents of trade even if they were ruthless conquerors. At the same time relationships between Europe and China were renewed, notably after the voyages of Marco Polo (1271–92).

During the Middle Ages, the Venetians and the Genoese controlled the bulk of the Mediterranean trade which connected to the major trading centers of Constantinople, Antioch and Alexandria. As European powers developed their maritime technologies

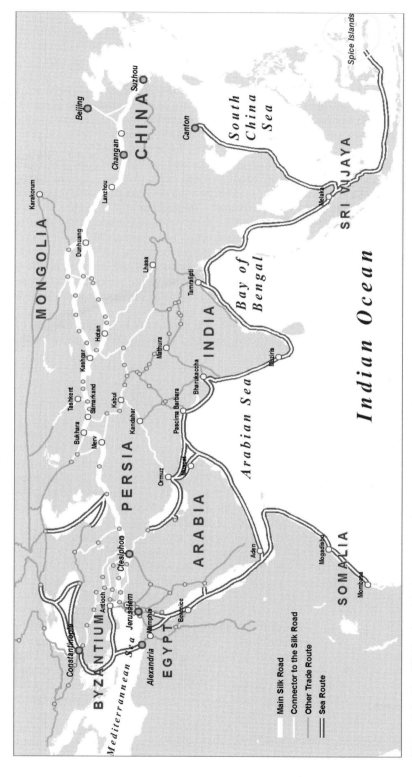

Figure 2.1 The Silk Road and Arab sea routes (eighth to fourteenth centuries)

from the fifteenth century, they successfully overthrew the Arab control of this lucrative trade route to replace it by their own. Ships that were able to transport commodities faster and cheaper marked the downfall of the Silk Road by the sixteenth century.

At that time, it was difficult to speak of an urban system, but rather of a set of relatively self-sufficient economic systems with very limited trade. The preponderance of city-states during this period can be explained a priori by transportation, in particular the difficulties of shipping goods (therefore to trade) from one place to another. Among the most notable exceptions to this were the Roman and Chinese Empires, which committed extraordinary efforts at building transportation networks and consequently maintained control over an extensive territory for a long time.

The transport system of the Roman Empire was a reflection of the geographical characteristics and constraints of the Mediterranean basin (Figure 2.2). The Mediterranean provided a central role to support trade between a network of coastal cities, which were the most important of the empire (Rome, Constantinople, Alexandria, Carthage, etc.). These cities were serviced by a road network permitting trade within their respective hinterlands. Little fluvial transportation took place since the major pan-European rivers, the Rhine and the Danube, were military frontiers, not the core, of the empire. The road served numerous functions, such as military movements, political control, cultural and economic (trade). To improve the traveling speed posthouses with fresh horses were laid every 15 kilometers along the route and lodgings for travelers could be found about every 40 kilometers.

The Appian Way (Via Appia), about 560 kilometers in length, was one of the first Roman roads (via) to be constructed (around 312 BC) under the initiative of emperor

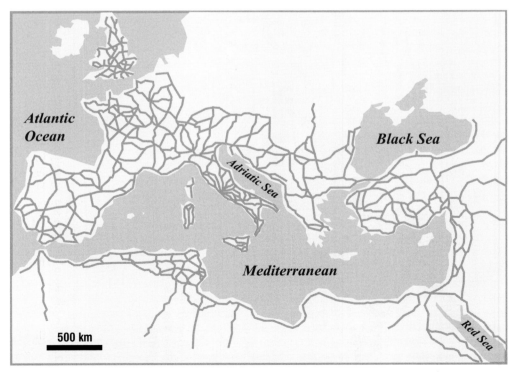

Figure 2.2 Roman road network, AD 200

Appius Claudius Caecus. As the empire grew, this system was expanded to cover 80,000 kilometers of first-class roads at the height of the Roman Empire (around AD 200). Most of the roads were constructed by soldiers, prisoners of war and slaves. The minimum requirement of a first-class road was a width of 5 meters and a drained stone surface. The Romans also built the world's first dual carriageway, Via Portuensis, between Rome and its port Ostia at the mouth of the Tiber. The Roman road network covered most of the conquered provinces, with Rome as the focal point (thus the saying "All roads lead to Rome"). At the center of Rome was located the *milliareum aureum* (the golden milestone), from which the Roman roads radiated. Maintenance was the responsibility of the inhabitants of the district through which the road passed by, but access was public. The system collapsed during the Middle Ages because of the lack of maintenance and plundering for construction material, but still, the remains of the Roman network provided transportation in Europe for a thousand years. Only small segments of this system are left today.

The economic importance and the geopolitics of transportation were recognized very early, notably for maritime transportation since, before the industrial revolution, it was the most convenient way to move freight and passengers around. Great commercial empires were established with maritime transportation. Initially, ships were propelled by rowers and sails were added around 2,500 BC as a complementary form of propulsion. By medieval times, an extensive maritime trade network, the highways of the time, centered along the navigable rivers, canals and coastal waters of Europe (and also China), was established. Shipping was extensive and sophisticated, using the English Channel, the North Sea, the Baltic and the Mediterranean where the most important cities were coastal or inland ports (London, Norwich, Königsberg, Hamburg, Bruges, Bordeaux, Lyon, Lisbon, Barcelona and Venice). Trade of bulk goods, such as grain, salt, wine, wool, timber and stone was taking place. By the fourteenth century, galleys were finally replaced by full-fledged sailships (the caravel and then the galleon) that were faster and required smaller crews. The year 1431 marked the beginning of European expansion with the discovery by the Portuguese of the North Atlantic circular wind pattern, better known as the trade winds. A similar pattern was also found on the Indian and Pacific Oceans with the monsoon winds.

The fall of Constantinople, the capital of the Byzantium Empire (Eastern Roman Empire), to the Turks in 1453 disrupted the traditional land trade route from Europe to Asia. Europe was forced to find alternate maritime routes. One alternative, followed by Columbus in 1492, was to sail to the west, and the other alternative, followed by Vasco de Gama in 1497, was to sail to the east. Columbus stumbled upon the American continent, while Gama found a maritime route to India using the Cape of Good Hope. These events were quickly followed by a wave of European exploration and colonization, initially by Spain and Portugal, the early maritime powers, then by Britain, France and the Netherlands. The traditional trade route to Asia no longer involved Italy (Venice) and Arabia, but involved direct maritime connections from ports such as Lisbon. European powers were able to master the seas with larger, better armed and more efficient sailing ships and thus were able to control international trade and colonization. By the early eighteenth century, most of the world's territories were controlled by Europe, providing wealth and markets to their thriving metropolises through a system of colonial trade (Figure 2.3).

A complex network of colonial trade was thus established over the North Atlantic Ocean. This network was partially the result of local conditions and of dominant wind patterns. It was discovered in the fifteenth century, notably after the voyages of Columbus, that there is a circular wind pattern over the North Atlantic. The eastward

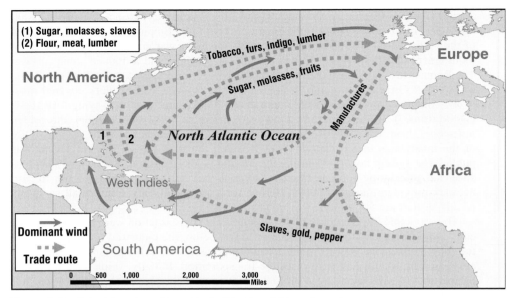

Figure 2.3 Colonial trade pattern, North Atlantic, eighteenth century

wind pattern, which blows on the southern part, came to be known as the "trade winds" since they enabled ships to cross the Atlantic. The westward wind pattern, blowing on the northern part, came to be known as the "westerlies".

Since sailing ships were highly constrained by dominant wind patterns, a trade system followed this pattern. Manufactured commodities were exported from Europe, some towards the African colonial centers, some towards the American colonies. This system also included the slave trade, mainly to Central and South American colonies (Brazil, West Indies). Tropical commodities (sugar, molasses) flowed to the American colonies and to Europe. North America also exported tobacco, furs, indigo (a dye) and lumber (for shipbuilding) to Europe. This system of trade collapsed in the nineteenth century with the introduction of steamships, the end of slavery and the independence of many of the colonies of the Americas.

Prior to the industrial revolution, the quantity of freight transported between nations was negligible by contemporary standards. For instance, during the Middle Ages, French imports via the Saint-Gothard Passage (between Italy and Switzerland) would not fill a freight train. The total amount of freight transported by the Venetian fleet, which dominated Mediterranean trade for centuries, would not fill a modern cargo ship. The volume, but not the speed, of trade improved under mercantilism (fifteenth to eighteenth centuries), notably for maritime transportation. In spite of all, distribution capacities were very limited and speeds slow. For example, a stage coach going through the English countryside in the sixteenth century had an average speed of two miles per hour; moving one ton of cargo 30 miles (50 km) inland in the United States by the late eighteenth century was as costly as moving it across the Atlantic. The inland transportation system was thus very limited, both for passengers and freight. By the late eighteenth century, canal systems started to emerge in Europe, initially in the

Netherlands and England. They permitted the beginning of large movements of bulk freight inland and expanded regional trade. Maritime and fluvial transportation were consequently the dominant modes of the pre-industrial era.

The industrial revolution and transportation (1800–70)

It was during the industrial revolution that massive modifications of transport systems occurred in two major phases, the first centered along the development of canal systems and the second along railways. This period marked the development of the steam engine that converted thermal energy into mechanical energy, and an important territorial expansion for maritime and railway transport systems followed. Much of the credit of developing the first efficient steam engine in 1765 is attributed to the British engineer Watt, although the first steam engines were used to pump water out of mines. It was then only a matter of time before the adaptation of the steam engine to locomotion. In 1769, the French engineer Cugnot built the first self-propelled steam vehicle (and was responsible for the first automobile accident ever recorded). The first mechanically propelled maritime vehicle was tested in 1790 by the American inventor Fitch as a mode of fluvial transportation on the Delaware River. This marked a new era in the mechanization of land and maritime transport systems alike.

From the perspective of land transportation, the early industrial revolution faced problems over bottlenecks, as inland distribution was unable to carry the growing quantities of raw materials and finished goods. Roads were commonly unpaved and could not be used to effectively carry heavy loads. Although improvements were made on road transport systems in the early seventeenth century, such as the Turnpike Trusts in Britain (1706) and the development of stagecoaches, this was not sufficient to accommodate the growing demands on freight transportation. The first coach services had speeds of about 5.5 miles per hour in the 1750s. By the 1820s turnpikes greatly improved overland transportation. However, roads were not profitable if used to haul anything except compact and valuable goods. Bulk products could be transported for about 100 miles, but in a slow and inefficient manner. In a horse-drawn era, road economics were clearly disadvantageous. For instance, four horses could pull a wagon weight of one ton 12 miles a day over an ordinary road and one-and-a-half tons 18 miles a day over a turnpike. Comparatively, four horses could draw a barge of 100 tons 24 miles a day on a canal.

From the 1760s a set of freight shipping canals were slowly built in emerging industrial cores such as England (e.g. Bridgewater Canal, 1761) and the United States (e.g. Erie Canal, 1825). These projects relied on a system of locks to overcome changes in elevation, thus linking different segments of fluvial systems into a comprehensive waterway system. Barges became increasingly used to moving goods at a scale and a cost that were not previously possible. Economies of scale and specialization, the foundation of modern industrial production systems, became increasingly applicable through fluvial canals. Physical obstacles made canal construction expensive, however, and the network was constrained. In 1830 there were about 2,000 miles of canals in Britain and by the end of the canal era in 1850, there were 4,250 miles of navigable waterways. The canal era was however short-lived as a new mode that would revolutionalize and transform inland transportation appeared in the second half of the nineteenth century.

Steam railway technology initially appeared in 1814 to haul coal. It was found that using a steam engine on smooth rails required less power and could handle heavier loads. The first commercial rail line linked Manchester to Liverpool in 1830 (a distance of 40 miles) and shortly after rail lines began to be laid throughout developed countries. The capital costs to build railway networks were enormous and often left to the private sector. They included rights of way, building, maintenance and operating costs. By the 1850s, railroad towns were being established and the railways were giving access to resources and markets of vast territories. Some 6,000 miles of railways were then operating in England and railways were quickly being constructed in Western Europe and North America. Railroads represented an inland transport system that was flexible in its spatial coverage and could carry heavy loads. As a result many canals fell into disrepair and were closed as they were no longer able to compete with rail services. In their initial phase of development, railways were a point-to-point process where major cities were linked one at a time by independent companies. Thus, the first railroad companies bore the name of the city pairs or the region they were servicing (e.g. the Camden and Amboy Railroad Company chartered in 1830). From the 1860s, integrated railway systems using standard gauges started to cohesively service whole nations with passenger and freight services. The journey between New York and Chicago was reduced from three weeks by stage coach to 72 hours by train. Many cities thus became closely interconnected. The transcontinental line between New York and San Francisco, completed in 1869, represented a remarkable achievement in territorial integration made only possible by rail. It reduced the journey across the continent (New York to San Francisco) from six months to one week, thus opening for the Eastern part of the United States a vast pool of resources and new agricultural regions. This was followed by Canada in 1886 (trans-Canada railway) and Russia in 1904 (trans-Siberian railway).

In terms of international transportation, the beginning of the nineteenth century saw the establishment of the first regular maritime routes linking harbors worldwide, especially over the North Atlantic between Europe and North America. Many of these long-distance routes were navigated by fast clipper ships, which dominated ocean trade until the late 1850s. Another significant improvement resided in the elaboration of accurate navigation charts where prevailing winds and sea currents could be used to the advantage of navigation. Composite ships (a mixture of wood and iron armature) then took over a large portion of the trade until about 1900, but they could not compete with steamships which had been continually improved since they were first introduced a hundred years before. Regarding steamship technology, 1807 marks the first successful use of a steamship, Fulton's *North River/Clermont*, on the Hudson River servicing New York and Albany. In 1820, the *Savannah* was the first steamship (used as auxiliary power) to cross the Atlantic, taking 29 days to link Liverpool to New York. The first regular services for transatlantic passenger transport by steamships was inaugurated in 1838, followed closely by the usage of the helix, instead of the paddle wheel, as a more efficient propeller (1840). The gradual improvement of steam engine technology slowly but surely permitted longer and safer voyages, enabling steamships to become the dominant mode of maritime transportation (Figure 2.4). Shipbuilding was also revolutionized by the usage of steel armatures (1860), enabling to escape the structural constraints of wood and iron armatures in terms of ship size. Iron armature ships were 30 to 40 percent lighter and had 15 percent more cargo capacity.

One of the main technical drawbacks of a steamship was the requirement of carrying coal in storage, which was done at the expense of the regular payload. As steam engine technology improved (better boiler and piston systems), more power could be generated by the same quantity of coal, implying that longer distances could be traveled. Thus, the break-even distance between sail and steam steadily improved from the 1850s. By the 1860s, transatlantic steamship services became cost-effective. By the 1870s, particularly in conjunction with the opening of the Suez Canal (1869), South

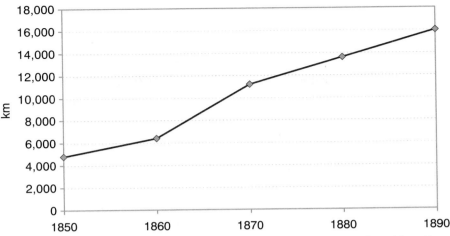

Figure 2.4 Break-even distance between sail and steam, 1850–90 (Source: adapted from Bernstein, 2008)

Asia became economically accessible. By the 1890s, steamship technology improved to enable long-distance voyages such as linking Great Britain with its Pacific Asian colonies. The use of coaling stations required the setting of intermediary refueling locations, namely Gibraltar, Cape Town, Suez, Malta and Singapore. This marked the downfall of sailing as a commercially viable form of freight transportation.

The main consequence of the industrial revolution was a specialization of transportation services and the establishment of large distribution networks of raw materials and energy.

Emergence of modern transportation systems (1870–1920)

By the end of the nineteenth century, international transportation undertook a new growth phase, especially with improvements in engine propulsion technology and a gradual shift from coal to oil in the 1870s. Although oil has been known for centuries for its combustion properties, its commercial use was only applied in the early nineteenth century. Inventors started experimenting with engines that could use the cheap new fuel. Oil increased the speed and the capacity of maritime transport. It also permitted to reduce the energy consumption of ships by a factor of 90 percent relatively to coal, the main source of energy for steam engines prior to this innovation. An equal size oil-powered ship could transport more freight than a coal-powered ship, reducing operation costs considerably and extending range. Also, coal refueling stages along trade routes could be bypassed. Global maritime circulation was also dramatically improved when infrastructures to reduce intercontinental distances, such as the Suez (1869) and the Panama (1914) Canals, were constructed (Figure 2.5).

The Panama Canal considerably shortens the maritime distances between the American East and West coasts by a factor of 13,000 km. Planned by the French but constructed by the British, the Suez Canal opened in 1869. It represents, along with the Panama Canal, one of the most significant maritime "shortcuts" ever built. It brought a new era of European influence in Pacific Asia by reducing the journey from Asia to Europe by about 6,000 km. The region became commercially accessible and colonial trade expanded as a result of increased interactions because of a reduced friction of

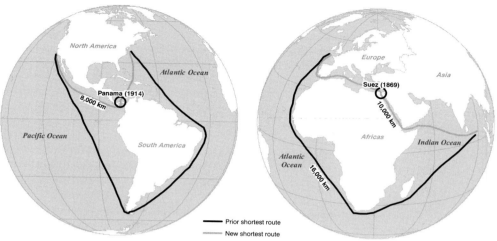

Figure 2.5 Geographical impacts of the Suez and Panama Canals

distance. Great Britain, the maritime power of the time, benefited substantially from this improved access. For instance, the Suez Canal shortened the distance on a maritime journey from London to Bombay by 41 percent and shortened the distance on the journey from London to Shanghai by 32 percent.

The increasing size of ships, the outcome of advances in shipbuilding, imposed massive investments in port infrastructures such as piers and docks to accommodate them. Ship size grew dramatically, from the largest tonnage of 3,800 gross registered tons (revenue making cargo space) in 1871 to 47,000 tons in 1914. Accordingly, ocean freight rates dropped by a factor of 70 percent from 1840 to 1910. The harbor, while integrating production and transshipping activities, became an industrial complex around which agglomerated activities using ponderous raw materials. From the 1880s, liner services linked major ports of the world, supporting the first regular international passenger transport services, until the 1950s when air transportation became the dominant mode. This period also marked the golden era of the development of the railway transport system as railway networks expanded tremendously and became the dominant land transport mode both for passengers and freight. As the speed and power of locomotives improved and as the market expanded, rail services became increasingly specialized with trains entirely devoted to passengers or freight. Rail systems reached a phase of maturity.

Another significant technological change of this era involved urban transportation, which until then solely relied on walking and different types of carriages (mainly horse drawn). The significant growth of the urban population favored the construction of the first public urban transport systems. Electric energy became widely used in the 1880s and considerably changed urban transport systems with the introduction of tramways (streetcars), notably in Western Europe and in the United States. They enabled the first forms of urban sprawl and the specialization of economic functions, notably by a wider separation between the place of work and residence. In large agglomerations, underground metro systems began to be constructed, London being the first in 1863. The bicycle, first shown at the Paris Exhibition of 1867, was also an important innovation which changed commuting in the late nineteenth century. Initially, the rich used it as a form of leisure, but it was rapidly adopted by the working class as a mode of transportation to the workplace. Today, the bicycle is much less used in developed countries (outside of recreational purposes), but it is still a major mode of transportation in developing countries, especially China.

This era also marked the first significant developments in telecommunications. The telegraph is considered to be the first efficient telecommunication device. In 1844, Samuel Morse built the first experimental telegraph line in the United States between Washington and Baltimore, opening a new era in the transmission of information. By 1852, more than 40,000 km of telegraph lines were in service in the United States. In 1866, the first successful transatlantic telegraph line marked the inauguration of an intercontinental telegraphic network. The growth of telecommunications is thus closely associated with the growth of railways and international shipping. Managing a rail transport system, especially at the continental level, became more efficient with telegraphic communication. In fact, continental rail and telegraphic networks were often laid concomitantly. Telecommunications were also a dominant factor behind the creation of standard time zones in 1884. From a multiplicity of local times, zones of constant time with Greenwich (England) as the reference were laid. This improved the scheduling of passenger and freight transportation at national levels. By 1895, every continent was linked by telegraph lines, a precursor of the global information network that would emerge in the late twentieth century. Business transactions became more efficient as production, management and consumption centers could interact with delays that were in hours instead of weeks and even months.

Transportation in the Fordist era (1920–70)

The Fordist era was epitomized by the adoption of the assembly line as the dominant form of industrial production, an innovation that benefited transportation substantially. The internal combustion engine, or four-stroke engine by Daimler (1889), which was a modified version of the diesel engine (1885), and the pneumatic tire (1885) by Dunlop made road vehicle operations faster and more comfortable. Compared with steam engines, internal combustion engines have a much higher efficiency and use a lighter fuel: petrol. Petrol, previously perceived as an unwanted by-product of the oil-refining process, which was seeking kerosene for illumination, became a convenient fuel. Initially, diesel engines were bulky, limiting their use to industrial and maritime propulsion, a purpose which they still fulfill today. The internal combustion engine permitted an extended flexibility of movements with fast, inexpensive and ubiquitous (door-to-door) transport modes such as automobiles, buses and trucks. Mass producing these vehicles changed considerably the industrial production system, notably by 1913 when Ford began the production of the Model T car using an assembly line. From 1913 to 1927, about 14 million Ford Model T cars were built, making it the second most produced car in history, behind the Volkswagen Beetle. The rapid diffusion of the automobile marked an increased demand for oil products and other raw materials such as steel and rubber.

Economies of scale also improved transportation in terms of capacity, which enabled to move low-cost bulk commodities such as minerals and grain over long distances. Oil tankers are a good example of the application of this principle to transport larger quantities of oil at a lower cost, especially after World War II when global demand surged. Maritime routes were thus expanded to include tanker routes, notably from the Middle East, the dominant global producer of oil. The very long distances concerned in the oil trade favored the construction of larger tankers. In the 1960s, tanker ships of 100,000 tons became available, to be supplanted by VLCCs (Very Large Crude Carriers) of 250,000 tons in the 1970s and by the ULCCs (Ultra Large Crude Carriers) of 550,000 tons at the end of the 1970s. A ship of 550,000 tons is able to transport 3.5 million tons of oil annually between the Persian Gulf and Western Europe.

Although the first balloon flight took place in 1783, due to the lack of propulsion no practical applications for air travel were realized until the twentieth century. The first propelled flight was made in 1903 by the Wright brothers and inaugurated the era of air transportation. The initial air transport services were targeted at mail since it was a type of freight that could be easily transported and initially proved to be more profitable than transporting passengers. The year 1919 marked the first commercial air transport service between England and France, but air transport suffered from limitations in terms of capacity and range. Several attempts were made at developing dirigible services, as the Atlantic was crossed by a Zeppelin dirigible in 1924. However, such technology was abandoned in 1937 after the *Hindenburg* accident, in which its hydrogen-filled reservoirs burned. The 1920s and 1930s saw the expansion of regional and national air transport services in Europe and the United States with successful propeller aircrafts such as the Douglas DC-3. The post-World War II period was however the turning point for air transportation as the range, capacity and speed of aircraft increased as well as the average income of the passengers. A growing number of people were thus able to afford the speed and convenience of air transportation. The year 1952 marks the beginning of commercial jet services with the Comet, but a design flaw grounded the plane the following year. In 1958, the first successful commercial jet plane, the Boeing 707, entered service and revolutionized international movements of passengers, marking the end of passenger transoceanic ships.

Basic telecommunications infrastructures, such as the telephone and the radio, were mass marketed during the Fordist era. However, the major change was the large diffusion of the automobile, especially from the 1950s when it became a truly mass consumption product and when the first major highway systems, such as the American Interstate, began to be built. No other modes of transportation have so drastically changed lifestyles and the structure of cities, notably for developed countries. It created sub-urbanization and expanded cities to areas larger than 100 km in diameter in some instances. In dense and productive regions, such as the Northeast of the United States, the urban system became structured and interconnected by transport networks to the point that it could be considered as one vast urban region: the Megalopolis.

A new context for transportation: the post-Fordist era (1970–)

Among the major changes in international transportation from the 1970s are the massive development of telecommunications, the globalization of trade, more efficient distribution systems and the considerable development of air transportation.

Telecommunications enabled growing information exchanges, especially for the financial and service sectors. After 1970, telecommunications successfully merged with information technologies. As such, telecommunication also became a medium of doing business in its own right, in addition to supporting and enhancing other transportation modes. The information highway became a reality as fiber optic cables gradually replaced copper wires, multiplying the capacity to transmit information between computers. This growth was however dwarfed by the tremendous growth in processing power of computers, which are now fundamental components of economic and social activities in developed countries. A network of satellite communication was also created to support the growing exchanges of information, especially for television images. Out of this wireless technology emerged local cellular networks which expanded and merged to cover whole cities, countries, regions and then continents. Telecommunications have reached the era of individual access, portability and global coverage.

In a post-Fordist system, the fragmentation of production, organizing an international division of labor, as well as the principle of "just-in-time" increased the quantity of freight moving at the local, regional and international levels. This in turn required increasing efforts to manage freight and reinforced the development of logistics, the science of physical distribution systems. Containers, main agents of the modern international transport system, enabled an increased flexibility of freight transport, mainly by reducing transhipment costs and delays. Handling a container requires about 25 times less labor than its equivalent in bulk freight resulting in a significant reduction in transhipment costs and time. They were introduced by the American entrepreneur Malcolm McLean who initially applied containerization to land transport but saw the opportunity of using container shipping as an alternative to acute road congestion in the early 1950s before the construction of the first Interstate highways. The initial attempts at containerization thus aimed at reducing maritime transhipment costs and time. Before containerization, a cargo ship could spend as much time in a port being loaded or unloaded than it did at sea. Later on, the true potential of containerization became clear when interfacing with other modes became an operational possibility, mainly between maritime, rail and road transportation.

The first containership (the *Ideal-X*, a converted T2 oil tanker) set sail in 1956 from New York to Houston and marked the beginning of the era of containerization. In 1960, the Port Authority of New York/New Jersey foreseeing the potential in container trade constructed the first specialized container terminal next to Port Newark: the Port Elizabeth Marine Terminal. The Sea-Land Company established the first regular

Plate 2.1 Panamax ship crossing the Miraflores Locks on the Panama Canal. *Credit*: Brian Slack

maritime container line in 1965 over the Atlantic between North America and Western Europe. By the early 1980s, container services with specialized ships (cellular containerships, first introduced in 1967) became a dominant aspect of international and regional transport systems, transforming the maritime industry. However, the size of those ships remained for 20 years constrained by the size of the Panama Canal, which *de facto* became the panamax standard (see plate 2.1). In 1988, the first post-Panamax containership was introduced, an indication of the will to further expand economies of scale in maritime container shipping.

Air and rail transportation experienced remarkable improvements in the late 1960s and early 1970s. The first commercial flight of the Boeing 747 between New York and London in 1969 marked an important landmark for international transportation (mainly for passengers, but freight became a significant function in the 1980s). This giant plane can transport around 400 passengers, depending on the configuration. It permitted a considerable reduction of air fares through economies of scale and opened intercontinental air transportation to the mass market. Attempts were also undertaken to establish faster-than-sound commercial services with the Concorde (1976; flying at 2,200 km/hr). However, such services proved to be financially unsound and no new supersonic commercial planes have been built since the 1970s. The Concorde was finally retired in 2003. At the regional level, the emergence of high-speed train networks provided fast and efficient inter-urban services, notably in France (1981; TGV; speeds up to 300 km/hr) and in Japan (1964; Shinkansen; speeds up to 275 km/hr).

Major industrial corporations making transportation equipment, such as car manufacturers, have become dominant players in the global economy. Even if the car is not an international transport mode, its diffusion has expanded global trade of vehicles, parts, raw materials and fuel (mainly oil). Car production, which used to be mainly concentrated in the United States, Japan and Germany, has become a global industry with a few key players part of well-integrated groups such as Ford, General Motors, Daimler Chrysler, Toyota and Mitsubishi. Along with oil conglomerates, they have pursued strategies aimed at the diffusion of the automobile as the main mode of individual transportation. This has led to growing mobility but also to congestion and waste of energy. As the twenty-first century begins, the automobile accounts for about 80 percent of the total oil consumption in developed countries.

The current period is also one of transport crises, mainly because of a dual dependency. First, transportation modes have a heavy dependence on fossil fuels and second, road transportation has assumed dominance. The oil crisis of the early 1970s, which saw a significant increase in fuel prices, induced innovations in transport modes, the reduction of energy consumption and the search for alternative sources of energy (electric car, adding ethanol to gasoline and fuel cells). However, from the mid-1980s to the end of the 1990s, oil prices declined and attenuated the importance of these initiatives. Again, oil prices surged in the beginning of the twenty-first century, placing alternative sources of energy on the agenda. Still, the reliance on fossil fuels continues with a particularly strong growth of motorization in developing countries.

Concept 2 – Transport and spatial organization

The spatial organization of transportation

Throughout history, transport networks have structured space at different scales. The fragmentation of production and consumption, the locational specificities of resources, labor and markets generate a wide array of flows of people, goods and information.

Transportation not only favors economic development but also has an impact on spatial organization. Space shapes transport as much as transport shapes space, which is a salient example of the reciprocity of transport and its geography. This reciprocity can be articulated over two points:

- This relationship concerns the **transport system** itself. Since the transport system is composed of nodes and links as well as the flows they are supporting the spatial organization of this system is a core defining component of the spatial structure. Even if streets are not the city, they are the means that are shaping its organization in terms of locations and relations. The same applies for maritime shipping networks, which are not international trade, but reflect the spatial organization of the global economy.
- This relationship concerns **activities** that are all dependent on transportation at one level or another. Since every single activity is based on a level of mobility, the relationship they have with transportation is reflected in their spatial organization. While a small retail activity is conditioned by local accessibility from which it draws its customers, a large manufacturing plant relies on accessibility to global freight distribution for its inputs as well as its outputs.

The more interdependent an economy is, the more important transportation becomes as a support and a factor shaping this interdependence. The relationship between transport and spatial organization can be considered from three major geographical scales: the global, the regional and the local (Figure 2.6). While the major nodes structuring spatial organization at the global level are gateways mainly supported by port, airport and telecommunication activities, at the local level, employment and commercial activities, which tend to be agglomerated, are the main structuring elements. Each of these scales is also characterized by specific links and relations ranging from locally based commuting to global trade flows.

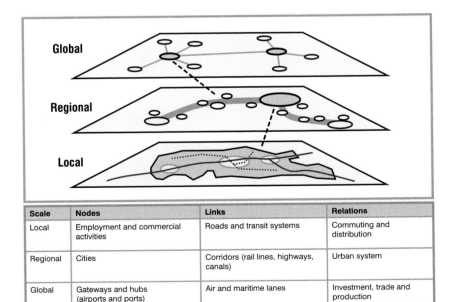

Scale	Nodes	Links	Relations
Local	Employment and commercial activities	Roads and transit systems	Commuting and distribution
Regional	Cities	Corridors (rail lines, highways, canals)	Urban system
Global	Gateways and hubs (airports and ports)	Air and maritime lanes	Investment, trade and production

Figure 2.6 Scales of spatial organization for transportation

Global spatial organization

At the global level, transportation supports and shapes economic specialization and productivity through international trade. Improvements in transport are expanding markets and development opportunities, but not uniformly. The inequalities of the global economy are reflected in its spatial organization and the structure of international transport systems. The patterns of globalization have created a growth in spatial flows (trade) and increased interdependencies. Telecommunications, maritime transport and air transport, because of their scale of service, support the majority of global flows. The nature and spatial structure of these flows can be considered from two major perspectives that seek to explain global differences in growth and accessibility:

- **Core/periphery**. This basic representation assumes that the global spatial organization favors a few core areas that grow faster than the periphery. Differential growth creates acute inequalities in levels of development. Transportation is thus perceived as a factor of polarization and unequal development. From this perspective, parts of the global economy are gaining, because they are more accessible, while others are marginalized and bound to dependency. However, this trend can be reversed if international transport costs are significantly reduced. This is evidenced by the substantial growth of many Pacific Asian countries that have opted for an export-oriented strategy which requires good access to global freight distribution. Consequently, the core/periphery relationship is flexible.
- **Poles**. Transportation is perceived as a factor of articulation in the global economy where the circulation of passengers and freight is regulated by poles corresponding to a high level of accumulation of transport infrastructures, distribution and economic activities. These poles are subject to centrifugal and centripetal forces that have favored geographical concentration of some activities and the dispersion of others. The global economy is thus based on the backbone of freight distribution, which in turn relies on networks established to support its flows and on nodes that are regulating the flows within networks. Networks, particularly those concerning maritime shipping and air transportation, are flexible entities that change with the ebb and flow of commerce while nodes are locations fixed within their own regional geography.

The global spatial organization is a priori conditioned by its nodality. Global flows are handled by gateways and hubs, each of which account for a significant share of the flows of people, freight and information.

Gateway. A location offering accessibility to a large system of circulation of freight and passengers. Gateways reap the advantage of a favorable physical location – such as highway junctions, the confluence of rivers, a good port site – and have been the object of a significant accumulation of transport infrastructures such as terminals and their links. A gateway is commonly an origin, a destination and a point of transit. It generally commands the entrance to and the exit from its catchment area. In other words, it is a pivotal point for the entrance and the exit to a region, a country or a continent and often requires intermodal transfers.

Hub. A central point for the collection, sorting, transhipment and distribution of goods for a particular area. This concept comes from a term used in air transport for passengers as well as for freight and describes collection and distribution through a single point such as the "hub-and-spoke" concept.

Gateways and hubs refer to a rather similar element of the spatial structure of flows. However, the concept of gateway tends to be more restrictive in its definition. While a hub is a central location in a transport system with many inbound and outbound connections of the same mode, a gateway commonly implies a shift from one mode to another (such as maritime/land) (Figure 2.7). Transport corridors are commonly linking gateways to the inland. Gateways also tend to be most stable in time as they often have emerged at the convergence on inland transport systems while the importance of a hub can change if transport companies decide to use another hub, which is common in the airline industry. Thus, a location can at the same time be a gateway and a hub.

Services follow a spatial trend which appears to be increasingly different than of production. As production disperses worldwide to lower cost locations, high level services increasingly concentrate into a relatively few large metropolitan areas, labeled as world cities. They are centers for financial services (banking, insurance), head offices of major multinational corporations and the seats of major governments. Thus, gateways and world cities may not necessarily correspond as locations. This is particularly the case for containerized traffic which is linked with new manufacturing clusters and the usage of intermediary hubs.

Regional spatial organization

Regions are commonly organized along an interdependent set of cities forming what is often referred to as an urban system. The key spatial foundation of an urban system is based on a series of market areas, which are a function of the level of activity of each center in relation with the friction of distance. The spatial structure of most regions can be subdivided into three basic components:

- A set of locations of **specialized industries** such as manufacturing and mining, which tend to group into agglomerations according to location factors such a raw materials, labor, markets, etc. They are often export-oriented industries from which a region derives the bulk of its basic growth.
- A set of **service industry locations**, including administration, finance, retail, wholesale and other similar services, which tend to agglomerate in a system of central places (cities) providing optimal accessibility to labor or potential customers.
- A pattern of **transport nodes and links**, such as road, railways, ports and airports, which services major centers of economic activity.

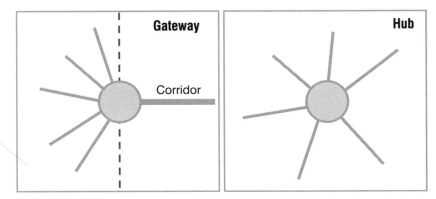

Figure 2.7 Gateway and hub

Jointly, these components define the spatial order of a region, mostly its organization in a hierarchy of relationships involving flows of people, freight and information. More or less well-defined urban systems spatially translate such development. Three conceptual categories of regional spatial organization can be observed:

- **Central places/urban systems** models try to find the relationships between the size, the number and the geographic distribution of cities in a region. Many variations of the regional spatial structure have been investigated by central place theory. The great majority of urban systems have a well-established hierarchy where a few centers dominate. Transportation is particularly important in such a representation as the organization of central places is based on minimizing the friction of distance. The territorial structure depicted by central place theory is the outcome of a region seeking the provision of services in a (transport) cost-effective way.
- **Growth poles** where economic development is the structural change caused by the growth of new propulsive industries that are the poles of growth. The location of these activities is the catalyst of the regional spatial organization. Growth poles first initiate, then diffuse, development. This concept attempts to be a general theory of the initiation and diffusion of development models. Growth gets distributed spatially within a regional urban system, but this process is uneven with the core benefiting first and the periphery eventually becomes integrated in a system of flows. In the growth poles theory transportation is a factor of accessibility which reinforces the importance of poles.
- **Transport corridors** represent an accumulation of flows and infrastructures of various modes and their development is linked with economic, infrastructural and technological processes. When these processes involve urban development, urbanization corridors are a system of cities oriented along an axis, commonly fluvial or a coastline. Corridors are also structured along articulation points that regulate the flows at the local, regional and global levels. Historically, urbanization was mainly organized by the communication capacities offered by fluvial and coastal maritime transportation. Many urban regions such as BosWash (Boston–Washington) or Tokaido (Tokyo–Osaka) share this spatial commonality.

Three geographical models relate to urbanization, transportation and corridors (Figure 2.8):

- The urban system and central places theory mainly considers cities as structurally independent entities that compete over overlapping market areas. Under the **location and accessibility model** (A) an urban region is considered as a hierarchy/order of services and functions and the corridor a structure organizing interactions within this hierarchy. Transport costs are considered a dominant factor in the organization of the spatial structure as the hinterland of each center is the outcome of the consumers' ability to access its range of goods and services. Because of higher levels of accessibility along the corridor, market areas are smaller and the extent of goods and services being offered are broader. Thus, differences in accessibility are the least significant along the corridor.
- The **specialization and interdependency model** (B) considers that some cities can have a level of interaction and that transportation could be not only a factor of market accessibility, but also of regional specialization and of comparative advantages. The Megalopolis concept introduced by Gottmann (1961) acknowledges the creation of large urban corridors structured by transportation infrastructures and terminals maintaining interactions. Accessibility and economies of scale, both in production

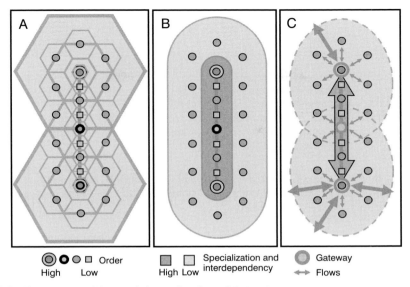

Figure 2.8 Transport corridors and the regional spatial structure

and consumption, are factors supporting the development of such entities where urban areas are increasingly specialized and interdependent. The main assumption is that the accessibility provided by the corridor reinforces territorial specialization and interdependency along its main axis, and consequently the reliance on a regional transport system.

- The **distribution/flow model** (C) is one where a major gateway of an urban region acts as the main interface between global, national and regional systems. Under such a paradigm, three core structural elements define a regional corridor: (1) Gateways regulating freight, passengers and information flows. (2) Transport corridors with a linear accumulation of transport infrastructures servicing a set of gateways. They provide for the physical capacity of distribution. (3) Flows, their spatial structure and the underlying activities of production, circulation and consumption.

Local spatial organization

Although transport is an important element in rural spatial organization, it is at the urban level that transportation has the most significant local spatial impact. Urbanization and transport are interrelated concepts. Every city relies on the need for mobility of passengers (residence, work, purchases and leisure) and freight (consumption goods, food, energy, construction materials and waste disposal) and where the main nodes are employment zones. Urban demographic and spatial evolution is translated in space by the breadth and amplitude of movements. Employment and attraction zones are the most important elements shaping the local urban spatial organization:

- **Employment zones**. The growing dissociation between the workplace and residence is largely due to the success of motorized transport, notably the private automobile. Employment zones located away from residential zones have contributed to an increase in number and length of commuting trips. Before suburbanization, public transit was wholly responsible for commuting. Today, the automobile supports the majority of these trips. This trend is particularly prevalent in highly populated, industrialized and urbanized zones, notably in North America and Western Europe, but motorization is also a dominant trend in developing countries.

- **Attraction zones**. Attraction zones linked to transport modes are areas to which a majority of the population travels for varied reasons such as shopping, professional services, education and leisure. As with central place theory, there is a certain hierarchy of services within an urban area ranging from the central business district offering a wide variety of specialized services to small local centers offering basic services such as groceries and personal banking.

The development of cities is conditioned by transport and several modes, from urban transit to the automobile, have contributed to the creation of urban landscapes. Three distinct phases can be noted:

- The **conventional/classic city**. Constructed for pedestrian interactions and constrained by them, the historic city was compact and limited in size. The emergence of the first public transit systems in the nineteenth century permitted the extension of the city into new neighborhoods. However, pedestrian movements still accounted for the great majority of movements and the local spatial organization remained compact. Many European and Asian cities still have a significant level of compactness today.
- **Suburbanization**. The advent of more efficient public transit systems and later on of the automobile permitted an increased separation of basic urban functions (residential, industrial and commercial) and the resulting spatial specialization. The rapid expansion of urban areas that resulted, especially in North America, created a new spatial organization, less cohesive than before but still relatively adjacent to the existing urban fabric. Although this process started in the early twentieth century, it accelerated after World War II.
- **Exurbanization**. Additional improvements in mobility favored urban expansion in the countryside where urban and rural activities are somewhat intermixed. Many cities became extended metropolitan regions, with a wide array of specialized functions including residential areas, commercial centers, industrial parks, logistics centers, recreational areas and high tech zones. These exurbanization developments have also been called "edge cities".

The automobile has clearly influenced the new spatial organization but other socio-economic factors have also shaped urban development such as gentrification and the increase in land values. The diffusion of the automobile has led to an urban explosion. The car has favored the mobility of individuals thus permitting a disorderly growth and an allocation of space between often conflicting urban functions (residential, industrial, commercial). Transport thus contributes to the local spatial organization; however, it must also adapt to urban morphologies. Transport networks and urban centers complement and condition each other.

Concept 3 – Transport and location

The importance of transport in location

In addition to being a factor of spatial organization, transportation is linked with the location of socio-economic activities, including retail, manufacturing and services. In a market economy, location is the outcome of a constrained choice where many issues are being considered, transportation being one of them. The goal is to find a suitable location that would maximize the economic returns for this activity. There is a long tradition within economic geography in developing location theories with a view to explain and predict the locational logic of economic activities by incorporating market, institutional and behavioral considerations. The majority of location theories have an explicit or implicit role attributed to transport. As there are no absolute rules dictating locational choices, the importance of transport can only be evaluated with varying degrees of accuracy. At best, the following observations concerning transportation modes and terminals and their importance for location can be made:

- **Ports and airports**. Convergence of related activities around terminals, particularly for ports since inland distribution costs tend to be high.
- **Roads and railroads**. A structuring and convergence effect that varies according to the level of accessibility. For rail transport, terminals also have a convergence effect.
- **Telecommunications**. No specific local influence, but the quality of regional and national telecommunication systems tends to ease transactions.

Globalization has been associated with significant changes in business operations and markets. Managing operations in such an environment has become increasingly complex, especially with the territorial extension of production and consumption. Manufacturing strategies tend to use different locations for each component of a product in order to optimize respective comparative advantages. Transport requirements have proportionally increased as well in order to organize the related flows. The requirement of faster long-distance transport services has propelled the importance of air transport, especially for freight. Air terminals have thus become a significant location factor for globally oriented activities, which tend to agglomerate in the vicinity. Additionally, the surge in long-distance trade has put logistical functions, namely transport terminals and distribution centers, at the forefront of locational considerations. Technological changes have also been linked with the relocation of industrial and even service activities. Global telecommunication facilities can favor the outsourcing of several services to lower cost locations, such as call centers in India.

Location factors

The location of economic activities is a priori dependent on the nature of the activity itself and on certain location factors such as the attributes of the site, the level of accessibility and the socio-economic environment. Location factors can be subdivided into three general functional categories (Figure 2.9):

- **Site**. Specific micro-geographical characteristics of the site, including the availability of land, basic utilities, the visibility (prestige), amenities (quality of life) and the nature and level of access to local transportation (such as the proximity to a highway). These factors have an important effect on the costs associated with a location.
- **Accessibility**. Includes a number of opportunity factors related to a location, mainly labor (wages, availability, level of qualification), materials (mainly for raw materials-dependent activities), energy, markets (local, regional and global) and accessibility to suppliers and customers (important for intermediate activities). These factors tend to have a meso (regional) connotation.

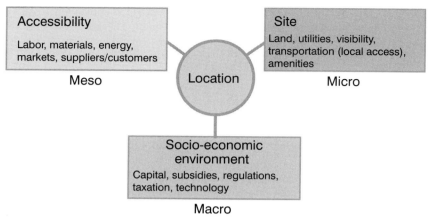

Figure 2.9 Basic location factors

- **Socio-economic environment**. Specific macro-geographical characteristics that tend to apply to jurisdictional units (nation, region, locality). They consider the availability of capital (investment, venture), varied subsidies, regulations, taxation and technology.

The role and importance of each factor depends on the nature of the activity whose locational behavior is being investigated. Although each type of economic activity has its own set of location factors, some general factors can be identified by major economic sector:

- **Primary economic activities**. Their dominant location factor is related to environmental endowments, such as natural resources. For instance, mining takes place where economically recoverable mineral deposits are found and agriculture is subject to environmental constraints such as soil fertility, precipitation and temperature. Primary activities are thus characterized by the most basic location factors but have a strong reliance on transportation since their locations rarely are close to centers of demand. Substantial investments in extraction and distribution infrastructures must thus be made before resources can be brought to markets. The capacity to transport raw materials plays a significant role in the possible development of extractive activities at a location.
- **Secondary economic activities**. Imply a complex web of location factors which, depending upon the industrial sector, relate to labor (cost and/or skill level), energy costs, capital, land, markets and/or proximity of suppliers. Location is thus an important cost factor (cost minimization). Considering the wide variety of industrial and manufacturing activities, understanding the rationale of each sector is a difficult task that has been subject to many investigations in economic geography. Globalization and recent developments in supply chain management and global production networks have made the situation even more complex with the presence of many intermediaries and significant locational changes.

- **Tertiary economic activities**. Involve activities that are most bound to market proximity, since the capacity to sell a product or service is the most important location requirement. As many of these activities are retail-oriented, consumer proximity (as well as their level of income) is essential and is directly related to sale levels. The main focus is to maximize sales revenues. Location is thus an important revenue factor (revenue maximization). The retail industry has significantly changed with the emergence of large retail stores that maximize sales through economies of scale and local accessibility. E-commerce also provides a new dynamic where information can easily be traded and where niche retailing markets can be developed in a situation of high product diversity.
- **Quaternary economic activities**. Imply activities not linked to environmental endowments or access to a market, but to high level services (banking, insurance), education, research and development; dominantly the high technology sector. With improvements in telecommunications, many of these activities can be located almost anywhere as demonstrated by the recent trend to locate call centers offshore. There are still some strong locational requirements for high technology activities that include proximity to large universities and research centers and to a pool of highly qualified workers (as well as cheap labor for supporting services), availability of venture capital, a high quality of life and access to excellent transportation and telecommunication facilities. However, as telecommunication infrastructures are becoming globally ubiquitous and accessible, such proximity is of lesser importance.

Each of these sectors thus has its own criteria, which vary in time and space. However, basic location strategies appear to be a cost minimization or a revenue maximization endeavor. Understanding location factors enables a better overview of the dynamics of the global economy and the associated territorial changes at the global, regional and local levels.

Accessibility and location

Since accessibility is dominantly the outcome of transportation activities, namely the capacity of infrastructures to support mobility, it presents the most significant influence of transportation on location. Hence, it appears that location (accessibility) and economic activities are intimately linked. Accessibility plays an important role by offering more customers through an expanded market area, by making distribution more efficient (in terms of costs and time) or by enabling more people to reach workplaces. While some transport systems have favored the dispersion of socio-economic activities (e.g. automobiles and suburbanization), others have favored their concentration (e.g. container terminals). All systems are bearers of spatial specialization and configuration. Among the main configuration forces are:

- **Transportation costs**. Refer to the benefits of a location that minimizes transport costs either for passengers or freight. This is at the core of classic industrial location theories where transport-dependent activities seek to minimize total transport costs. With the expansion of transport infrastructures, shifts in manufacturing, new economic activities such as high technology, logistical management and an overall decline in transport costs, cost minimization is no longer a substantial consideration in location. However, transport costs cannot be easily dismissed and must be considered in a wider context where the quality and reliability of transport is of growing importance. It has been demonstrated that travel time, instead of distance, is the determining factor behind commuting ranges, a notion that increasingly applies to freight distribution.

- **Agglomeration economies**. Refer to the benefits of having activities locate (cluster) next to one another, such as the use of common infrastructures and services. Clustering continues to be a powerful force in location as the reduction in transport costs favored the agglomeration of retail, manufacturing and distribution activities at specific locations. For instance, shopping malls are based on agglomeration economies, offering customers a wide variety of goods and services in a single location. Distribution activities, even unrelated, have also a tendency to cluster. The development of special economic zones, many export-oriented, also benefit from the clustering effect.
- **Economies of density**. Somewhat related to economies of agglomeration, but focus on spatial coverage and proximity. For instance, a retailer can achieve several types of cost savings by locating its stores in proximity to one another. Such a structure reduces logistics and delivery costs by sharing a distribution center. Other advantages may include the possibility to relocate part of the workforce between nearby facilities and having shared advertising. In such a circumstance, the locational strategies are based on proximity to existing facilities, even if this implies the selection of sub-optimal locations.

Because of the level of accessibility they provide, new transport infrastructures influence the setting of economic activities. It becomes a particularly strong effect when new infrastructure are added to an undeveloped (or underdeveloped) site and thus locational decisions tend to be simpler and unhindered by the existing spatial structure. The locational effects on activities are not always automatic or evident. They are important however when infrastructure is accompanied by social, economic and urban transformations of space. New infrastructures therefore play a catalytic role, because they are capable of transforming space.

Concept 4 – Future transportation

Past trends and uncertain future

Where are the flying cars? Where are the supersonic passenger jets?

In 200 years since the beginning of mechanized transportation, the capacity, speed, efficiency and geographical coverage of transport systems has improved dramatically. These processes can be summarized as follows:

- Each mode, due to its geographical and technical specificities, was characterized by **different technologies** and **different rates of innovation and diffusion**. A transport innovation can thus be an additive/competitive force where a new technology expands or makes an existing mode more efficient and competitive. It can also be a destructive force when a new technology marks the obsolescence and the demise of an existing mode often through a paradigm shift.
- Technological innovation was linked with **faster and more efficient transport systems**. This process implies a space–time convergence where a greater amount of space can be exchanged with lesser amount of time. The comparative advantages of space can thus be more efficiently used.
- Technological evolution in the transport sector has been linked with the **phases of economic development** of the world economy. Transportation and economic development are consequently interlinked as one cannot occur without the other.

The growth of transport systems, as the case of the United States exemplifies, went through a series of waves of introduction, growth, maturity and decline as massive investments in infrastructures and development of the system took place (Figure 2.10). Each time there is a substitution from one mode to another, moving to a higher level of speed (and sometimes efficiency). A paradigm shift represents an event that marks the prominence of transport systems, often characterized by the completion of a significant infrastructure project which starts to impact economic and spatial systems. A peak year is when the system is about to reach maturity and experience a slowdown in its growth.

The development of the canal system lasted 30 years (ΔT, time for the transport system to grow from 10 percent to 90 percent of its full extent) and peaked around 1836. Its paradigm shift took place in 1825 with the opening of the Erie Canal and its maturity and decline was caused by the emergence of a more flexible and efficient inland transport system: rail. The growth of rail in the later part of the nineteenth century took off with the completion of the transcontinental railway in 1869 marking its paradigm shift. By the late nineteenth and early twentieth centuries most of the American territory was serviced by rail. Road transportation emerged in the beginning of the twentieth century, especially after the introduction of the Model T in 1913. The growth of the road transport system marked the maturity and downfall of rail trans- port, especially for short and medium distances. Surprisingly, the development of the Interstate highway system marked the maturity of the road transport system as national trade was increasingly taking place along major high-capacity road corridors between major metropolitan areas, lessening the need for regional road construction. The latest

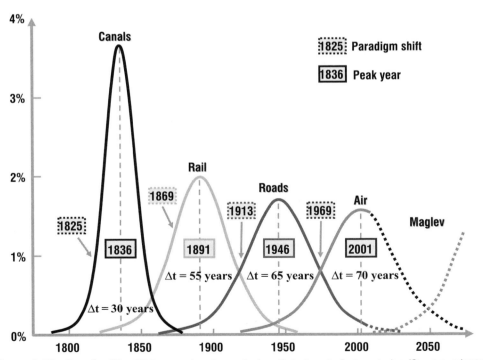

Figure 2.10 Growth of the US transport system, nineteenth to twenty-first centuries (Source: adapted from Ausubel et al., 1998)

wave of development is related to air transportation, which peaked around 2001 and is expected to last about 70 years. A key event that marked the dominance of air transport was the introduction of the Boeing 747 in 1969 which opened air travel to the masses. The next transport technology is likely to be Maglev (magnetic levitation), but little is known about its potential impacts and which event would trigger a paradigm shift.

One of the pitfalls in discussing future trends is looking at the future as an extrapolation of the past. It is assumed that the future will involve a technology that already exists, but simply operating on an extended scale beyond what is currently possible. The parameters of such an extrapolation commonly involve a greater speed, mass availability, a higher capacity and/or better accessibility, all of which imply similar or lower costs. Popular literature (such as popular mechanics or popular science) of the first half of the twentieth century is abundant with extrapolations and speculations, some spectacular, about what transportation technology would look like in the (their) "future".

A common failure about predictions is their incapacity at anticipating the paradigm shifts brought by new technologies as well as economic and social conditions. Another failure relates to the expectation of a massive diffusion of a new technology with profound economic and social impacts, and this over a short period of time (the "silver bullet effect"). This rarely takes place as most innovations go through a cycle of introduction, adoption, growth, peak and then obsolescence, which can take several years, if not decades. Even in the telecommunications sector, which accounts for the fastest diffusion levels, the adoption of a technology takes place over a decade. Any discussion about the future of transportation must start with the realization that much of what is being presented as plausible is unlikely to become a reality, more so if the extrapolation goes several decades into the future. Thus, as much as someone would have been unable at the beginning of the twentieth century to even dream of what transportation would look like half a century later (e.g. air transportation and the automobile), we may be facing the same limitations at the beginning of the twenty-first century. However, since substantial technological innovations took place in the twentieth century, we are likely better placed to evaluate which technological trends will emerge in the near future.

Technological trends

Since the introduction of commercial jet planes, high-speed train networks and the container in the late 1960s, no significant technological changes have impacted passengers and freight transport systems. The early twenty-first century is an era of car and truck dependency, which tends to constrain the development of alternative modes of transportation, as most of the technical improvements aim at ensuring the dominance of oil as a source of energy. However, with dwindling oil reserves, the end of the dominance of the internal combustion engine is approaching. As oil production is expected to peak by 2008–10 and then decline, energy prices are expected to soar, triggering the most important technological transition in transportation since the automobile. In such an environment the most promising technologies are:

- **Automated transport systems**. Refers to a set of alternatives to improve the speed, efficiency, safety and reliability of movements, by relying upon complete or partial automation of vehicles, transhipment and control. These systems could involve the improvement of existing modes such as automated highway systems, and the creation of new modes and new transhipment systems such as for public transit and freight transportation. The goal of such initiatives is mainly to efficiently use existing infrastructures through information technologies. Much gains still remain to be achieved through the management of existing infrastructures and vehicles.

- **Alternative modes**. There is a range of modes that could replace but will more likely complement existing modes, particularly for passenger transportation. One such technology is maglev, short for magnetic levitation, which has the advantage of having no friction with its support and no moving parts, enabling to reach operational speeds of 500–600 km per hour (higher speeds are possible if the train circulates in a low pressure tube). This represents an alternative for passengers and freight land movements in the range of 75 to 1,000 km. Maglev improves from the existing technology of high-speed train networks which are limited to speeds of 300 km per hour. In fact, maglev is the first fundamental innovation in railway transportation since the industrial revolution. The first commercial maglev system opened in Shanghai in 2003 and has an operational speed of about 440 km per hour.

- **Alternative fuels**. This mainly concerns existing modes but the sources of fuel, or the engine technology, are modified. For instance, hybrid vehicles involve the use of two types of motor technologies, commonly an internal combustion engine and an electric motor. Simplistically, breaking is used to recharge a battery, which then can be used to power the electric motor. Although the gasoline appears to be the most prevalent fuel choice, diesel has a high potential since it can also be made from coal or organic fuels. Diesel as a fuel can thus be part of a lower petroleum dependency energy strategy. Hybrid engines have often been perceived as a transitional technology to cope with higher energy prices. There is also a possibility of greater reliance on biofuels as an additive (and possibly a supplement) to petroleum, but their impacts on food production must be carefully assessed. Far more reaching in terms of energy transition are fuel cells, which involve an electric generator using the catalytic conversion of hydrogen and oxygen. The electricity generated can be used for many purposes, such as supplying an electric motor. Current technological prospects do not foresee high output fuel cells, indicating they are applicable only to light vehicles, notably cars, or to small power systems. Nevertheless, fuel cells represent a low environmental impact alternative to generate energy and fuel cell cars are expected to reach mass production by 2010. Additional challenges in the use of fuel cells involve hydrogen storage (especially in a vehicle) as well as establishing a distribution system to supply consumers.

Still, anticipating future transport trends is very hazardous since technology is a factor that historically created paradigm shifts and is likely to do so again with unforeseen consequences. For instance, one of the major concerns about future transportation for London, England, in the late nineteenth century, was that by the mid-twentieth century the amount of horse manure generated by transport activities would become unmanageable.

Economic and regulatory trends

Future transportation systems face concerns related to energy, the environment, safety and security. They are either going to be developed to accommodate additional demands for mobility or to offer alternatives (or a transition) to existing demand. An important challenge relies on the balance between market forces and public policy, as both have a role to play in the transition.

A fundamental component of future transport systems, freight and passenger alike, is that they must provide increased flexibility and adaptability to changing market circumstances (origins, destinations, costs, speed, etc.) while complying with an array of environmental, safety and security regulations. This cannot be effectively planned and governments have consistently been poor managers and slow to understand technological changes, often impeding them through regulations and preferences to specific modes or to specific technologies. For instance, because of biofuel policies aimed at ethanol production using corn, the unintended consequence was a surge on global food prices as more agricultural land was devoted for energy production instead of food production.

It is thus likely that future transport systems will be the outcome of private initiatives with the market (transport demand) the ultimate judge about the true potential of a new transport technology. Economic history has shown that the market will always try to find and adopt the most efficient form of transportation available. Some transport systems or technologies have become obsolete and have been replaced by others that are more efficient and cost-effective based upon the prevailing input conditions such as labor, energy and commodities. This fundamental behavior is likely to endure in future transportation systems.

Method 1 – The notion of accessibility

Definition

Accessibility is a key element to transport geography, and to geography in general, since it is a direct expression of mobility either in terms of people, freight or information. Well-developed and efficient transportation systems offer high levels of accessibility (if the impacts of congestion are excluded), while less-developed ones have lower levels of accessibility. Thus accessibility is linked with an array of economic and social opportunities.

> **Accessibility** is defined as the measure of the capacity of a location to be reached by, or to reach different locations. Therefore, the capacity and the structure of transport infrastructure are key elements in the determination of accessibility.

All places are not equal because some are more accessible than others, which implies inequalities. The notion of accessibility consequently relies on two core concepts:

- The first is **location** where the relativity of places is estimated in relation to transport infrastructures, since they offer the means to support movements.
- The second is **distance**, which is derived from the connectivity between locations. Connectivity can only exist when there is a possibility to link two locations through transportation. It expresses the friction of space (or deterrence) and the location which has the least friction relative to others is likely to be the most accessible. Commonly, distance is expressed in units such as in kilometers or in time, but variables such as cost or energy spent can also be used.

There are two spatial categories applicable to accessibility problems, which are interdependent:

- The first type is known as **topological accessibility** and is related to measuring accessibility in a system of nodes and paths (a transportation network). It is assumed that accessibility is a measurable attribute significant only to specific elements of a transportation system, such as terminals (airports, ports or subway stations).

- The second type is known as **contiguous accessibility** and involves measuring accessibility over a surface. Under such conditions, accessibility is a measurable attribute of every location, as space is considered in a contiguous manner.

Last, accessibility is a good indicator of the underlying spatial structure since it takes into consideration location as well as the inequality conferred by distance (Figure 2.11). Due to different spatial structures, two different locations of the same importance will have different accessibilities. On example A of Figure 2.11 representing a spatial structure where locations are uniformly distributed, locations 1 and 2 have different accessibilities, with location 1 being the most accessible. As distance (Euclidean) increases, location 1 has access to a larger number of locations than location 2. To access all locations, location 2 would require a longer traveled distance (roughly twice) than location 1. This is particularly the case as the spatial structure changes to one concentrated around location 1 (example B). In this case, the number of locations that can be reached by location 1 climbs rapidly and then eventually peaks. The third example (C) has a spatial structure with roughly two foci. Although the number of locations that can be reached from location 2 initially climbs faster than for location 1, location 1 catches up and is actually the most accessible, but by a lesser margin.

Connectivity

The most basic measure of accessibility involves network connectivity where a network is represented as a connectivity matrix ($C1$), which expresses the connectivity

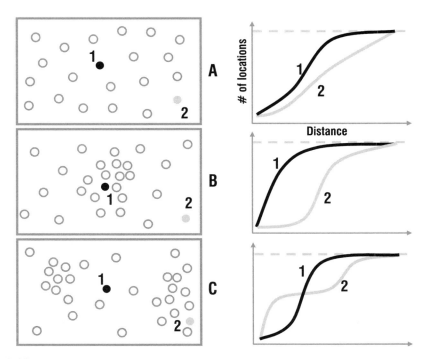

Figure 2.11 Accessibility and spatial structure

of each node with its adjacent nodes. The number of columns and rows in this matrix is equal to the number of nodes in the network and a value of 1 is given for each cell where there is a connected pair and a value of 0 for each cell where there is an unconnected pair. The summation of this matrix provides a very basic measure of accessibility, also known as the degree of a node:

$$C1 = \sum_{j}^{n} c_{ij}$$

- $C1$ = degree of a node.
- c_{ij} = connectivity between node i and node j (either 1 or 0).
- n = number of nodes.

The network on Figure 2.12 can be represented as a connectivity matrix, which is rather simple to construct. The size of the connectivity matrix involves a number of rows and cells equivalent to the number of nodes in the network. Since the network on Figure 2.12 has five nodes, its connectivity matrix is a five-by-five grid. Each cell representing a connection between two nodes receives a value of 1 (e.g. cell B–A). Each cell that does not represent a connection gets a value of 0 (e.g. cell D–E). If all connections in the network are bi-directional (a movement is possible from node C to node D and vice versa), the connectivity matrix is transposable. Adding up a row or a column gives the degree of a node. Node C is obviously the most connected since it has the highest summation of connectivity compared to all other nodes. However, this assumption may not hold true on a more complex network because of a larger number of indirect paths which are not considered in the connectivity matrix. The connectivity matrix does not take into account all the possible indirect paths between nodes. Under such circumstances, two nodes could have the same degree, but may have different accessibilities.

Geographic and potential accessibility

From the accessibility measure developed so far, it is possible to derive two simple and highly practical measures, defined as geographic and potential accessibility. Geographic accessibility considers that the accessibility of a location is the summation of all distances between other locations divided by the number of locations. The lower its value, the more a location is accessible.

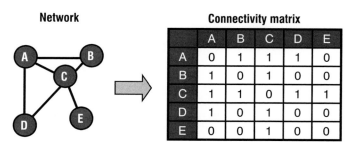

Figure 2.12 Connectivity matrix

$$A(G) = \sum_{i}^{n} \left(\sum_{j}^{n} d_{ij} \right) \bigg/ n$$

- $A(G)$ = geographical accessibility matrix.
- d_{ij} = shortest path distance between location i and j.
- n = number of locations.

In this measure of accessibility, the most accessible place has the lowest summation of distances. As shown on Figure 2.13, the construction of a geographic accessibility matrix, $A(G)$, is a rather simple undertaking. First, build a matrix containing the shortest distance between the nodes (node A to node E), here labeled as the L matrix. Second, build the geographic accessibility matrix $A(G)$ with the summation of rows and columns divided by the number of locations in the network. The summation values are the same for columns and rows since this is a transposable matrix. The most accessible place is node C, since it has the lowest summation of distances.

Although geographic accessibility can be solved using a spreadsheet (or manually for simpler problems), geographic information systems have proven to be a very useful and flexible tool to measure accessibility, notably over a surface simplified as a matrix (raster representation). This can be done by generating a distance grid for each place and then summing all the grids to form the total summation of distances (Shimbel) grid. The cell having the lowest value is thus the most accessible place.

Potential accessibility is a more complex measure than geographic accessibility, since it includes simultaneously the concept of distance weighted by the attributes of a place. All places are not equal and thus some are more important than others. Potential accessibility can be measured as follows:

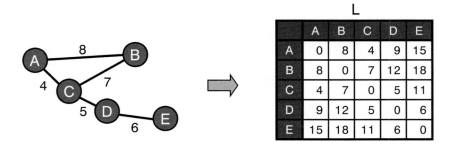

Figure 2.13 Geographic accessibility

$$A(P) = \sum_{i}^{n} P_i + \sum_{j}^{n} P_j/d_{ij}$$

- $A(P)$ = potential accessibility matrix.
- d_{ij} = distance between place i and j (derived from valued graph matrix).
- P_j = attributes of place j, such as its population, retailing surface, parking spaces, etc.
- n = number of locations.

The potential accessibility matrix is not transposable since locations do not have the same attributes, which brings the underlying notions of emissiveness and attractiveness:

- **Emissiveness** is the capacity to leave a location, the sum of the values of a row in the $A(P)$ matrix.
- **Attractiveness** is the capacity to reach a location, the sum of the values of a column in the $A(P)$ matrix.

By considering the same shortest distance matrix (L) as on Figure 2.13 and the population matrix P, the potential accessibility matrix, $P(G)$, can be calculated (Figure 2.14). The value of all corresponding cells (A–A, B–B, etc.) equals the value of their respective attributes (P). The value of all non-corresponding cells equals their attribute divided by the corresponding cell in the L matrix. The higher the value, the more a location is accessible, node C being the most accessible. The matrix being non-transposable, the summation of rows is different from the summation of columns, bringing forward the issue of attractiveness and emissiveness. Node C has more emissiveness than attractiveness (2525.7 versus 2121.3), while node B has more attractiveness than emissiveness (1358.7 versus 1266.1). Likewise, a geographic information system can be used to measure potential accessibility, notably over a surface.

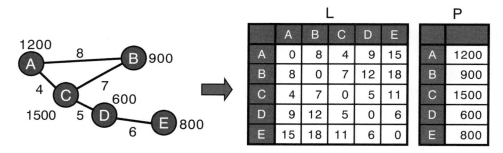

L

	A	B	C	D	E
A	0	8	4	9	15
B	8	0	7	12	18
C	4	7	0	5	11
D	9	12	5	0	6
E	15	18	11	6	0

P

A	1200
B	900
C	1500
D	600
E	800

P (G)

i\j	A	B	C	D	E	Σi
A	1200.0	150.0	300.0	133.3	80.0	1863.3
B	112.5	900.0	128.6	75.0	50.0	1266.1
C	375.0	214.3	1500.0	300.0	136.4	2525.7
D	66.6	50.0	120.0	600.0	100.0	936.6
E	53.3	44.4	72.7	133.3	800.0	1103.7
Σj	1807.4	1358.7	2121.3	1241.6	1166.4	7695.4

Figure 2.14 Potential accessibility

Method 2 – Network data models

Nature and utility

Graph theory gives a topological and mathematical representation of the nature and structure of transportation networks. However, graph theory can be expanded for the analysis of real world transport networks by encoding them in an information system. In the process, a digital representation of the network is created, which can then be used for a variety of purposes such as managing deliveries or planning the construction of transport infrastructure. This digital representation is highly complex, since transportation data is often multimodal, can span several local, national and international jurisdictions and has different logical views depending on the particular user. In addition, while transport infrastructures are relatively stable components, vehicles are very dynamic elements.

It is thus becoming increasingly relevant to use a data model where a transportation network can be encoded, stored, retrieved, modified, analyzed and displayed. Obviously, geographic information systems have received a lot of attention over this issue since they are among the best tools to store and use network data models. Network data models are an implicit part of many GIS, if not an entire GIS package of its own. There are four basic application areas of network data models:

- **Topology**. The core purpose of a network data model is to provide an accurate representation of a network as a set of links and nodes. Topology is the arrangement of nodes and links in a network. Of particular relevance are the representations of location, direction and connectivity. Even if graph theory aims at the abstraction of transportation networks, the topology of a network data model should be as close as possible to the real world structure it represents. This is especially true for the usage of network data models in a GIS.

Figure 2.15 represents the basic topology of an urban transport network composed of linked nodes. It has been encoded into a network data model to represent the

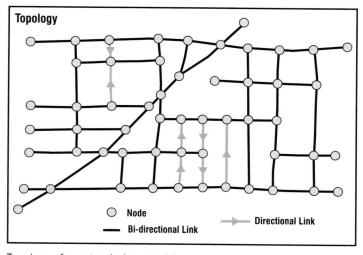

Figure 2.15 Topology of a network data model

reality as close as possible, both topologically and geographically. Topologically, each node has been encoded with the connectivity it permits, such as if a left turn is possible or not (although this attribute is not displayed here). Further, a direction has been encoded in each link (directional or bi-directional) to represent one-ways. Geographically, each node is located at a coordinate which matches, within a tolerated accuracy, the actual intersection it represents. In addition, the links between each node have been decomposed into several segments (not implicitly shown) to respect the positional accuracy of the road they represent.

● **Cartography**. Allows the visualization of a transport network for the purpose of reckoning and simple navigation and serves to indicate the existence of a network. Different elements of the network can have a symbolism defined by some of their attributes. For instance, a highway link may be symbolized as a thick line with a label such as its number, while a street may be symbolized as an unlabeled simple line. The symbolized network can also be combined with other features such as landmarks to provide a better level of orientation to the user. This is commonly the case for road maps used by the general public.

By using attributes encoded in the network data model, such as road type, each segment can be displayed to reflect its importance. For instance, the cartographic representation of a network data model on Figure 2.16 displays three road classes (highway, main street and street) differently. Descriptive labels for the most important elements and directional signs for one-ways have also been added. To enrich the cartographic message, additional layers of information have been added, namely landmarks (City Hall, Central Park and a college campus). Nodal attributes can also have a cartographic utility, such as displaying whether an intersection has traffic lights.

● **Geocoding**. Transportation network models can be used to derive a precise location, notably through a linear referencing system. For instance, the great majority of addresses are defined according to a number and a street. If address information is embedded

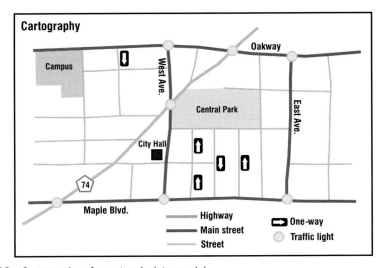

Figure 2.16 Cartography of a network data model

in the attributes of a network data model, it becomes possible to use this network for geocoding and pinpoint the location of an address, or any location along the network, with reasonable accuracy.

Geocoding is possible if a linear referencing system is embedded in a network data model. One of the most common linear referencing systems is the address system, where each link has a corresponding street name and address range. The address range of Figure 2.17 illustrates even (right side) and odd (left side) addresses, very common attributes in most network data models such as TIGER (Topologically Integrated Geographic Encoding and Referencing; developed by the US Census Bureau). For instance, finding the approximate location of the address "197 East Ave." would first imply querying the network data model to find all the links that have "East Ave." as a name attribute. Then, the appropriate address range is found and the location interpolated. "197" corresponds to the 191–209 address range, located on the left side of East Ave. Its approximate location would be at 1/3 [1 − (209 − 197)/ (209 − 191)] of the length of the link that has the 191–209 address range. The same procedure can be applied to the "188 East Ave." address, which in this case would be located at 1/4 of the length of the link that has the 172–210 address range.

- **Routing and assignment.** Network data models may be used to find optimal paths and assign flows with capacity constraints in a network. While routing is concerned by the specific behavior of a limited number of vehicles, traffic assignment is mainly concerned by the system-wide behavior of traffic in a transport network. This requires a topology in which the relationship of each link with other intersecting segments is explicitly specified. Impedance measures (e.g. distance) are also attributed to each link and will have an impact on the chosen path or on how flows are assigned in the network. Routing and traffic assignment at the continental level is generally simple since small variations in impedance are of limited consequences. Routing and traffic assignment in an urban area is much more complex as it must consider stop signs, traffic lights and congestion in determining the impedance of a route.

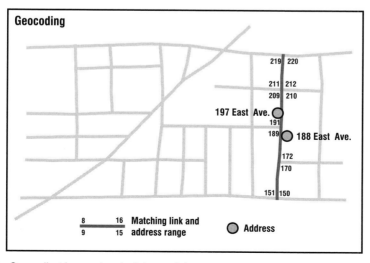

Figure 2.17 Geocoding in a network data model

Routing in a network data model can be simulated if impedance is attributed to links and nodes. For links, impedance is often characterized by travel time, while turn penalties are often used to characterize impedance at nodes, which is how difficult (if possible) it is to turn in one direction, as opposed to another. The network in Figure 2.18 represents a typical routing "traveling salesperson" type of problem. Starting and ending at a warehouse, a delivery truck has a set of deliveries and pickups to perform. The locations of those pickup and delivery points could have been derived from address matching (geocoding). Considering link and node (turn penalties) impedance attributes that are encoded in the network data model, it is possible to plot an optimal route minimizing travel time that would satisfy basic constraints related to the start and end points, pickup and delivery points, as well as link and turn penalty impedances.

Basic representation

Constructing the geometry of a network depends on the mode and the scale being investigated. For urban road networks, information can be extracted from aerial photographs or topographic maps. Air transport networks are derived from airport locations (nodes) and scheduled flights between them (links). Two fundamental tables are required in the basic representation of a network data model that can be stored in a database:

- **Node table**. This table contains at least three fields: one to store a unique identifier and the others to store the node's X and Y coordinates. Although these coordinates can be defined by any Cartesian reference system, longitudes and latitudes would ensure an easy portability to a GIS.
- **Link table**. This table also contains at least three fields: one to store a unique identifier, one to store the node of origin and one to store the node of destination. A fourth field can be used to state if the link is unidirectional or not.

Once those two tables are relationally linked, a basic network topology can be constructed and all the indexes and measures of graph theory can be calculated. Attributes

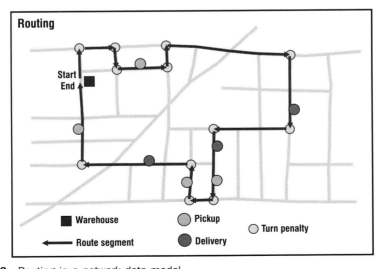

Figure 2.18 Routing in a network data model

such as the connectivity and the Shimbel matrix can also easily be derived from the link table. This basic representation enables to define the topology of networks as structured by graph theory.

A network can be represented by using two tables, one defining nodes and the other defining links (Figure 2.19). The three core elements (fields) of a nodes table are a unique identifier and locational attributes in a coordinate system, such as latitude and longitude values. On Figure 2.19, coordinates are in decimal degrees, meaning that the location of these nodes can directly be imported to a GIS. Additional attributes can also be included in this table.

The links table has four core elements (fields). The first one is a unique identifier for each link, the next two are the nodes of origin and destination of the link and the fourth is a directional tag indicating if the link is unidirectional or not. Another alternative would be to assume that all links are unidirectional and define each of them implicitly. This would require the addition of three new records if the directional tag field is not used (C–D, E–D and E–C). However, this would involve serious redundancies on a complex network. As for the nodes table, additional attributes can be included, such as name, number of lanes, maximum speed, etc.

Both the nodes and links tables have little value if they are considered individually, as a network is the combination of the information contained on both tables. A way to combine these tables is by building a relational join between them. In Figure 2.19, a relational join can be established between the [From] and [To] fields of the links table with the [ID] field of the nodes table. The resulting relational database contains the basic topological elements of the network.

Many efforts have been made to create comprehensive transportation network databases to address a wide variety of transportation problems ranging from public

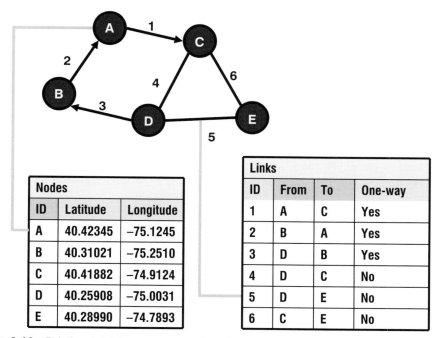

Figure 2.19 Relational database representation of a simple network

transit to package distribution. Initially, these efforts were undertaken within transportation network optimization packages (e.g. EMME/2, TransCAD) which created topologically sound representations. Many of these representations were however geographically inaccurate and had limited visual and geocoding capabilities. Using a network data model for the purposes of cartography, geocoding and routing requires further developments.

Layer-based approach

Most conventional GIS data models separate information in layers, each representing a different class of geographical elements symbolized as points, lines and polygons in the majority of cases. As such, a network data model must be constructed with the limitation of having points and lines in two separate layers; thus the layer-based approach. Further, an important requirement is that the geometry of the network matches the reality as closely as possible since these networks are often part of a geographic information system where an accurate location and visualization is a requisite. This has commonly resulted in the fragmentation of each logical link into a multitude of segments, with most of the nodes of these segments mere intermediate cosmetic elements. The topology of such network data models is not well defined, and has to be inferred. However, these network data models benefit from the attribute linking capabilities of the spatial database models they are derived from. Among the most significant attributes that can be attached to network layers are:

- **Classification and labeling**. Each segment can be classified into categories such as its function (street, highway, railway, etc.), importance (number of lanes) and type (paved, non-paved). Also, a complex labeling structure can be established with prefixes, proper names and suffixes.
- **Linear referencing system**. Several systems to locate elements along a segment have been established. One of the most common is the address system where each segment is provided with an address range. Through linear interpolation, a specific location can be derived (geocoding).
- **Segment travel costs**. Can consider a vast array of impedance measures. Among the most common is the length of the segment, a typical travel time or a speed limit. Congestion can also be assessed, either as a specific value of impedance or as a mathematical function.
- **Direction**. To avoid unnecessary and often unrealistic duplication of links, especially at the street level, a directional attribute can be included in the attribute table.
- **Overcrossing and undercrossing**. Since the great majority of layer-based network models are planar, they are ill-designed to deal with non-planar representations. A provision must be made in the attribute table to identify segments that are overcrossing or undercrossing a segment they intersect with.
- **Turn penalties**. An important attribute to ensure accurate routing within a network. Each intersection has different turn constraints and possibilities. Commonly in road transportation, a right turn is assumed to have a lesser penalty than a left turn.

The TIGER model is a notable example of a layer-based structure which has been widely accepted. TIGER was developed by the US Census Bureau to store street information constructed for the 1990 census. It contains complete geographic coordinates in a line-based structure. The most important attributes include street name and address information, offering an efficient linear referencing system for geocoding. The layer-based approach is consequently good to solve the cartography and geocoding issues.

However, it is ill-suited to comprehensively address routing and assignment transport problems.

Object-oriented approach

The object-oriented approach represents the latest development in spatial data models. It assumes that each geographical feature is an object having a set of properties and a set of relationships with other objects. As such, a transportation network is an object composed of other objects, namely nodes and links. Since topology is one of the core concepts defining transportation networks, relationships expressing it are embedded in object-oriented representations. The basic elements of an object-oriented transportation network data model are:

- **Classes**. They categorize objects in a specific taxonomy, which has a proper set of properties and relationships. The two basic classes of a network are obviously nodes and links, but each of these classes can be subdivided into subclasses. For instance, a link can be subdivided as a road link, a rail link, a walkway, etc.
- **Properties**. They refer to a set of measurable characteristics that are associated with a specific class. For instance, the properties of a road class could be its length, number of lanes, name, surface, speed limit, etc.
- **Relationships**. They describe the type of logical relations objects have with one another. Instance (is–a) and membership (is–in) are among the most common relations. For example, a street is an instance of the road class, which itself is an instance of a transport infrastructure. A specific road segment can be considered part of a specific transport system through a membership relation. From these relations inheritance can be derived, where the characteristics of one object can be passed to another. Using the previous example, it is logical to derive that a street is a transport infrastructure, thus the object street inherits the properties of the object transport infrastructure.

By their structure, especially with their embedded topology, an object-oriented transport network data model would be effective to solve the routing issue in transport. However, object-oriented data models are still in the design phase with proposals such as UNETRANS (Unified NEtwork-TRANSportation data model) hoping to become accepted standards. The potential of the object-oriented approach for GIS remains to be seen as well as the amount of effort required to convert or adapt existing transport network databases, which are mainly layer-based, into the new representational structure.

CASE STUDY UPS and the management of distribution networks

The company

United Parcel Service (UPS) is an enterprise specializing in the collection and the routing of parcels throughout the world. It represents an excellent example of a corporation actively involved using network strategies to manage its distribution system. In 2007, UPS generated incomes around $50 billion and employed 425,000 people, 358,000 of them in the United States. Its service area covers 200 nations and it handles 4.0 billion parcels per year; around 15.8 million per day, of which 2 million are carried by air transport. It is estimated that UPS delivers more than

6 percent of the American gross domestic product and 2 percent of global GDP each and every day.

The infrastructures of UPS are extensive and include 2,400 distribution centers, 93,000 vehicles and a fleet of airplanes, making it the second largest freight airline in the world and the ninth largest airline in terms of revenue. UPS has also an extensive information system specifically adapted to the needs of parcel collection.

UPS was established in 1907, in Seattle, under the name American Messenger Co., to support the need for private messenger and delivery services. Since phones and vehicles were not as common as they are today, message couriers were quite useful for an urban population mainly walking or using crowded public transit. The company started as an enterprise specializing in the routing of parcels from department stores. One of the main factors that explain the success of the enterprise is the early adoption of a management strategy based on the consolidation of freight. It implies the combining of packages addressed to a certain neighborhood onto one delivery vehicle to optimize transport costs.

By the 1930s, the company expanded to Oakland and then California and took the name it is known by today. It inaugurated United Air Express, offering package air delivery throughout the West Coast. The consolidation system was still the key infrastructure for efficient delivery. This service was also expanded to New York City area, as UPS's service was still mainly intra-urban. From the 1940s to the 1960s, many elements favored the growth of the company: the shortage of fuel and rubber, caused by World War II, considerably reduced the usage of personal cars. The post-World War II expansion of suburbs in many metropolitan areas, where people needed extra delivery services especially where large shopping malls opened, also provided for growth. Simultaneously, the consolidation of the service economy expanded the demand for parcel services.

A major change for the company occurred in the 1950s when UPS became a common carrier, receiving the right to deliver packages between any civic address within the territory this right was granted. However, it was not until 1975 that UPS was granted the right to be a common carrier for the 48 contiguous states and was able to offer second-day deliveries throughout the United States. Shortly after, UPS expanded from coast to coast and began to consolidate and expand its international services, initially in Canada and then Western Europe (Germany). By 1987, UPS was servicing almost every address in North America, Western Europe and Japan. This was done mainly by the establishment of high-throughput distribution centers forming major air hubs. Since 1988, UPS operates its own airline, UPS Airline. From the hub, UPS delivers to more than 391 national airports and over 219 international ones. By 2001, UPS was offering direct air freight services to China. This totals about 1,000 flights per day.

The 1990s also represented an important stage in the logistics industry, namely through the growing number of transactions occurring online. The growth of Amazon-type commercial activities has been accompanied with a surge of parcels being shipped. Further, customers are able to track the location of their parcels throughout the distribution system.

The system

The UPS system is mostly aimed at servicing businesses since 80 percent of the traffic handled is business to business. To be effective, UPS relied on the efficiency of its distribution system. Reliability and efficiency are key issues in the establishment and management of freight distribution systems leaning on parcels. Optimal locations for the hubs are sought, as well as the possible delivery routes, to avoid unnecessary movements, congestion and assure timely deliveries. Every single parcel has to go through the UPS network regardless of its destination. It could be bound for the other side of the planet or addressed to the neighbor; the parcel will have to go through the distribution system, which has a hub-and-spoke structure. This distribution system involves three major functions (Figure 2.20):

- **Consolidation.** The first step obviously involves the collection of parcels by trucks assigned to specific routes. To optimize the driver's effectiveness, traffic trends and road conditions are continuously monitored to ensure that the optimal path is taken. The parcels are then assembled at the closest distribution center.
- **Distribution.** The distribution function works on a hub-to-hub basis; depending on the distance involved, the mode used between hubs will either be trucking or air. Commonly, trucks are used for distances less than 400 miles (600 km). The main air hub is Louisville, Kentucky, which handles over 100 flights a day. In 2002, a new distribution center of 4 million square foot, called UPS Worldport, opened at the Louisville International Airport. The main land hub is Willow Springs in Chicago, which is the largest distribution center in the United States (Plate 2.2).
- **Fragmentation.** This step is the inverse of consolidation as parcels have to be delivered to each individual destination. Commonly, fragmentation is combined with consolidation as a delivery truck route can be integrated with a pickup route. This can be achieved only with a high level of control on the logistical chain.

Furthermore, UPS is investing in transport technology research. Innovations such as alternative fuels and electric vehicles are among their extensive testing programs, reduction in fuel consumption being the main concern. Engineers and geographers study optimal roads and driving speeds to enhance efficiency and reduce costs. In technology, UPS is also working on computer software to simplify shipping red tape, optimize routing strategies and facilitate package tracking. The system also enables customers to locate and track their parcel directly from the Internet.

UPS is a textbook example of intensive research in transport geography. Optimal routing networks are essential to assure efficient delivery in only 24 hours throughout the world. Rigorous planning can also save considerable amounts in transport costs such as fuel, wages, vehicle maintenance, etc. Strategies such as the consolidation principle and the hub network strategy are very important and useful in transport geography analysis.

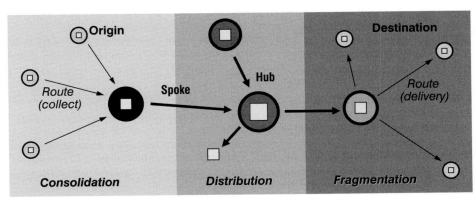

Figure 2.20 The hub-and-spoke structure of UPS

Plate 2.2 UPS Willow Springs Distribution Center, Chicago. *Credit:* Jean-Paul Rodrigue

Bibliography

Ausubel, J.H. and C. Marchetti (2001) "The Evolution of Transportation", *Industrial Physicist*, April/May, pp. 20–4. http://www.aip.org/tip/INPHFA/vol-7/iss-2/p20.pdf.

Ausubel, J.H., C. Marchetti and P. Meyer (1998) "Toward Green Mobility: The Evolution of Transport", *European Review*, 6(2), 137–56.

Bernstein, W.J. (2008) *A Splendid Exchange: How Trade Shaped the World*, New York: Atlantic Monthly Press.

Gottmann, J. (1961) *Megalopolis: The Urbanized Northeastern Seaboard of the United States*, New York: Twentieth Century Fund.

Henderson, J.V., Z. Shalizi and A.J. Venables (2000) *Geography and Development*, http://econ.lse.ac.uk/staff/ajv/vhzstv3.pdf.

Miller, H.J. and S.L. Shaw (2001) "GIS-T Data Models", in *Geographic Information Systems for Transportation: Principles and Applications*, Oxford: Oxford University Press. http://www.gisvisionmag.com/Book/miller_shaw.pdf.

Rioux, J.-P. (1989) *La révolution industrielle, 1780–1880*, Paris: Éditions du Seuil.

Sletmo, G.K. (1989) "Shipping's Fourth Wave: Ship Management and Vernon's Trade Cycles", *Maritime Policy and Management*, 16(4), 293–303.

Taaffe, E.J., H.L. Gauthier and M.E. O'Kelly (1996) *Geography of Transportation*, 2nd edn, Upper Saddle River, NJ: Prentice Hall.

Victoria Transport Policy Institute, "Defining, Evaluating and Improving Accessibility", *Transport Demand Management Encyclopedia*, http://www.vtpi.org/tdm/tdm84.htm.

Williams, A. (1992) "Transport and the Future", in B.S. Hoyle and R.D. Knowles (eds) *Modern Transport Geography*, London: Belhaven Press, pp. 257–70.

Zeiler, M. (1999) *Modeling our World: The ESRI Guide to Geodatabase Design*, Redlands, CA: ESRI Press.

3 Transportation and the economy

Transport systems are closely related to socio-economic changes. The mobility of people and freight and levels of territorial accessibility are at the core of this relationship. Economic opportunities are likely to arise where transportation infrastructures are able to answer mobility needs and ensure access to markets and resources. From the industrial revolution in the nineteenth century to globalization and economic integration processes of the late twentieth and early twenty-first centuries, regions of the world have been affected differently by economic development. International, regional and local transportation systems alike have become fundamental components of economic activities. A growing share of the wealth is thus linked to trade and distribution. However, even if transportation has positive impacts on socio-economic systems, there are also negative consequences such as congestion, accidents and mobility gaps.

Transportation is also a commercial activity derived from operational attributes such as transportation costs, capacity, efficiency, reliability and speed. Transportation systems are evolving within a complex set of relationships between transport supply, mainly the operational capacity of the network, and transport demand, the mobility requirements of an economy.

Concept 1 – Transportation and economic development

The economic importance of transportation

The transport sector is an important component of the economy, impacting on development and the welfare of populations. When transport systems are efficient, they provide economic and social opportunities and benefits that impact throughout the economy. When transport systems are deficient, they can have an economic cost in terms of reduced or missed opportunities. Transport also carries an important social and environmental load, which cannot be neglected. From a general standpoint, the economic impacts of transportation can be direct and indirect:

- **Direct impacts** relate to accessibility change where transport enables larger markets and enables to save time and costs.
- **Indirect impacts** relate to the economic multiplier effect where the price of commodities goods or services drop and/or their variety increases.

Table 3.1 shows a wide range of economic benefits conveyed by transportation systems, some direct (income related) and some indirect (accessibility related), impacting transport supply and demand both at the microeconomic (sector-wise) and macroeconomic (whole economy) levels.

Table 3.1 Economic benefits of transportation

Direct transport supply	Direct transport demand	Indirect microeconomic	Indirect macroeconomic
Income from transport operations (fares and salaries) Access to wider distribution markets and niches	Improved accessibility Time and cost savings Productivity gains Division of labor Access to a wider range of suppliers and consumers Economies of scale	Rent income Lower price of commodities Higher supply of commodities	Formation of distribution networks Attraction and accumulation of economic activities Increased competitiveness Growth of consumption Fulfilling mobility needs

The impacts of transportation are not always intended, and can have unforeseen consequences such as congestion. Mobility is one of the most fundamental and important characteristics of economic activity as it satisfies the basic need of going from one location to another, a need shared by passengers, freight and information. All economies do not share the same level of mobility as most are in a different stage in the transition. Economies that possess greater mobility are often those with better opportunities to develop than those suffering from scarce mobility. Reduced mobility impedes development while greater mobility is a catalyst for development. Mobility is thus a reliable indicator of development.

Providing this mobility is an industry that offers services to its customers, employs people and pays wages, invests capital and generates income. The economic importance of the transportation industry can thus be assessed from a macroeconomic and microeconomic perspective:

- At the **macroeconomic level** (the importance of transportation for a whole economy) transportation and the mobility it confers are linked to a level of output, employment and income within a national economy. In many developed countries, transportation accounts for between 6 percent and 12 percent of GDP.
- At the **microeconomic level** (the importance of transportation for specific parts of the economy) transportation is linked to producer, consumer and production costs. The importance of specific transport activities and infrastructure can thus be assessed for each sector of the economy. Transportation accounts on average for between 10 percent and 15 percent of household expenditures while it accounts for around 4 percent of the costs of each unit of output in manufacturing.

Transportation links together the factors of production in a complex web of relationships between producers and consumers. The outcome is a more efficient division of production by an exploitation of geographical comparative advantages, as well as the means to develop economies of scale and scope. The productivity of space, capital and labor is thus enhanced with the efficiency of distribution. It is acknowledged that economic growth is increasingly linked with transport development. The following impacts can be assessed:

- **Networks**. Setting of routes enabling new or existing interactions between economic entities.
- **Performance**. Improvement in cost and time attributes for existing passenger and freight movements.
- **Reliability**. Improvement in time performance, notably in terms of punctuality, as well as reduced loss or damage.
- **Market size**. Access to a wider market base where economies of scale in production, distribution and consumption can be achieved.
- **Productivity**. Increases in productivity from access to a larger and more diverse base of inputs (raw materials, parts, energy or labor) and broader markets for diverse outputs (intermediate and finished goods).

Transportation and economic development

Transportation developments that have taken place since the beginning of the industrial revolution have been linked to growing economic opportunities. At each stage of human societal development, a particular transport mode has been developed or adapted. However, it has been observed that throughout history no single transport has been solely responsible for economic growth. Instead, modes have been linked with the direction and the geographical setting in which growth was taking place. For instance, major flows of international migration that occurred since the eighteenth century were linked with the expansion of international and continental transport systems. Transport has played a catalytic role in these migrations, transforming the economic and social geography of many nations. Concomitantly, transportation has been a tool of territorial control and exploitation, particularly during the colonial era where resource-based transport systems supported the extraction of commodities in the developing world.

Each transport mode and technology is linked to a set of economic opportunities, notably in terms of market areas, types of commodities that can be transported (including passengers) and economies of scale (Figure 3.1). All these issues are related to a

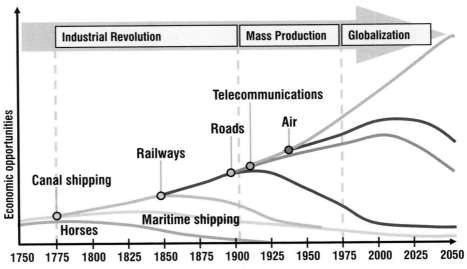

Figure 3.1 Cumulative modal contribution to economic opportunities

scale and level of commercial geography. Prior to the industrial revolution, economic opportunities were limited by the low capacity to move commodities over long distances, as most activities were very localized in scale and scope. The industrial revolution unleashed greater economic opportunities, initially with the development of inland canal systems, steamships and then railway systems. Passenger and freight transportation expanded as well as production and consumption while new markets and resources became available. In many instances, the development of a transportation mode built on the opportunities developed by another, such as maritime and canal shipping. In other situations, the growth of a new mode of transportation favored the decline of others, such as the collapse of many inland canal networks in the late nineteenth century because of rail competition.

The development of the mass production system at the beginning of the twentieth century increasingly relied on the commercial opportunities introduced by road trans-portation, particularly the automobile. Later in the twentieth century, globalization became a possibility with the joint synergy of maritime transportation, roadways, railways, air and telecommunications. Economic opportunities became global in scale and scope, particularly because of the capacity to maintain an intricate network of trade and trans-actions through transport systems. More recently, new opportunities arose with the convergence of telecommunications and information technologies, supporting a higher level of management of production, consumption and distribution. It is expected that such a process, building upon the advantages conferred by other transportation modes, will account for a significant share of economic opportunities in the first half of the twenty-first century.

While some regions benefit from the development of transport systems, others are often marginalized by a set of conditions in which inadequate transportation plays a role. Transport by itself is not a sufficient condition for development; however, the lack of transport infrastructures can be seen as a constraining factor on development. Investment in transport infrastructures is thus seen as a tool of regional development, particularly in developing countries and for the road sector. The relationship between transportation and economic development is thus difficult to formally establish and has been debated for many years. The complexity lies in the variety of possible impacts:

- **Timing of the development** varies as the impacts of transportation can either precede, occur during or take place after economic development. The lag, concomitant and lead impacts make it difficult to separate the specific contributions of transport to development, therefore. Each case study appears to be specific to a set of timing circumstances that are difficult to replicate elsewhere.
- **Types of impacts** vary considerably. The spectrum of impacts ranges from the positive through the permissive to the negative. In some cases transportation impacts can promote, in others they may hinder economic development in a region. In many cases, few, if any, direct linkages can be clearly established.

Cycles of economic development provide a revealing conceptual perspective about how transport systems evolve in time and space as they include the timing and the nature of the transport impact on economic development. Transport, as a technology, typically follows a path of experimentation, introduction, adoption and diffusion and, finally, obsolescence, each of which has an impact on economic development. In addi-tion, transport modes and infrastructures are depreciating assets that constantly require maintenance and upgrades. At some point, their useful lifespan is exceeded and the vehicle must be retired or the infrastructure rebuilt. Thus, transport investments for their amortization must consider the lifespan of the concerned mode or infrastructure.

In general, transport technology can be linked to five major waves of economic development where a specific mode or system emerged:

- **Seaports**. Linked with the early stages of European expansion from the sixteenth to the eighteenth centuries. They supported the development of international trade through colonial empires, but were constrained by limited inland access.
- **Rivers and canals**. The first stage of the industrial revolution in the late eighteenth and early nineteenth centuries was linked to the development of canal systems in Western Europe and North America, mainly to transport heavy goods. This permitted the development of rudimentary and constrained inland distribution systems.
- **Railways**. The second stage of industrial revolution in the nineteenth century was intimately linked to the development and implementation of rail systems enabling a more flexible inland transportation system.
- **Roads**. The twentieth century saw the development of road transportation systems and automobile manufacturing. Individual transportation became a commodity available to the masses, especially after World War II. This process was reinforced by the development of highway systems.
- **Airways and information**. The later part of the twentieth century saw the development of global air and telecommunication networks in conjunction with the globalization of economic activities. New organization, control and maintenance capacities were made possible. Electronic communications have become consistent with transport functions, especially in the rapidly developing realm of logistics and supply chain management.

Technological innovation and economic growth are closely related and can be articulated within the concept of cycles or waves. Each wave represents a diffusion phase of technological innovations, creating entirely new industrial sectors and thus opportunities for investment and growth. Five waves have been identified so far (Figure 3.2):

- **1st wave (1785–1845)**. Depended on innovations such as water power, textiles and iron. The beginning of the industrial revolution in England was mainly focused on simple commodities such as clothes and tools. The conventional maritime technology relying on sailships was perfected, supporting the creation of large colonial/trading empires, mainly by Great Britain, France, the Netherlands and Spain. Significant inland waterway systems were also constructed.
- **2nd wave (1845–1900)**. Involved the massive application of coal as a source of energy, mainly through the steam engine. This induced the development of rail transport systems, opening new markets and giving access to a wider array of resources. The

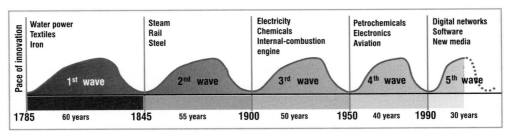

Figure 3.2 Long wave cycles of innovation

steamship had a similar impact for maritime transportation and permitted expanded commercial opportunities.

- **3rd wave (1900–50).** Electrification was a major economic change as it permitted the usage of a variety of machines and appliances. This permitted the development of urban transit systems (subways and tramways). Another significant improvement was the internal combustion engine, around which the whole automotive industry was created and permitted the motorization of mobility.
- **4th wave (1950–90).** The post-World War II period represented significant industrial changes such as plastics (petrochemicals) and electronics (television). The jet engine expanded the aviation industry towards the mass market and mobility could be realized globally.
- **5th wave (1990–2020?).** The current wave mainly relies on information systems, which have tremendously modified the transactional environment with new methods of communication and more efficient management of production and distribution systems (logistics). This has spawned new industries, mainly computer manufacturing and software programming, but more recently e-commerce as information processing converged with telecommunications.

As time progressed, the lapse between each wave got shorter. For instance, the first wave lasted 60 years while the fourth wave lasted 40 years. This reflects a growing capacity for innovation and the capacity of economic systems to derive wealth from it. Innovations are no longer the result of individual efforts, but are organized and concerted actions whose results are rapidly diffused. It is thus expected that the fifth wave will last about 30 years.

Contemporary trends have underlined that economic development has become less dependent on relations with the environment (resources) and more dependent on relations across space. While resources remain the foundation of economic activities, the commodification of the economy has been linked with higher levels of material flows of all kinds. Concomitantly, resources, capital and even labor have shown increasing levels of mobility. This is particularly the case for multinational firms that can benefit from transport improvements in two significant markets:

- **Commodity market.** Improvement in the efficiency with which firms have access to raw materials and parts as well as to their respective customers. Thus, transportation expands opportunities to acquire and sell a variety of commodities necessary for industrial and manufacturing systems.
- **Labor market.** Improvement in the access to labor and a reduction in access costs, mainly by improved commuting (local scale) or the use of lower cost labor (global scale).

Transport as a factor of production

Transportation is an economic factor of production of goods and services. It provides market accessibility by linking producers and consumers. An efficient transport system with modern infrastructures favors many economic changes, most of them positive. The major impacts of transport on economic processes can be categorized as follows:

- **Geographic specialization.** Improvements in transportation and communication favor a process of geographical specialization that increases productivity and spatial interactions (Figure 3.3). An economic entity tends to produce goods and services with the most appropriate combination of capital, labor, and raw materials. A given

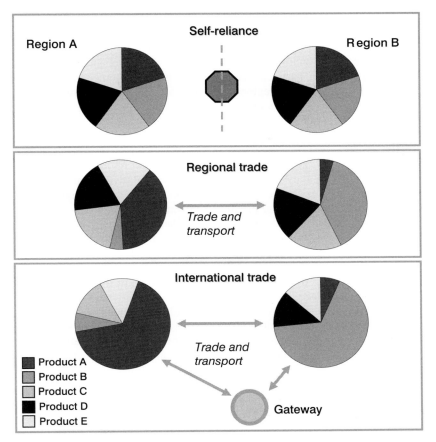

Figure 3.3 Economic production and specialization

area will thus tend to specialize in the production of goods and services for which it has the greatest advantages (or the least disadvantages) compared to other areas, as long as appropriate transport is available for trade. Through geographic specialization supported by efficient transportation, the economic productivity is promoted. This process is known in economic theory as comparative advantages.

- **Large-scale production**. An efficient transport system offering cost, time and reliability advantages permits goods to be transported further. This facilitates mass production through economies of scale because more markets can be accessed. The concept of "just-in-time" has further expanded the productivity of production and distribution. Thus, the more efficient transportation becomes, the larger the markets that can be serviced and the larger the scale of production.

- **Increased competition**. When transport is efficient, the potential market for a given product (or service) increases, and so does competition. A wider array of goods and services becomes available to consumers through competition which tends to reduce costs and promote quality and innovation.

- **Increased land value**. Land which is adjacent or serviced by good transport services generally has greater value due to the utility it confers to many activities. In some cases, the opposite can be true. Land located near airports and highways, near noise and pollution sources, will thus suffer from corresponding diminishing land value.

The evolution of transport systems has a set of impacts on the regional economy in terms of spatial specialization. Figure 3.3 represents a simplified example of how changes in transport may impact on the specialization of regional economies which a priori have similar environmental endowments. In a situation of self-reliance there is no apparent transport link between two regions. They are isolated from one another and must satisfy their own needs. Each region thus tends to be similar in terms of economic output. While regions have different environmental endowments, they must still provide for every basic necessity such as food. Quantities produced depend on the demand and the industrial capacity.

With a transport link between two regions, specialization can take place. Each region develops its respective potential; product A for the first region and product B for the second, assuming that they respectively have a comparative advantage for these two products. If product A is cheaper to produce in the first region, it becomes more efficient to lessen the production of other products and concentrate on product A. Similarly, the second region can concentrate on product B. Therefore, the first region can delegate more resource for the production of product A and can then sell the surplus (minus local consumption) to the second region. The key of this specialization becomes the difference between transport costs and production costs of a product. If the unit costs savings resulting from specialization exceed the unit transports costs, then specialization can take place.

Regional specialization is greatly expanded with international trade. By having access to a larger market through a gateway – namely a seaport – regions 1 and 2 can specialize even more in the production they have respective comparative advantages in. They can even cease production in a specific array of products, which are now imported. Under such circumstances, the reliance on transportation increases, even if its relative costs may be declining.

Transport also contributes to economic development through job creation and its derived economic activities. Accordingly, a large number of direct (freighters, managers, shippers) and indirect (insurance, packaging, handling, travel agencies, transit operators) employment are associated with transport. Consumers take economic decisions on products, markets, costs, location and prices, which are themselves based on transport services, their availability, costs and capacity.

Socio-economic impacts

While many of the economic impacts of transportation are positive, there are also significant negative impacts that are assumed by individuals or by society in one way or another. Among the most significant are:

● **Mobility gaps.** Since mobility is one of the fundamental components of the economic benefits of transportation, its variations are likely to have substantial impacts on the opportunities of individuals. Mobility needs do not always coincide due to several factors, namely the lack of income, lack of time, lack of means and the lack of access. People's mobility and transport demands thus depend on their socio-economic situation. The higher the income, the higher the mobility, which may give rise to substantial mobility gaps between different population groups. Gender gaps exist in mobility as women tend to have lower incomes. Mobility gaps are particularly prevalent for long-distance travel. With the development of air transport, a segment of the global population has achieved a very high level of mobility for their business and leisure activities, while the great majority of the global population has little mobility. This issue is expected to become more acute as the population of many of the most advanced countries is aging rapidly, which implies that access to mobility will not be an income issue but an age issue. By 2020, about 10 percent of the global population (719 million) will be over 65, while by 2050 it will be 16 percent (1,492 million).

- **Costs differences**. Locations that have low levels of accessibility tend to have higher costs for many goods (sometimes basic necessities such as food) as most have to be imported, often over long distances. The resulting higher transport costs inhibit the competitiveness of such locations and limits opportunities. Consumers and industries will pay higher prices, impacting on their welfare (disposable income) and competitiveness.
- **Congestion**. With the increased use of transport systems, it has become increasingly common for parts of the network to be used above design capacity. Congestion is the outcome of such a situation with its associated costs, delays and waste of energy. Distribution systems that rely upon on-time deliveries are particularly susceptible to congestion.
- **Accidents**. The use of transport modes and infrastructure is never entirely safe. Every motorized vehicle contains an element of danger and nuisance. Due to human errors and various forms of physical failures (mechanical or infrastructural), injuries, damages and even death occur. Accidents tend to be proportional to the intensity of use of transport infrastructures, which means the more traffic the higher the probability for an accident to occur. They have important socio-economic impacts including healthcare, insurance, damage to property and loss of life. The respective level of safety depends on the mode of transport and the speed at which an accident occurs. No mode is completely safe but the road remains the most dangerous medium for transportation, accounting for 90 percent of all transport accidents on average. China has one of the highest car accident death rates in the world, with more than 110,000 fatalities per year (300 per day), a factor mainly due to recent growth in vehicle ownership.

The emission of pollutants related to transport activities has a wide range of environmental consequences that have to be assumed by society (Chapter 8 provides a comprehensive overview about transport and the environment), more specifically on four elements:

- **Air quality**. Atmospheric emissions from pollutants produced by transportation, especially by the internal combustion engine, are associated with air pollution, acid rain and arguably, global climate change. Some pollutants (NO_x, CO, O_3, volatile organic compounds (VOCs), etc.) can produce respiratory troubles and aggravate cardiovascular illnesses. In urban regions, about 50 percent of all air pollution emanates from automobile traffic.
- **Noise**. A major irritant, noise can impact on human health and most often human welfare. Noise can be manifested in three levels depending on emissions intensity: psychological disturbances (perturbations, displeasure), functional disturbances (sleep disorders, loss of work productivity, speech interference) or physiological disturbances (health issues such as fatigue, and hearing damage). Noise and vibration associated with trains, trucks and planes in the vicinity of airports are major irritants.
- **Water quality**. Accidental and nominal runoff of pollutants from transport such as oil spills, are sources of contamination for both surface water and groundwater.
- **Land take**. Transport is a large consumer of space when all of its supporting infrastructure and equipment are considered. Furthermore, the planning associated with these structures does not always consider aesthetic values as is often the case

in the construction of urban highways. These visual impacts have adverse consequences on the quality of life of nearby residents.

Concept 2 – Transportation and commercial geography

Trade and commercial geography

Economic systems are based on trade and transactions since specialization and
efficiency requires interdependency. People trade their labor for a wage while corporations trade their output for capital. Trade is the transmission of a possession in return
for a counterpart, generally money, which is often defined as a medium of exchange.
This exchange involves a transaction and its associated flows of capital, information,
commodities, parts or finished products. All this necessitates the understanding of commercial geography.

> **Commercial geography** investigates the spatial characteristics of trade and
> transactions in terms of their cause, nature, origin and destination. It leans on the
> analysis of contracts and transactions. From a simple commercial transaction
> involving an individual purchasing a product at a store, to the complex network
> of transactions maintained between a multinational corporation and its suppliers,
> the scale and scope of commercial geography varies significantly.

As each transaction involves movements of people, freight and information, there
is a close relationship between the sphere of transactions (the geographical setting of
transactions) and the sphere of circulation (the geographical setting of movements) (figure
3.4). This implies transaction costs and transportation costs. The main transaction costs
are: (1) Search and information costs: costs related to finding the appropriate goods on
the market, who has them and at what price. (2) Negotiation costs: costs involved
in reaching an agreement with the other party to the transaction, the contract being
the outcome. (3) Policing and enforcement costs: costs related to ensuring that both
parties respect the terms of the contract and, if not the case, taking legal actions to
correct the situation.

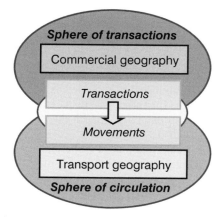

Figure 3.4 Commercial and transport geography

Trade, in terms of its origins and destinations, has a spatial logic. It reflects the economic, social and industrial structure of the concerned markets, but also implies other factors such as transport costs, distance, political ties, exchange rates and the reciprocal economic advantages proponents get from trade. For trade to occur, several conditions must be met:

- **Availability**. Commodities, from coal to computer chips, must be available for trade and there must be a demand for these commodities. In other terms, a surplus must exist at one location and a demand in another. A surplus can often be a simple matter of investment in production capabilities, such as building an assembly plant, or can be constrained by complex environmental factors like the availability of resources such as fossil fuels, minerals and agricultural products.
- **Transferability**. There are three major impediments to transferability, namely policy barriers (tariffs, custom inspections, quotas), geographical barriers (time, distance) and transportation barriers (the simple capacity to move the outcome of a transaction). Transport infrastructures in allowing commodities to be moved from their origins to their destinations favor the transferability of goods. Distance often plays an important role in trade, as does the capacity of infrastructures to route and to tranship goods.
- **Transactional capacity**. It must be legally possible to make a transaction. This implies the recognition of a currency for trading and legislations that define the environment in which transactions take place, such as taxation. In the context of a global economy, the transactional environment is very complex but is important in facilitating trade at the regional, national and international levels.

Once these conditions are met, trade is possible and the outcome of a transaction results in a flow. Three particular issues relate to the concept of flow:

- **Value**. Flows have a negotiated value and are settled in a common currency. The American dollar, which has become the major global currency, is used to settle and/or measure many international transactions. Further, nations must maintain reserves of foreign currencies to settle their transactions and the relationship between the inbound and outbound flows of capital is known as the balance of payments. Although nations try to maintain a stable balance of payments, this is rarely the case.
- **Volume**. Flows have a physical characteristic, mainly involving a mass. The weight of flows is a significant variable when trade involves raw materials such as petroleum or minerals. However, in the case of consumption goods, weight has little significance relative to the value of the commodities being traded. With containerization, a new unit of volume has been introduced: the TEU (Twenty-foot Equivalent Unit), which can be used to assess trade flows.
- **Scale**. Flows have a range which varies significantly based on the nature of a transaction. While retailing transactions tend to occur at a local scale, transactions related to the operations of a multinational corporation are global in scale.

Trends in commercial geography

The contemporary commercial setting is marked by increasing free trade and profound technological, industrial and geopolitical changes. The liberalization of trade, as confirmed by the implementation of the World Trade Organization, has given a strong impetus and a positive trend in the growth rate of world trade and industrial production. However, in a true free trade environment, regulatory agencies would not be required. But in spite of attempts at deregulation, transactions and trade are prone to disputes, litigations and perceived imbalances concerning who benefits the most. Although these issues mainly apply to international trade, there are also situations where trade is constrained between the provinces/states of a nation.

In spite of globalization, much trade is still dominantly regional. An overview of world trade flows indicate that trade within regions is more significant than trade between regions, but long-distance trade is steadily growing. Figures indicate the increasing share of East Asia, especially China, in world trade both in terms of exports and imports. Flows of merchandise have also been accompanied by a substantial growth in foreign direct investments. There is thus a remarkable reallocation of production capacities following changes in comparative advantages around the world. This trend goes in tandem with mergers and acquisitions of enterprises that are increasingly global in scope. The analysis of international trade thus reveals the need to adopt different strategies to adapt to this new trading environment. As production is being relocated, there is a continuous shift in emphasis in the structure of export and import of world economies.

Major changes have occurred in the organization of production. There is a noticeable increase in the division of labor concerning the design, planning and assembly in the manufacturing process of the global economy. Interlocking partnerships in the structure of manufacturing have increased the trade of parts and the supply of production equipment around the world. One-third of all trade takes place among parent companies and their foreign affiliates. A part of this dynamism resides in the adoption of standards, a process which began in the late nineteenth century to promote mass production. It permitted the rapid development of many sectors of activity, including railways, electricity, the automobile and the telecommunications industry more recently (Internet, electronic data interchange). In the realm of globalization of economic activities, the International Organization for Standardization developed the ISO norms that serve as comparison between various enterprises around the world. These norms are applicable to the manufacturing and services industries and are a necessary tool for growth.

Another significant force of change in commercial geography implies the growth of personal consumption, although this is not taking place uniformly. The bulk of consumption remains concentrated in a limited number of countries, with the G7 countries alone accounting for about two-thirds of the global gross domestic product. As a result, commercial geography is influenced by the market size, the consumption level of an economy (often measured in GDP per capita), and also by the growth potential of different regions of the world. Economic growth taking place in East and Southeast Asia has been one of the most significant forces shaping changes in the contemporary commercial environment. The commodification of the economy has led to significant growth in retail and wholesale and the associated movements of freight.

Commercialization of the transport industry

The liberalization of trade was accompanied by a growth of transportation since transactions involve movements of freight, capital, people and information. Developments in the transport sector are matched by global and regional interdependence and competition. Transportation, like commodities, goods and services, is traded, sometimes openly and subject to full market forces, but more often subject to a form of public control (regulation) or ownership. The core component of a transport-related transaction involves its costs that either have to be negotiated between the provider of the service and the user or are subject to some arbitrary decree (price fixing such as public transit). Since transportation can be perceived as a service to people, freight or information, its commercialization, how it is brought to the market, is an important dimension of its dynamics. Transport service providers tend to be private entities, particularly in the global freight sector. Local passenger transportation providers (transit) tend to be publicly owned.

The extension of the operational scale of freight distribution ensures that a production system reaches its optimal market potential, namely by a combination of strategies related to the exploitation of comparative advantages and a wider market base (Figure 3.5). Although an optimal market size can never be attained due to regulations preventing monopolies and differences in consumer preferences, the trend to ensure maximal market exposure is unmistakable. The emergence of global brands and global production networks clearly underlines this. Within freight distribution, four distinct cyclic phases of extension and functional integration can be identified:

- **Introduction**. Initially, a transport system is introduced to service a specific opportunity in an isolated context. The technology is often "proprietary" and incompatible with other transport systems. Since they are not interconnected, this does not represent much of an issue.
- **Expansion and interconnection**. As the marketability and the development potential of a transport system become apparent, a phase of expansion and interconnection occurs. The size of the market serviced by these transport systems consequently increases as they become adopted in new locations. At some point, independently developed transport systems connect. This connection is however often subject to a function of transhipment between two incompatible transport systems.
- **Standardization and integration**. This phase often involves the emergence of a fully developed distribution system servicing vast national markets. The major challenge to be addressed involves a standardization of modes and processes, further expanding the commercial potential of the supply chains concerned. Modal flows are moving more efficiently over the entire network and are able to move from one mode to the other through intermodal integration. A process of mergers and acquisitions often accompanies this phase for the purpose of rationalization and market expansion.
- **Integrated demand**. The most advanced stage of extension of a distribution system involves a system fully able to answer freight mobility needs under a variety of

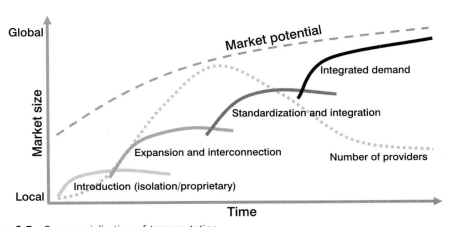

Figure 3.5 Commercialization of transportation

circumstances, either predicted or unpredicted demand. As this system tends to be global, it commonly operates close to market potential. In such a setting, a distribution system expresses an integrated demand where the distribution capabilities are tuned to the demand in an interdependent system.

Each of these phases tends to be sequential and related to an historical process of transport development. For instance, up to the mid-nineteenth century, most distribution systems were isolated and developed independently from one another. Even global maritime transport was fragmented by national flags and trading systems. As regional transport systems grew in the second half of the nineteenth century, they gradually interconnected, but moving from one system to another required a form of transhipment. By the early twentieth century, most national transport systems were integrated, but interconnection between modes was difficult. The next challenge resided in the development of intermodal transportation, accelerated by containerization and information technologies.

One important component of the commercialization of transportation concerns investments in infrastructure, modes and terminals, as well as marketing and financing. Investments are performed either to expand the geographical extent and/or the capacity of a transport system or to maintain its operating conditions. The public and private sectors have contributed to the funding of transport investments depending on economic, social and strategic interests. For obvious reasons, the private sector seeks transport investments that promise economic returns while the public sector often invests for social and strategic reasons. In many cases private transport providers have difficulty in acting independently to formulate and implement their transport investments. Various levels of government are often lobbied by transport firms for financial and/or regulatory assistance in projects that are presented as of public interest and benefit. The consolidation of regional markets and the resulting increase in transborder traffic has led transport firms to seek global alliances and greater market liberalization in the transport and communication sector as a means to attract investments and to improve their productivity.

Deregulation and divestiture policy in the transport industry has led governments to withdraw from the management, operations and ownership of national carriers, ports and airports. This has given rise to a major reorganization of the international and national transport sectors, with the emergence of transnational transport corporations that are governing the global flow of air, maritime and land trade and the management of airports, ports and railyards (Table 3.2).

Concept 3 – Transportation costs

Transportation costs and rates

Transport systems face requirements to increase their capacity and to reduce the costs of movements. All users (e.g. individuals, enterprises, institutions, governments, etc.) have to negotiate or bid for the transfer of goods, people, information and capital because supplies, distribution systems, tariffs, salaries, locations, marketing techniques as well as fuel costs are changing constantly. There are also costs involved in gathering information, negotiating, and enforcing contracts and transactions, which are often referred to as the cost of doing business. Trade involves transactions costs that all agents attempt to reduce since transaction costs account for a growing share of the resources consumed by the economy.

Table 3.2 Major commercial actors in freight distribution

Transport sector	Function
Maritime shipping companies	Control long-distance segments of the global freight distribution linking major markets. Highly capital intensive industry.
Global port operators	Control important intermodal infrastructures (terminals) within the world's largest container ports. Have strong linkages with maritime shipping companies.
Port authorities	Manage and plan port infrastructures. Tend to lease the operation of terminals. Important intermediaries for regional distribution (hinterland).
Maritime lock and canal operators	Ensure the operation of strategic passages in global and national distribution. This mainly includes the Panama and Suez Canals and the St. Lawrence Seaway.
Rail and rail terminal operators	Strategic inland freight carriers transporting a wide array of raw materials and commodities. Responsible for many of the transhipments between rail and road, particularly for containerized freight.
Trucking industry	Controls vast and diverse assets that include critical segments of freight distribution in all economic sectors.
Third-party logistics providers	Important managerial and organizational skills within supply chains. Often act as brokers between transport customers and service providers.
Air freight transport companies and air freight terminals	Important assets for the rapid distribution of high value-added freight.
Distribution centers	A crucial element of modern supply chains. Perform tasks such as packaging, labeling and the consolidation of shipments to customers.

Frequently, enterprises and individuals must take decisions about how to route passengers or freight through the transport system. This choice has been considerably expanded in the context of the production of lighter and high value consumer goods, such as electronics, and less bulky production techniques. It is not uncommon for transport costs to account for 20 percent of the total cost of a product. Thus, the choice of a transportation mode to route people and freight within origins and destinations becomes important and depends on a number of factors such as the nature of the goods, the available infrastructures, origins and destinations, technology and particularly their respective distances. Jointly, they define transportation costs.

Transport costs are a monetary measure of what the transport provider must pay to produce transportation services. They come as fixed (infrastructure) and variable (operating) costs, depending on a variety of conditions related to geography, infrastructure, administrative barriers and energy, and on how passengers and freight are carried. Three major components, related to transactions, shipments and the friction of distance, impact on transport costs.

As shown on Figure 3.6, a movement between location A and B involves three cost components in the assessment of its transport cost. Transaction costs relate to resolving the setting of the movement, including legal costs and insurance. For international

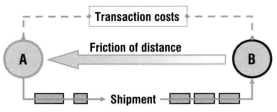

Figure 3.6 Components of transport costs

movements they can be significant as issues related to currency exchange as well as custom duties have to be considered. The friction of distance represents how many units of space can be traded per unit of cost, which indicates how much effort (time, energy, etc.) must be made to ensure that a movement takes place. Distance is considered to be the simplest attribute for such a purpose. Shipment implies the physical characteristics of the transportation process, including the mode used, the frequency, as well as economies of scale that can be realized by consolidating shipments.

Transport costs have significant impacts on the structure of economic activities as well as on international trade. Empirical evidence underlines that raising transport costs by 10 percent reduces trade volumes by more than 20 percent. In a competitive environment where transportation is a service that can be bid on, transport costs are influenced by the respective rates of transport companies and the portion of the transport costs charged to users.

> **Rates** are the price of transportation services paid by their users. They are the negotiated monetary cost of moving a passenger or a unit of freight between a specific origin and destination. Rates are often visible to consumers since transport providers must provide this information to secure transactions. They may not necessarily express the real transport costs.

The difference between costs and rates either results in a loss or deficit from the service provider. Considering the components of transport costs previously discussed, rate setting is a complex undertaking subject to constant change. For public transit, rates are often fixed and the result of a political decision where a share of the total costs is subsidized by society. The goal is to provide an affordable mobility to the largest possible segment of the population even if this implies a recurring deficit (public transit systems rarely make any profit). For freight transportation and many forms of passenger transportation (e.g. air transportation) rates are subject to a competitive pressure. This means that the rate will be adjusted according to the demand and the supply. They either reflect costs directly involved with shipping (cost-of-service) or are determined by the value of the commodity (value-of-service).

Costs and time components

Among the most significant conditions affecting transport costs and thus transport rates are:

- **Geography**. Its impacts mainly involve distance and accessibility. Distance is commonly the most basic condition affecting transport costs. The more it is difficult to trade space for a cost, the more the friction of distance is important. The friction of distance can be expressed in terms of length, time, economic costs or the amount of energy used. It varies greatly according to the type of transportation mode involved and the efficiency of specific transport routes. Landlocked countries tend to have higher transport costs, often twice as much, as they do not have direct access to maritime transportation.

There are four major categories of friction of distance functions (Figure 3.7): (1) No effects of distance. Economic activities on which distance has no effects are uncommon. However, the distance–cost function of telecommunication networks and the virtual space of the Internet have such a cost structure. Telephone calls can be included in this category as well as postal fees and public transit fares. All those activities generally have a fixed cost which is not related to distance, but often to a service zone. Once a new zone is entered (such as international phone calls) a new cost structure applies. (2) Linear effects of distance. Transport costs are increasing proportionally to distance. Fuel consumption can be included in this category since it is a direct function of the distance traveled. For reasons of simplicity, a step-wise approach is often used to establish transport rates. (3) Non-linear effects of distance. Freight distribution costs are growing in a non-linear fashion with distance from the distribution center. This mainly involves the costs of returning back empty. Inversely, intercontinental air transportation costs may be considered, which are not much higher than continental air transportation costs. (4) Multimodal transport chain. Is a combination of linehaul and terminal costs. Transhipment costs at terminals (e.g. ports and airports) which, without involving a distance, increase the friction of distance as effort must be spent to load or unload.

- **Type of product**. Many products require packaging, special handling, or are bulky or perishable. Coal is obviously a commodity that is easier to transport than fruit or fresh flowers as it requires rudimentary storage facilities and can be transhipped using rudimentary equipment. Insurance costs are also to be considered and are

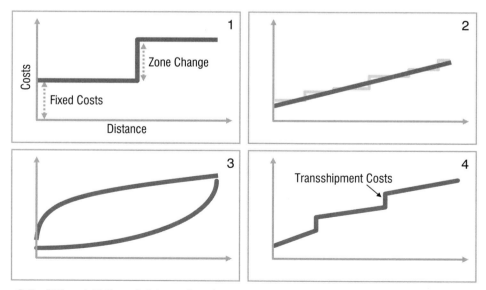

Figure 3.7   Different friction of distance functions

commonly a function of the value to weight ratio and the risk associated with the movement. As such, different economic sectors incur different transport costs as they each have their own transport intensity. For passengers, comfort and amenities must be provided, especially if long-distance travel is involved.

- **Economies of scale.** Another condition affecting transport costs is related to economies of scale or the possibilities to apply them as the larger the quantities transported, the lower the unit cost. Bulk commodities such as energy (coal, oil), minerals and grains are highly suitable to obtain lower unit transport costs if they are transported in large quantities. For instance, moving a barrel of oil over 4,000 km would cost $1 on a 150,000 deadweight tons tanker ship and $3 on a 50,000 deadweight tons tanker ship. A similar trend also applies to container shipping with larger containerships involving lower unit costs.
- **Energy.** Transport activities are large consumers of energy, especially oil. About 60 percent of all global oil consumption is attributed to transport activities. Transport typically accounts for about 25 percent of all the energy consumption of an economy. The costs of several energy-intensive transport modes, such as air transport, are particularly susceptible to fluctuations in energy prices.
- **Trade imbalances.** Imbalances between imports and exports have impacts on transport costs. This is especially the case for container transportation since trade imbalances imply the repositioning of empty containers that have to be taken into account in the total transport costs. Consequently, if a trade balance is strongly negative (more imports than exports), transport costs for imports tend to be higher than for exports. Significant transport rate imbalances have emerged along major trade routes. The same condition applies at the national and local levels where freight flows are often unidirectional, implying empty movements.
- **Infrastructures.** The efficiency and capacity of transport modes and terminals have a direct impact on transport costs. Poor infrastructures imply higher transport costs, delays and negative economic consequences. More developed transport systems tend to have lower transport costs since they are more reliable and can handle more movements.
- **Mode.** Different modes are characterized by different transport costs, since each has its own capacity limitations and operational conditions. When two or more modes are directly competing for the same market, the outcome often results in lower transport costs.
- **Competition and regulation.** Concerns the complex competitive and regulatory environment in which transportation takes place. Transport services taking place over highly competitive segments tend to be of lower cost than on segments with limited competition (oligopoly or monopoly). International competition has favored concentration in many segments of the transport industry, namely maritime and air modes. Regulations, such as tariffs, cabotage laws, labor, security and safety impose additional transport costs.

The transport time component (Figure 3.8) is also an important consideration as it is associated with the service factor of transportation costs, particularly since logistics concomitantly involves cost and time management. The major time related elements are:

- **Transport time.** Concerns the real duration of a transport, which tends to be easily understood since commonly it is a proportional function of distance. Geographical constraints such as weather or technical limitations such as operational speed have a direct impact on transport time. Transport time on road is technically limited

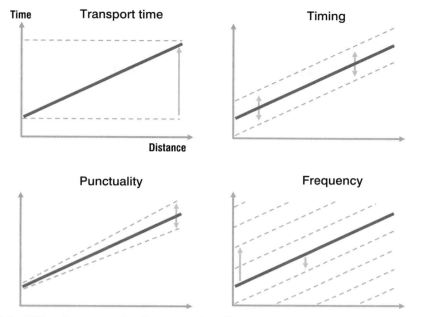

Figure 3.8 Different components of transport time (Source: adapted from Woxenius, 2006)

to legal speed limits. For maritime and air, the limitation mainly concerns fuel economy and design speed. Although rail can accommodate a variety of speeds, tight schedules impose limited variations in operational speeds.

- **Order time**. Almost all transport requires a form of advance preparation, mainly to secure a capacity, an itinerary and a rate. In some cases, the order time is short and a matter of queuing on a first-come first-served basis, while in others orders have to be secured months in advance.
- **Timing**. Involves the usage of a specific departure time, which depending on the mode can have a level of flexibility. While for air and rail travel timing is commonly tight due to fixed schedules and access to intermodal capacity (such as a gate and a takeoff time) commuters and trucking have more flexibility. If there is congestion either at the origin, destination or in-between, trucking companies may elect to modify their schedule accordingly (earlier or later delivery).
- **Punctuality**. The ability to keep a specified schedule, which can be represented as an average deviation from a scheduled arrival time. The longer the distance, the more likely are potential disruptions to affect punctuality. Some movements may have a level of tolerance to disruptions in punctuality while others, such as heading to a business meeting or flows in a just-in-time supply chain, have limited tolerance.
- **Frequency**. The number of departures for a specific time range. The higher the frequency, the better the level of service. However, a high frequency ties up a larger quantity of vehicles. Distance is also a factor for lower frequency since transport demand tends to decline accordingly. Combining long-distance travel and frequency is an expensive undertaking for transport providers as a greater number of vehicles must be assigned to a specific route, as the case of maritime container shipping indicates.

For instance, a maritime shipper may offer a container transport service between a number of North American and Pacific Asian ports. It may take 12 days to service two ports across the Pacific (transport time) and a port call is done every two days (frequency). In order to secure a slot on a ship, a freight forwarder must call at least five days in advance (order time). For a specific port terminal, a ship arrives at 8 a.m. and leaves at 5 p.m. (timing) with the average delay being two hours (punctuality).

Types of transport costs

Mobility tends to be influenced by transport costs. Empirical evidence for passenger vehicle use underlines the relationship between annual vehicle mileage and fuel costs, implying that the higher fuel costs are, the lower the mileage. At the international level, doubling of transport costs can reduce trade flows by more than 80 percent. The more affordable mobility is, the more frequent the movements and the more likely they will take place over longer distances. A wide variety of transport costs can be considered:

- **Freight on board (FOB)**. Is a transport rate where the price of a good is the combination of the factory costs and the shipping costs from the factory to the consumer. In the case of FOB, the consumer pays for the freight transport costs. Consequently, the price of a commodity will vary according to transportation costs and distance.
- **Costs–Insurance–Freight (CIF)**. Is a transport rate that considers the price of the good, insurance costs and transport costs. It implies a uniform delivered price for all customers everywhere, with no spatially variable shipping price. The average shipping price is built into the price of a good. The CIF cost structure can be expanded to include several rate zones, such as one for local, another for the nation and another for exports.
- **Terminal costs**. Costs that are related to loading, transhipment and unloading. Two major terminal costs can be considered: loading and unloading at the origin and destination, which are unavoidable, and intermediate (transhipment) costs that can be avoided.
- **Linehaul costs**. Costs that are a function of the distance over which a unit of freight or passenger is carried. Weight is also a cost function when freight is involved. They include labor and fuel and commonly exclude transhipment costs.
- **Capital costs**. Costs applying to the physical assets of transportation mainly infrastructures, terminals and vehicles. They include the purchase or major enhancement of fixed assets, which can often be a one-time event. Since physical assets tend to depreciate over time, capital investments are required on a regular basis for maintenance.

With an FOB cost structure, customers located nearby will have a lower overall cost than customers that are further away (Figure 3.9). Under the CIF cost structure, every consumer is charged the same price, which commonly reflects the average transport cost. Customers located close to production are "subsidizing" the costs paid by customers located further away. This price structure is common for consumer goods.

Real freight rates can be complicated to calculate for a transport company, especially when there are numerous customers. A common answer to this problem is to establish a set of geographic zones where freight rates are equal (Figure 3.10). The rate is commonly set through the CIF principle where the closest customers in a zone are partially subsidizing the furthest customers. For instance, under a zonal rate system a customer located at D1 pays the same rate as a customer located at D2. Under a distance-based system, the customer at D1 would have paid a lower rate than a customer located at D2. Many transit systems also use a zonal rate structure.

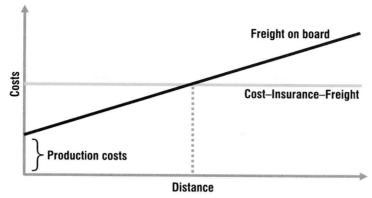

Figure 3.9 FOB and CIF transport costs

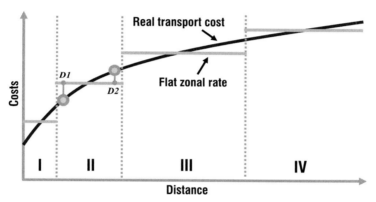

Figure 3.10 Zonal freight rates

Transport providers make a variety of decisions based on their cost structure, a function of all the above types of transport costs. The role of transport companies has sensibly increased in the general context of global commercial geography. However, the nature of this role is changing as a result of the reduction of transport costs but growing infrastructure costs, mainly due to greater flows and competition for land. Each transport sector must consider variations in the importance of different transport costs. While operating costs are high for air transport, terminal costs are significant for maritime transport.

Technological changes and their associated decline in transport costs have weakened the links between transport modes and their terminals. There is less emphasis on heavy industries and more importance given to manufacturing and transport services (e.g. warehousing and distribution). Indeed, new functions are being grafted on to transport activities that henceforward facilitate logistics and manufacturing processes. Relations between terminal operators and carriers have thus become crucial, notably in containerized traffic. They are needed to overcome the physical and time constraints of transhipment, notably at ports.

The requirements of international trade gave rise to the development of specialized and intermediary firms providing transport services. These are firms that do not physically transport goods, but are required to facilitate the grouping, storage and handling of freight as well as the complex paperwork and financial and legal transactions involved in international trade. Examples included freight forwarders, customs brokers, warehousing, insurance agents and banking, etc. Recently, there has been a trend to consolidate these different intermediate functions, and a growing proportion of global trade is now being organized by multinational corporations that are offering door-to-door logistics services.

Concept 4 – Transport supply and demand

The supply and demand for transportation

What are the differences between a Boeing 747, an oil tanker, a car and a bicycle? Many indeed, but they each share the common goal of fulfilling a derived transport demand, and they thus all fill the purpose of supporting mobility. Transportation is a service that must be utilized immediately and thus cannot be stored. Mobility must occur over transport infrastructures, providing a transport supply. In several instances, transport demand is answered in the simplest means possible, notably by walking. However, in some cases elaborate and expensive infrastructures and modes are required to provide mobility, such as for international air transportation.

An economic system including numerous activities located in different areas generates movements that must be supported by the transport system. Without movements infrastructures would be useless and without infrastructures movements could not occur, or would not occur in a cost-efficient manner. This interdependency can be considered according to two concepts: transport supply and transport demand (Figure 3.11):

- **Transport supply.** The expression of the capacity of transportation infrastructures and modes, generally over a geographically defined transport system and for a specific period of time. Therefore, supply is expressed in terms of infrastructures (capacity), services (frequency) and networks. The number of passengers, volume (for liquids or containerized traffic) or mass (for freight) that can be transported per unit of time and space is commonly used to quantify transport supply.
- **Transport demand.** The expression of the transport needs, even if those needs are satisfied, fully, partially or not at all. Similar to transport supply, it is expressed in terms of number of people, volume or tons per unit of time and space.

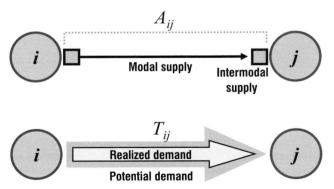

Figure 3.11 Transport supply and demand

Transport supply is generally expressed by A_{ij}; transport supply between location i and j. Indirectly it combines modal supply, the capacity of a mode to support traffic, and intermodal supply, the capacity to tranship traffic from one mode to the other. Transport demand is represented by T_{ij}, which expresses the transport demand between location i and j. The potential transport demand would be the amount of traffic if transport costs were negligible. The realized transport demand, a subset of the potential transport demand, is the traffic that actually takes place, namely in view of costs between the origins and the destinations.

There is a simple statistical way to measure transport supply and demand for passengers or freight:

The **passenger-km** (or passenger-mile) is a common measure expressing the realized passenger transport demand as it compares a transported quantity of passengers with a distance over which it gets carried. The ton-km (or ton-mile) is a common measure expressing the realized freight transport demand. Although both the passenger-km and ton-km are most commonly used to measure realized demand, the measure can equally apply for transport supply.

For instance, the transport supply of a Boeing 747-400 flight between New York and London would be 426 passengers over 5,500 kilometers (with a transit time of about five hours). This implies a transport supply of 2,343,000 passenger-km. In reality, there could be a demand of 450 passengers for that flight, or of 2,465,000 passenger-km, even if the actual capacity could only be 426 passengers (if a Boeing 747-400 is used). In this case the realized demand would be 426 passengers over 5,500 kilometers out of a potential demand of 450 passengers, implying a system where demand is at 105 percent of capacity.

Transport demand is generated by the economy, which is composed of persons, institutions and industries and which generates movements of people and freight. When these movements are expressed in space they create a pattern, which reflects mobility and accessibility. The location of resources, factories, distribution centers and markets is obviously related to freight movements. Transport demand can vary under two circumstances that are often concomitant: the quantity of passengers or freight increases or the distance over which these passengers or freight are carried increases. Geographical considerations and transport costs account for significant variations in the composition of freight transport demand between countries. For the movements of passengers, the location of residential, commercial and industrial areas tells a lot about the generation and attraction of movements.

The realized transport demand, expressed in passenger-km or ton-km, can increase for two reasons (Figure 3.12). The first is obviously that more passengers or freight are being carried. This is an outcome of growth in population, production, consumption and income. The second is a growth in the average distance over which passengers or freight are being carried. Industrial relocation, economic specialization (factors linked with globalization) and suburbanization are relevant factors behind this trend. These two factors often occur concomitantly: more passengers and freight being carried over longer distances.

Supply and demand functions

Transport supply can be simplified by a set of functions representing what are the main variables influencing the capacity of transport systems. These variables are different for each mode (Table 3.3). For road, rail and telecommunications, transport supply is

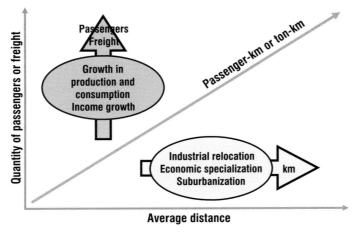

Figure 3.12 Growth factors in transport demand

Table 3.3 Factors behind freight transport demand

Economy	General derived demand impact. Linked with GDP. Function of the structure of the economy in terms of resources, goods and services.
Industrial location	Effect on ton-km and on modal choice.
Spatial structure	Effect on ton-km. Function of international trade structure. Containerization and intermodal transportation.
International agreements	Both concerning trade and transportation. Economic specialization. Increased transborder traffic. Simplified custom procedures.
JIT practices and warehousing	Decreased inventories. More shipments. Smaller line hauls. Shift to faster and more reliable modes. Use of third-party logistics providers.
Strategic alliances	Between carriers, shippers and often producers and retailers. Lower distribution costs.
Packaging and recycling	Increased transportability of products. Lower freight density. Reverse distribution.
Regulation and deregulation	Increased competition, level of service and lower costs. Growth of intermodal transportation.
Fuel costs, taxes and subsidies	Large and volatile cost components, specifically for energy intensive modes. Preferred mode or carrier.
Infrastructure and congestion	Efficiency, operating costs and reliability.
Safety and environmental policies	Operating speed, conditions and costs. Capacity and weight limits.
Technology	Containerization, double-stacking, automation and robotics, handling and interchange systems and automated terminals. Information systems. Lower costs, increased efficiency and reliability and new opportunities.

Source: adapted from Cambridge Systematics, 1996

often dependent on the capacity of the routes and vehicles (modal supply), while for air and maritime transportation, transport supply is strongly influenced by the capacity of the terminals (intermodal supply).

- **Modal supply**. The supply of one mode influences the supply of others, such as for roads where different modes compete for the same infrastructure, especially in congested areas. For instance, transport supply for cars and trucks is inversely proportional since they share the same road infrastructure.
- **Intermodal supply**. Transport supply is also dependent on the transhipment capacity of intermodal infrastructures. For instance, the maximum number of flights per day between Montreal and Toronto cannot be superior to the daily capacity of the airports of Montreal and Toronto, even though the Montreal–Toronto air corridor has potentially a very high capacity.

Transport demand tends to be expressed at specific times that are related to economic and social activity patterns. In many cases, transport demand is stable and recurrent, which allows a good approximation in planning services. In other cases, transport demand is unstable and uncertain, which makes it difficult to offer an adequate level of service. For instance, commuting is a recurring and predictable pattern of movements, while emergency response vehicles such as ambulances deal with an unpredictable demand. Transport demand functions vary according to the nature of what is to be transported:

- **Passengers**. For the road and air transport of passengers, demand is a function of demographic attributes of the population such as income, age, standard of living, race and sex, as well as modal preferences.
- **Freight**. For freight transportation, the demand is a function of the nature and the importance of economic activities (GDP, commercial surface, number of tons of ore extracted, etc.) and of modal preferences. Freight transportation demand is more complex to evaluate than passengers.
- **Information**. For telecommunications, the demand can be a function of several criteria including the population (telephone calls) and the volume of financial activities (stock exchange). The standard of living and education levels are also factors to be considered.

Supply/demand relationships

Relationships between transport supply and demand continually change, but they are mutually interrelated. From a conventional economic perspective, transport supply and demand interact until an equilibrium is reached between the quantity of transportation the market is willing to use at a given price and the quantity being supplied for that price level.

Many transport systems behave in accordance with supply and demand, which are influenced by cost variations. On Figure 3.13 the demand curve assumes that if transport costs are high, demand is low as the consumers of a transport service (either freight or passengers) are less likely to use it. If transport costs are low, the demand would be high as users would get more services for the same cost. The supply curve behaves inversely. If costs are high, transport providers would be willing to supply high quantities of services since high profits are likely to arise under such circumstances. If costs are low, the quantity of transport services would be low as many providers would see little benefit in operating at a loss.

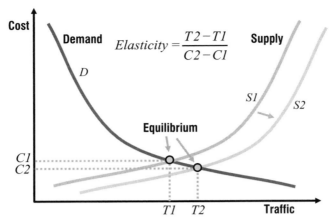

Figure 3.13 Classic transport demand/supply function

The equilibrium point represents a compromise between what users are willing to pay and what providers are willing to offer. Under such circumstances, an amount of traffic $T1$ would flow at an operating cost $C1$. If because of an improvement a larger amount of service is possible for the same cost (the supply curve moves from $S1$ to $S2$), a new equilibrium will be reached with a quantity of traffic $T2$ at a price $C2$. Elasticity refers to the variation of the demand in accordance with the variation of the price: the higher it is, the more the traffic in a transport system is influenced by costs variations.

However, several considerations are specific to the transport sector which complexify supply/demand relationships:

● **Entry costs**. The costs incurred to operate at least one vehicle in a transport system. In some sectors, notably maritime, rail and air transportation, entry costs are very high, while in others such as trucking, they are very low. High entry costs imply that transport companies will consider seriously the additional demand before adding new capacity or new infrastructures (or venturing in a new service). In a situation of low entry costs, the market sees companies coming in or dropping, fluctuating with the demand. When entry costs are high, the emergence of a new player is uncommon while dropping out is often a dramatic event linked to a large bankruptcy. Consequently, transport activities with high entry costs tend to be oligopolistic while transport activities with low entry costs tend to have many competitors.
● **Public sector**. Few other sectors of the economy have seen such a high level of public involvement than transportation, which creates many disruptions in conventional price mechanisms. The provision of transport infrastructures, especially roads, was massively funded by governments, namely for the sake of national accessibility and regional equity. Transit systems are also heavily subsidized, namely to provide accessibility to urban populations and more specifically to the poorest segment judged to be deprived in mobility. As a consequence, transport costs are often considered as partially subsidized. Government control (and direct ownership) was also significant for several modes, such as rail and air transportation in a number of countries. Recent years have however been characterized by less governmental involvement and deregulation.

- **Elasticity**. The variation of demand in response to a variation of cost. For example, an elasticity of −0.5 for vehicle use with respect to vehicle operating costs means that an increase of 1 percent in operating costs would imply a 0.5 percent reduction in vehicle mileage or trips. Variations of transport costs have different consequences for different modes, but transport demand has a tendency to be inelastic. While commuting tends to be inelastic in terms of costs, it is elastic in terms of time. This fact is underlined by empirical evidence that shows that drivers are marginally influenced by variations in the price of fuel in their commuting behavior, especially in highly motorized societies. Since work is a major, if not the only, source of income, commuting can simply not be forfeited under any circumstances short of being cost prohibitive. For economic sectors where freight costs are a small component of the total production costs, variations in transport costs have limited consequences on the demand. Activities that confer limited economic benefits tend to have high elasticities (Figure 3.14). Social and recreation-oriented movements are commonly those whose users have the least cost tolerance. Consequently, as transport costs increase, recreational movements experience the fastest decline. For air transportation, especially the tourism sector, price variations have significant impacts on the demand.

As transport demand is a derived demand from individuals, groups and industries it can be desegregated into a series of partial demands fulfilled by the adaptation and evolution of transport techniques, vehicles and infrastructures to changing needs. Moreover, the growing complexity of economies and societies linked with technological changes force the transport industry to make constant changes. This leads to growing congestion, a reduction in transport safety, a degradation of transport infrastructures and growing concerns on environmental impacts.

Generally, transport demand is variable in time and space whereas transport supply is fixed. When demand is lower than supply, transit times are stable and predictable, since the infrastructures are able to support the demand. When transport demand exceeds supply for a period in time, there is congestion with significant increases in transit times and higher levels of unpredictability. A growth of the transport demand

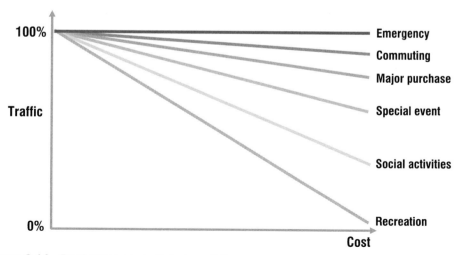

Figure 3.14 Road transport elasticity by activity

increases the load factor of a transport network until transport supply is reached. Speed and transit times drop afterwards. The same journey can thus have different durations according to the time of the day.

Method 1 – The transportation problem (linear programming)

Linear inequalities

Mathematical programming is a set of techniques used to determine the optimal solution to allocation problems. Linear programming, the most widely used of these techniques, is a method that helps to attain a desired objective, such as minimizing costs or maximizing profits, subject to constraints on the amount of commodities required or the resources available. The term linear implies proportionality; programming is used in the mathematical sense of selecting an optimum allocation of resources. In transport geography, linear programming is used to optimize flow patterns in which optimality implies the minimization of distance or transportation costs subject to certain constraints.

The following example is used to demonstrate the methodology. A government aid agency requires quantities of materials to assist in disaster relief of a region devastated by a major earthquake. The aid agency has two suppliers A and B, both of which are depots. Supplier A can deliver 6 tons of food, 2 tons of medical materials and 2 tons of shelter supplies. Supplier B can supply 2 tons of food, 8 tons of medical materials and 2 tons of shelter supplies. The minimum requirements of aid materials for the devastated region consist of 120 tons of food, 160 tons of medical materials and 80 tons of shelter supplies. See Table 3.4.

First the problem is stated in algebraic form. Let the minimum requirements consist of x loads for supplier A and y loads for supplier B. Thus:

For food:	$6x + 2y \geq 120$
For medical materials:	$2x + 8y \geq 160$
For shelter supplies:	$2x + 2y \geq 80$

Moreover, x and y cannot be negative, so the following conditions are added:

$x \geq 0$
$y \geq 0$

The above set of five inequalities is called the constraints of the problem. A particular point (x, y) satisfying all the constraints is called a feasible point. The set of all points satisfying the constraints is the constraint set. Geometrically the constraint set represents a region of the x, y plane called the feasible region. The constraint set is graphed in Figure 3.15 (shaded region)

Table 3.4 Amount of materials required from two suppliers

	Supplier A	Supplier B	Minimum requirements
Food	6	2	120
Medical materials	2	8	160
Shelter supplies	2	2	80

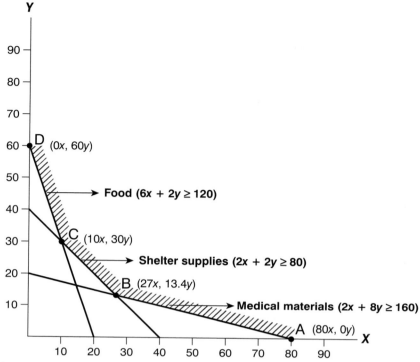

Figure 3.15 Linear inequalities

Notice that the region is unbounded and includes segments AB, BC and CD and the axis *Ax* and *Dy*. The vertexes A, B, C, and D are feasible points. There are four points where the inequalities cut each other on the *x* and *y* axes. For points A, B, C and D, the coordinates are given in Table 3.5.

A vertex of the constraint set must be: (1) the intersection of a pair of constraint lines and (2) a feasible point. This constraint set or feasible region represents a convex region in the *x*, *y* plane. Any point in this region is a feasible solution, namely that it satisfies the inequalities.

Linear programming

In practical problems, the best, or optimal, solution is sought. Typically, the objective is to minimize costs or to maximize profits when a set of activities is carried out. In this simple problem, the minimum requirements of material for the aid agency at the

Table 3.5 Coordinates of the feasible region

Point	x coordinate	y coordinate
Point A	80	0
Point B	27	13.4
Point C	10	30
Point D	0	60

lowest transportation cost are to be met. Let the distance between Supplier A and the agency be 60 km and from Supplier B be 30 km. The problem can then be modeled as:

Minimize: $Z = 60x + 30y$

Subject to the constraints:

$6x + 2y \geq 120$
$2x + 8y \geq 160$
$2x + 2y \geq 80$
$x \geq 0$
$y \geq 0$

Z, a linear function of x and y, is called the objective function; x and y are the activity variables to be determined. The model of the problem in terms of linear constraints and linear objective is called a linear program. Program must be taken in the sense of model or plan and not a computer program.

Optimal solution

Before solving a linear program, there is a need to consider the objective function more closely. The objective function is not represented by an area on the graph, but by an infinite number of parallel lines. The dashed parallel lines in Figure 3.16 represents the

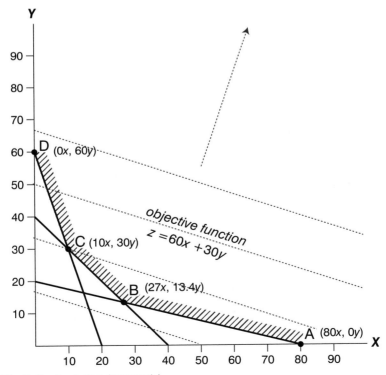

Figure 3.16 Optimal solution (geometric)

equations $Z = 60x + 30y$ for several values of Z. The further these lines shift to the right, the greater the value of Z they represent.

Taking the coordinates of the feasible region from Table 3.5, the distance value $60x$ and $30y$ of the objective function ($Z = 60x + 30y$) can be substituted into the equations as shown in Table 3.6.

The optimum (minimum cost) value of Z occurs when Z passes the point C where the lines $6x + 2y = 60$ and $2x + 2y = 40$ intersect. Solving these equations gives $x = 10$ and $y = 30$ and so the minimum transport value of $Z = 1,500$ load-kilometers. See Tables 3.7 and 3.8.

Advanced linear programming

Linear programming is the simplest and most widely used method of mathematical programming. The procedure explained above may be used to determine an optimal solution to problems involving the physical transportation of items. Linear programming can be used in selecting the shortest paths for distributing commodities from a variety of dispatch points to numerous destinations with a view to minimize total transportation costs. The method is extremely useful in evaluating the impact of changes in a transportation network such as road extension on various location problems.

Table 3.6 Calculating the optimal solution

Point	x coordinate	y coordinate	Load-kilometers
Point A	80	0	$(60 \times 80) + (0) = 4,800$
Point B	27	13.4	$(60 \times 27) + (30 \times 13.4) = 2,022$
Point C	**10**	**30**	$\mathbf{(60 \times 10) + (30 \times 30) = 1,500}$
Point D	0	60	$(0) + (30 \times 60) = 1,800$

Table 3.7 Optimal solution (algebraic)

Material	Food	Medical materials	Shelter supplies	Loads
	$6x + 2y \geq 120$	$2x + 8y \geq 160$	$2x + 2y \geq 80$	
Supplier A (x)	6×10 tons = 60 tons	2×10 tons = 20 tons	2×10 tons = 20 tons	10 loads
Supplier B (y)	2×30 tons = 60 tons	8×30 tons = 240 tons	2×30 tons = 60 tons	30 loads
Total	120 tons	260 tons	80 tons	40 loads

Table 3.8 Optimal solution (summary)

Material	Food	Medical materials	Shelter supplies	Loads
Supplier A	60 tons	20 tons	20 tons	10 loads
Supplier B	60 tons	240 tons	60 tons	30 loads
Total	120 tons	260 tons	80 tons	40 loads

Standard linear programming assumes simple transportation problems in which supplies and demands and the costs of flows between pairs of location are known. But transportation problems are more complicated. They often relate to problems involving modal split, transhipment and several allocations where decisions are taken at different stages because they involve assembling components. These transportation problems can also involve probability constraints and queuing theory. To solve these problems, more complex linear programming methods to transportation problems have been developed to meet variable costs, uncertain demand and even non-linear functions. As a result, linear programming and its extension represents one of the most important tools for transport geographers in the analysis of flows in a network.

Method 2 – Market area analysis

Market size and shape

Each economic activity possesses a location, but the various demands (raw materials, labor, parts, services, etc.) and flows it generates also have a spatial dimension called a market area.

> **Market area** The surface over which a demand or supply offered at a specific location is expressed. For a factory it includes the areas where its products are shipped; for a retail store it is the tributary area from which it draws its customers.

Transportation is particularly important in market area analysis because it impacts on the location of the activities as well as their accessibility. The size of a market area is a function of its threshold and range:

- **Market threshold.** Minimum demand necessary to support an economic activity such as a service. Since each demand has a distinct location, a threshold has a direct spatial dimension. The size of a market has a direct relationship with its threshold.
- **Market range.** The maximum distance each unit of demand is willing to travel to reach a service or the maximum distance a product can be shipped to a customer. The range is a function of transport costs, time or convenience in view of intervening opportunities. To be profitable, a market must have a range higher than its threshold.

Figure 3.17 considers a fairly uniform distribution of customers on an isotropic plain and a single market where goods and services may be purchased. If each customer is willing to purchase one unit per day and the market needs to sell 11 units per day to cover its costs (production or acquisition), then the threshold of the market would be the circle of distance $D(T)$ from the market. However, 29 customers per day, including customers 1 and 2, patronize the market, of which an extra 18 are beyond the threshold distance $D(T)$. They contribute directly to the profitability of the market. The market range of all these customers is below distance $D(R)$. Beyond this range, customers are unwilling to go to the market, such as customer 3. There are different thresholds according to the variety of products or services that can be offered on a market. A threshold may be as low as 250 people for a convenience store or as high as 150,000 people for a theater. If the demand falls below the threshold level, the activity will run at a loss and will eventually fail. If the demand increases above the minimum, the activity will increase its profits, which may also lead to increased

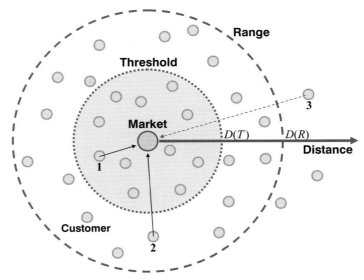

Figure 3.17 Market threshold and range

competition from new service activities. The frequency of use of goods or services is important in assessing the extent of the market threshold, which is often linked to the level of income. A movie theater needing 500 visitors per night will require a threshold population of around 150,000 if the average number of visits is one per year. But, if the average number of visits is three per year, the population threshold drops to 50,000; then three movie theaters instead of one can be supported by the same population.

In the case of a single market area, its shape in an isotropic plain is a simple concentric circle having the market range as radius. Since the purpose of commercial activities is to service all the available demand, when possible, and that the range of many activities is limited, more than one location is required to service an area. For such a purpose, a hexagonal-shaped structure of market areas represents the optimal market shape under a condition of isotropy. This shape can be modified by non-isotropic conditions mainly related to variations in density and accessibility.

The left part of Figure 3.18 represents the standard hexagonal shape of a set of markets under isotropic conditions. Each market has the same market area and is evenly spaced. This theoretical condition cannot obviously be found in reality. The two

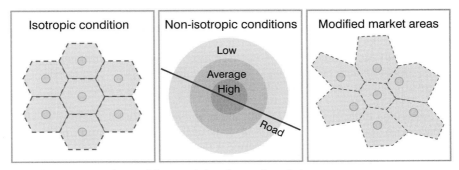

Figure 3.18 Non-isotropic conditions and the shape of market areas

most important non-isotropic conditions impacting on the shape of market areas are differences in density and accessibility. The middle part represents conditions where there is a concentric gradient of population density (from low to high) and a highway crossing through. Their possible outcome on the shape of market areas is portrayed on the right part of the figure.

Economic definition of a market area

A market depends on the relationship between supply and demand. It acts as a price fixing mechanism for goods and services. Demand is the quantity of a good or service that consumers are willing to buy at a given price. It is high if the price of a commodity is low, while in the opposite situation – a high price – demand would be low. Outside market price, demand can generally be influenced by the following factors:

- **Utility**. While goods and services that are necessities (such as food) do not see much fluctuation in demand, the demand for items deemed of lesser utility (even frivolous) would vary according to income and economic cycles.
- **Income level**. Income, especially disposable income, is directly proportional with consumption. A population with a high income level has much more purchasing power than a population with a low income.
- **Inflation**. Involves an increase in the money supply in relation to the availability of assets, commodities, goods and services. Commonly the outcome of an indirect confiscation of wealth through an over-issuance of currency by central banks and governments. Although it directly influences prices, inflation is outside the supply–demand relationship and decreases purchasing power, if wages are not increased accordingly.
- **Taxation**. Sale and value-added taxes can have an inhibiting effect on sales of goods and services as they add to the production costs and claim a share of consumers' income.
- **Savings**. The quantity of capital available in savings can provide a potential to acquire consumption goods. Also, people may restrain from consuming if saving is a priority, namely in periods of economic hardship. The wide availability of credit in a fiat currency system has considerably skewed the relationships between savings and consumption as it promotes current consumption levels, but at the expense of future consumption.

Supply is the amount of goods or services that firms or individuals are able to produce, taking account of a selling price. Outside price, supply can generally be influenced by the following factors:

- **Profits**. Even if the sales of a product are limited, if profits are high an activity providing goods or services may be satisfied with this situation. This is particularly the case for luxury goods. If profits are low, an activity can cease, thus lowering the supply.
- **Competition**. Competition is one of the most important mechanisms for establishing prices. Where competition is absent (an oligopoly), or where there is too much (over-competition), prices artificially influence supply and demand.

According to the market principle, supply and demand are determined by the price, which is an equilibrium between both. It is often called equilibrium price or market price. This price is a compromise between the desire of firms to sell their goods and services at the highest price possible and the desire of consumers to buy goods and services at the lowest possible price.

For many economists, the market is a point where goods and services are exchanged and does not have a specific location, since it is simply an abstraction of the relationships between supply and demand. It is important to nuance in this reasoning since most of the time consumers must move in order to acquire a good or service. The producer must also ship a commodity to a place where the consumer can buy it, be it at the store or at his/her residence (in the case of online shopping). The concept of distance thus must be considered concomitantly with the concept of market. In those conditions, the real price includes the market price plus the transport price from the market to the location of final consumption.

Competition over market areas

Competition involves similar activities trying to attract customers. Although the core foundation of competition for a comparable good or service is price, there are several spatial strategies that impact the price element. The two most common are:

- **Market coverage**. Activities offering the same service will occupy locations with a view to offering goods or services to the whole area. This aspect is well explained by the central place theory and applies well for sectors where spatial market saturation is a growth strategy (fast food, coffee shops, etc.). The range of each location will be a function of customer density, transport costs and the location of other competitors.
- **Range expansion**. Existing locations try to expand their ranges in order to attract more customers. Economies of scale resulting in larger retail activities are a trend in that direction, namely the emergence of shopping malls. Taken individually, each store would have a limited range. However, as a group they tend to attract additional customers from wider ranges for many reasons. First, a complementarity of goods or services is offered. A customer would thus find it convenient to be able to buy clothes, shoes and personal care products at the same location. Second, a diversity of the same goods or services is offered even if they compete between one other. Third, other related amenities are provided such as safety, food, indoor walking space, entertainment and also parking space.

Making market area competition models operational has been the object of numerous approaches. The early work of Hotelling (1929) with his principle of market competition, created the foundations of market area analysis by considering factors such as retail location and distance decay. Hotelling investigated how sellers would choose locations along a linear market. He assumed that the product was uniform so customers would buy from the nearest seller and that the friction of distance was linear and isotropic. The total price to the customer is thus the market price plus the transport price (time or effort spent to go to the market). Under such circumstances, two competitors will select locations A and B for optimal market coverage (Figure 3.19). With P1 being the market price, the market boundary would be F1 (point of cost indifference) since right of F1, customers would get a lower price at location B instead of location A and left of F1, customers would get a lower price at location A. If for any reasons, location A is able to lower the market price from P1 to P2, then its market area would expand at the expense of location B, from F1 to F2.

Later, factors such as market size were taken into consideration (Reilly's law) permitting to build complex market areas. Since market areas are often non-monopolistic, this factor was included with market areas becoming ranges of probabilities that customers will attend specific locations (Huff's law). Although market areas are

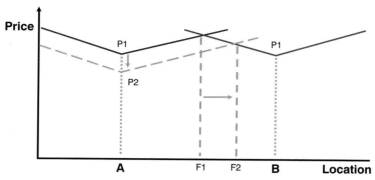

Figure 3.19 Hotelling principle of market competition

particularly relevant for retail analysis, the methodology also applies to time-dependent activities, such as freight distribution.

The purpose of Reilly's law of retail gravitation (1931) is to find a point of indifference between two locations, so the trading area of each can be determined. This point is assumed to be a function of the distance between two locations pondered by their respective size (population often used for this purpose). One location can thus be more attractive than another. For instance, on Figure 3.20 two locations are 75 km apart. According to the Hotelling principle, the point of indifference should be halfway in between (35 km). However, since location A has a larger population, it is assumed that it will draw more customers. Under such circumstances, the point of indifference is 45.9 km away from location A. Huff's retail model (1963) assumes that customers have a choice to patronize a location in view of other alternatives and thus a market area is expressed as probabilities (unless there are no other alternative locations). The point of indifference becomes the point of equal probability that a customer will patronize one location or another. On Figure 3.20, a customer has a greater chance (0.71) to patronize location A at the midpoint than to patronize location B (0.29). The advantage of Huff's retail model is that it leaves room for customer choice.

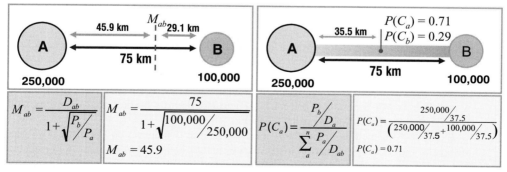

Figure 3.20 Reilly's and Huff's laws

GIS and market areas analysis

GIS have become useful tools to evaluate market areas, especially in retailing. With basic data, such as a list of customers and their addresses (or ZIP codes), it is relatively easy to evaluate market areas with a reasonable level of accuracy, a task that would have been much more complex beforehand. With GIS, market area analysis left the realm of abstraction to become a practical tool used by retailers and service providers. The market area is a polygon which can be measured and used to perform operations such as intersection (zones of spatial competition) or union (area serviced). Among the major methods of using a GIS to evaluate market areas are (Figure 3.21):

- **Concentric circles**. The simplest method since it assumes an isotropic effect of distance in all directions. The radius represents the maximum distance a customer is willing to travel. It is useful to have a rough overview of the situation when limited information is available. Buffer creation, a common GIS procedure, associates each concentric circle with a distance (or a time value). They can include the threshold and the range of a store. On Figure 3.21 three concentric circles have been created with a 5-minute distance increment each. Five minutes is assumed to be the threshold of this activity, while the range is estimated to be 10 minutes.
- **Share by polygon**. When data is available at the zonal level, such as the ZIP code, the market area can be expressed as a share of the market for each zone. Share by polygon can be estimated in many ways. For instance, it can be an aggregation of individual customers within a geographical unit of reference (ZIP code, census bloc, etc.) or a statistical calculation based on a set of representative variables, such as distance, population, income and age. On Figure 3.21 a market share was calculated for each unit and then classified according to a graduated color expressing the level of membership to the store's market area.
- **Star map**. Composed of straight lines between each customer and locations. It is an indication of the extent and the shape of the market area and particularly relevant for distribution systems where relationships between distribution centers and their customers is shown. It depicts a market area as a set of customers connected to a store. Qualitative and quantitative attributes can be attached to each vector.

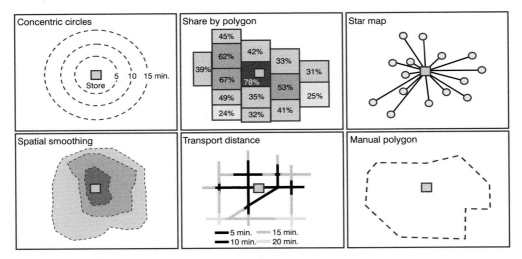

Figure 3.21 GIS methods to estimate market areas

- **Spatial smoothing**. A trend surface based on the location of actual customers. The higher the density of customers (the importance of each customer can be weighted), the higher the membership to a market area. Spatial smoothing is the outcome of statistical modeling that interpolates data from a known set of points (customers) to a continuous surface. The density of customers thus becomes a statistical surface expressing the market area. On Figure 3.21, spatial smoothing created a statistical surface expressing three levels of membership to the store's market area (high, average and low).
- **Transport distance**. Particularly useful for retailing or any activity that depends on consumer accessibility or timed deliveries. A measure of transport distance, often driving time in minutes, is calculated on road segments radiating from the facility location. It takes into consideration transport distance, often quite different from Euclidean distance, as well as the different capacities of road segments (number of lanes, driving speed, turn penalties, etc.). A new layer is created where each former road vector is segmented according to distance/time decay through a routing procedure that originates from the store. On Figure 3.21, roads are segmented according to 5-minute driving time increments from the store location.
- **Manual polygon**. Based on local knowledge, common sense and judgment. It may implicitly consider other methods. Manual polygons are created with tracing where the analyst evaluates a market area from a set of assumptions, often based on specific expertise and empirical knowledge about that market. For instance, the analyst may empirically know that for various reasons few customers may come from a nearby neighborhood, excluding it from the market area. It may also been known that few customers come from further away than a specific street, making that street a boundary for the market area.

CASE STUDY The financing of transportation infrastructure

Private participation in transport infrastructure

Transportation infrastructure, like several infrastructure classes, has a significant level of public involvement ranging from direct ownership and management to a regulatory framework that defines operational standards. This is notably the outcome of a tradition where transportation, particularly roads, was seen as a public good to be free of access and not subject to market forces. A similar trend applied to port and airport infrastructures that were placed under the management of public authorities. Although rail freight has essentially been a private endeavor in the United States, it was significantly regulated by the Interstate Commerce Commission in terms of fares and level of service. Rail terminals are mostly managed by private rail operators, while the warehousing/distribution industry is almost completely private. Like many civil engineering sectors, the private sector can be involved in transportation project delivery, which can include design and construction, project management, such as maintenance and operations, and project financing, namely raising capital.

The trend towards greater private involvement in the transportation sector initially started with the privatization (or deregulation) in the 1980s of existing transportation firms. New relationships started to be established with financial institutions since public funding and subsidies were substantially reduced and new competitors entered the market. Then, many transportation firms were able to expand through mergers and acquisitions into new networks and markets. Some, particularly in the maritime and terminal operation sectors, became large multinational enterprises controlling substantial assets and revenues.

The transport finance sector involves two major groups (Figure 3.22):

- **Providers**. Concern the major actors that can be tapped to finance transport infrastructure. Various levels of government are the conventional source as well as private lenders (e.g. bond issuers) that simply provide capital. Private investors, namely terminal operators, are a relatively new source of financing commonly taking direct involvement in the management of transport infrastructure and equipment. All actors, particularly the private sector, expect a form of return on their capital investments.
- **Recipients**. Investments in transport infrastructure, once completed, eventually impact an array of recipients. The most obvious concern the users of the infrastructure who contribute to transport finance mainly by the usage fees (e.g. tolls) they provide. There are also others that contribute or benefit indirectly. Beneficiaries are actors that even if not using the infrastructure directly will derive a benefit. For instance, a new terminal (or additional traffic at an existing terminal) will benefit the regional economy with additional employment, a larger taxation base and a greater demand for a range of goods and services. The general public, particularly if governments are involved in the financing, will contribute indirectly through taxes.

As the freight transport sector became increasingly efficient and profitable it received the attention of large equity firms in search of returns on capital investment. A new wave of mergers and acquisitions is taking place at the global and national levels as equity firms see transport terminals as an asset class that has an intrinsic value (real estate), an operational value (rent, income) as well as providing a form of diversification:

- **Asset**. Globalization and the growth of international trade have made many terminal assets more valuable since they are key elements in establishing and maintaining global supply chains. Terminals occupy premium locations conferring accessibility to either maritime, rail or road transport systems. These locations, such as waterfronts, are rare and cannot easily (if at all) be substituted for other locations. Traffic growth is commonly linked with valuation growth of a transport infrastructure since the same amount of land generates a higher income. Thus, terminals and some transport infrastructure are seen as fairly liquid assets with an anticipation that they will gain in value.

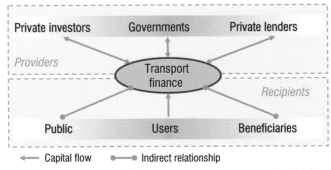

Figure 3.22 Actors in transport finance (Source: adapted from Russell, 2008)

- **Source of income**. In addition to be an asset, intermodal terminals also guarantee a source of income linked with the traffic volume they handle. They have a constant revenue stream with a fairly limited seasonality (unlike many bulk terminals), which make terminals particularly attractive in light of the substantial traffic growth that most terminal facilities have experienced. Traffic growth expectations result in income growth expectations.
- **Diversification**. Intermodal terminals offer a form of functional and geographical asset diversification for a holding company and help lower risks. Terminals represent an asset class on their own. They also offer a potential of geographical diversification as holding terminals at different locations help mitigate risks linked with a specific regional or national market.

Causes and forms of public divesture

With the growing inability of governments to manage and fund transport infrastructure, recent decades have seen deregulation and more active private participation. Many factors have placed pressures on public officials to consider the privatization of transport infrastructure, including terminals:

- **Fiscal problems**. The level of government expenses in a variety of social welfare practices is a growing burden on public finances, leaving limited options but divesture. Current fiscal trends clearly underline that all levels of governments have limited if any margin and that accumulated deficits have led to unsustainable debt levels. Since transport infrastructures are assets of substantial value, they are commonly a target for privatization. This is also known as "monetization" where a government seeks a large lump sum by selling or leasing an infrastructure for budgetary relief.
- **High operating costs**. Mainly due to managerial and labor costs issues, the operating costs of public transport infrastructure, including maintenance, tend to be higher than their private counterparts. Private interests tend to have a better control of technical and financial risks, are able to meet construction and operational guidelines, as well as providing a higher quality of services to users. Operating deficits thus must be covered by public funds, namely through cross-subsidies. Otherwise, users are paying a higher cost than a privately managed system. High operating costs are thus a significant incentive to privatize.
- **Cross-subsidies**. Several transport infrastructures are subsidized by revenues from other streams since their operating costs cannot be compensated by existing revenue. For instance, public transport systems are subsidized in part by revenues coming from fuel taxes or tolls. Privatization can thus be a strategy to end cross-subsidizing by taping private capital markets instead of relying on public debt. The subsidies can either be reallocated to fund other projects (or pay existing debt) or removed altogether, thus reducing taxation levels.
- **Equalization**. Since public investments are often a political process facing pressures from different constituents to receive their "fair share", many investments come with "strings attached" in terms of budget allocation. An infrastructure investment in one region must often be compensated with a comparable investment in another region or project, even if this investment may not be necessary. This tends to significantly increase the general cost of public infrastructure investments, particularly if equalization creates non-revenue generating projects. Thus, privatization removes the equalization process for capital allocation as private enterprises are less bound to such a forced redistribution.

One of the core goals of privatization concerns the derived efficiency gains compared to the transaction costs of the process. Efficiency gains involve a higher output level with the same or fewer input units, implying a more productive use of the infrastructure. Transaction costs are the costs related to the exchange (from public to private ownership) and could involve various buyouts, such as compensation for existing public workers. For public infrastructure, they tend to be very high and involve delays due to the regulatory changes of the transaction.

Privatization and financing models

Once privatization is considered, an important issue concerns which form it will take. There are several options ranging from a complete sale of the infrastructure to a management contract where the public sector retains ownership and a share of the revenues. Three forms of privatization are particularly dominant:

- **Sale or concession agreement (lease) of existing facilities**. Divesture is part of a political agenda which began with deregulation. As discussed before, budget relief is sought because of mismanagement; the public sector is essentially forced to sell or lease some of its infrastructures. For a sale, the infrastructure is transferred on a freehold basis with the requirement that it will be used for its initial purpose (unless another agreement was negotiated). For a concession agreement, it commonly takes the form of a long-term lease with the requirement that the concessionaire maintains, upgrades and builds infrastructure and equipment.
- **Concessions for new projects**. Tap new sources of capital outside conventional public funding. It can take place in the context of fiscal restraints or as a way to experiment with a more limited form of privatization since existing assets remain untouched. It also confers the advantage of getting the latest technical and managerial expertise for the infrastructure project.
- **Management contract**. While ownership remains public, management is given to a private operator, commonly through a bidding process. This strategy has been particularly popular in the terminal operation business as many rail and maritime terminals are managed by private operators who do not own the facilities but have long-term leases. The outcome commonly involves efficiency improvements.

Concessions are a simple and fair strategy involving a bidding process, which underlines the importance to have it take place in a transparent and open way. This is particularly relevant as retirement funds, sovereign wealth funds, investment banks and other financial institutions are increasingly involved in the funding of transportation infrastructure. A lack of transparency can be perceived negatively by the general public and can transform a simple transaction into a complex political process. Since some concessions are set over long time periods (50–75 years), they bring the issue of changing market conditions that may force a renegotiation of the contract. It is next to impossible to foresee long-term market changes and traffic levels, so a provision for renegotiation should be provided. Again, this renegotiation can be subject to controversy and public debate, particularly if performed in an un-transparent manner.

Private/public partnerships

A public/private partnership (PPP) is a contractual agreement between a public agency (federal, state or municipal) and a private sector entity that allows for the design, building, operation or financing of transport infrastructure. PPPs thus confer a wide range of options in terms of capital allocation and respective levels of participation. They can simply cover the standard design/build contracting process common in many road projects or involve innovative approaches where a private operator takes charge of the construction and management of a transport infrastructure over a long-term concession. The main forms of PPP include (Figure 3.23):

- **Design–Bid–Build**. In the first stage, a contract is awarded to an engineering design firm to set a clear guideline in terms of the potential costs, materials and equipment required to complete a public works project. Then private contractors are invited to bid on the proposed specifications, which are reviewed by the public entity. The winning contractor then undertakes the construction phase and once completed, management and maintenance will be performed by the public sector. All steps are financed by the public sector.
- **Private contract fee services**. A common contract structure where the public sector transfers the responsibility of specific services, such as operation and maintenance of public infrastructures, to the private sector. There exists a variety of private firms that have specialized in providing services to transport infrastructure, particularly in terms of maintenance, repairs and upgrades.
- **Design–Build**. Similar to the design–bid–build partnership with the exception that they are combined in a single contract. As usual, the public sector owns the infrastructure as well as bearing the responsibility for its financing, operating and maintenance.
- **Build–Operate–Transfer**. While the public sector is responsible for the financing of the infrastructure, a private entity provides for construction and operation. It is also known as a "turnkey" PPP since, after a specified amount of time, the public sector takes over the infrastructure. It can be decided to extend the operation contract to the same operator or put it up for bid.
- **Design–Build–Finance–Operate**. The responsibilities for designing, building, financing, and operating the infrastructure fall in the hands of the private sector, but ownership remains public. There is however some flexibility in the PPP as the respective shares of the financing could come from a pool of public and private interests. Flexibility can also be in terms of the nature of the financing, which can be capital or in kind. The expectation is that the contracted debt used to finance transport infrastructure will be recovered by future revenues, which implies that user fees will be applied and that debt (such as bonds) is leveraged by future revenues.

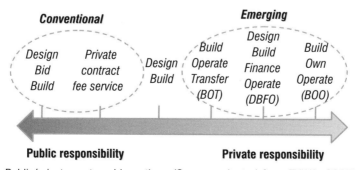

Figure 3.23 Public/private partnership options (Source: adapted from FHWA, 2007)

- **Build–Own–Operate**. The design, development, financing, building, operation and maintenance of an infrastructure fall completely under the responsibility of the private sector and this for the duration of the concession, which is dominantly long term. Public sector involvement is limited to the general regulatory framework and assuring compliance to the terms of the contract.

PPPs take place in situations where stakeholders alone cannot clearly evaluate the respective advantages of the investment and find it too risky to finance. The public sector thus helps leverage the position of the private sector, which commonly results in a better allocation of resources than if they had done so independently. While the public perception tends to relate PPP to toll roads, the reality places these initiatives in every segment of the transportation industry from modes to terminals. PPPs take a particular dimension in the freight sector as freight transportation is much in the realm of the private sector with public interests mainly covering the regulatory framework. The most significant infrastructure assets are related to freight transport terminals, particularly ports and rail, a reason why they are dominantly owned or operated by private interests, which makes public involvement problematic. There is thus a conventional approach to PPP which is gradually being supplemented by an emerging framework where private entities are taking a higher level of responsibility.

Bibliography

Berry, B.J.L. (1967) *Geography of Market Centers and Retail Distribution*, Englewood Cliffs, NJ: Prentice Hall.

Berry, B.J.L. (1991) *Long-wave Rhythms in Economic Development and Political Behavior*, Baltimore, MD: Johns Hopkins University Press.

Cambridge Systematics (1996) *Quick Response Freight Manual*, Federal Highway Administration, Office of Planning and Environmental Technical Support Services for Planning Research.

European Conference of Ministers of Transport (2001) *Transport and Economic Development*, Paris: OECD. http://www1.oecd.org/publications/e-book/7502101E.PDF.

Federal Highway Administration (FHWA) (2007) *Financing Freight Improvements*, Publication #FHWA-HOP-06-108, US Department of Transportation, http://ops.fhwa.dot.gov/freight/publications/freightfinancing/index.htm.

Goodbody Economic Consultants (2003) *Transport and Regional Development*, Ballsbridge, Dublin, http://www.irishspatialstrategy.ie/docs/pdf/Transport and Regional Development.pdf.

Gottmann, J. (1961) *Megalopolis: The Urbanized Northeast Seaboard of the United States*, New York: Twentieth Century Fund.

Hall, P. (1984) *The World Cities*, 3rd edn, New York: St. Martin's Press.

Harrington, J.W. and B. Warf (1995) *Industrial Location: Principles, Practice and Policy*, London: Routledge.

Henderson, J.V., Z. Shalizi and A.J. Venables (2000) *Geography and Development*, http://econ.lse.ac.uk/staff/ajv/vhzstv3.pdf.

ICF Consulting & HLB Decision-Economics (2002) *Economic Effects of Transportation: The Freight Story*, http://www.ops.fhwa.dot.gov/freight/.

Isard, W. (1956) *Location and Space-Economy: A General Theory Relating to Industrial Location, Market Areas, Land Use, Trade, and Urban Structure*, Cambridge, MA: MIT Press.

Llewelyn-Davies (2004) *Transport and City Competitiveness – Literature Review*, Department for Transport, http://www.dft.gov.uk/162259/163944/Transport_and_City_Competit1.pdf.

McQuaid, R.W., M. Greig, A. Smyth and J. Cooper (2004) *The Importance of Transport in Business' Location Decisions*, Department for Transport, http://www.dft.gov.uk/stellent/groups/dft_science/documents/pdf/dft_science_pdf_027294.pdf.

Perroux, F. (1955) "Note sur la Notion de Pôle de Croissance", *Economie Appliquée*, 7, 307–20.

Pred, A. (1977) *City Systems in Advanced Economies: Past Growth, Present Processes and Future Development Options*, New York: Wiley.

Preston, R.E. (1985) "Christaller's Neglected Contribution to the Study of the Evolution of Central Places", *Progress in Human Geography*, 9, 177–93.

Weisbrod, Glen (2007) *Models to Predict the Economic Development Impact of Transportation Projects: Historical Experience and New Applications*, Working Paper, Economic Development Research Group, http://www.edrgroup.com/edr1/bm~doc/models-to-predict-the-eco.pdf.

Woxenius, J. (2006) "Temporal Elements in the Spatial Extension of Production Networks", *Growth and Change*, 37(4), 526–49.

4 Transportation modes

Transportation modes are an essential component of transport systems since they are the means by which mobility is supported. Geographers consider a wide range of modes that may be grouped into three broad categories based on the medium they exploit: land, water and air. Each mode has its own requirements and features, and is adapted to serve the specific demands of freight and passenger traffic. This gives rise to marked differences in the ways the modes are deployed and utilized in different parts of the world. Recently, there is a trend towards integrating the modes through intermodality and linking the modes ever more closely into production and distribution activities. At the same time, however, passenger and freight activity is becoming increasingly separated across most modes.

Concept 1 – A diversity of modes

Transport modes are the means by which people and freight achieve mobility. They fall into one of three basic types, depending on what surface they travel over – land (road, rail and pipelines), water (shipping) and air. Each mode is characterized by a set of technical, operational and commercial characteristics (see Figure 4.1). Because of their operational characteristics, several freight transportation modes have different capacities and levels of efficiency. While the truck is certainly the mode which has the least capacity, it has a level of flexibility (speed and door-to-door services) unmatched by rail and fluvial transportation.

Road transportation

Road development accelerated in the first half of the twentieth century. By the 1920s, the first all-weather transcontinental highway, the Lincoln Highway, spanned over 5,300 km between New York and San Francisco. The Germans were the first to build the modern highway (autobahn) in 1932, with specifications such as restricted access, overpasses and road separation that would eventually become common characteristics of highway systems. The post-World War II era represented a period of rapid expansion of road transportation networks worldwide. The most remarkable achievement is without doubt the American Interstate highway system initiated in 1956. Its strategic purpose was to provide a national road system servicing the American economy and also able to support troop movements and act as air strips in case of an emergency. About 56,000 km was built from the 1950s to the 1970s, but between 1975 and 2006 only 15,000 km were added to the system, underlining growing construction costs and diminishing returns. Overall, about 70,000 km of four-lane and six-lane highways were constructed, linking all major American cities, coast to coast. A similar project took place in Canada with the Trans-Canada highway completed in 1962. By the 1970s, every

Vehicle	Capacity	Truck equivalency
Barge	1,500 tons 52,500 bushels 453,600 gallons	57.7 (865.4 for 15 barges in tow)
Hopper car	100 tons 3,500 bushels 30,240 gallons	3.8
100-car train unit	10,000 tons 350,000 bushels 3,024,000 gallons	384.6
Semi-trailer truck	26 tons 910 bushels 7,865 gallons 9,000 for a tanker truck	1
Panamax containership	5,000 TEU	2,116
VLCC	300,000 tons 2 million barrels of oil	9,330
747-400F	124 tons	5

Figure 4.1 Performance comparison for selected freight modes

modern nation had constructed a national highway system, which in the case of Western Europe resulted in a pan-European system. This trend now takes place in many industrializing countries. For instance, China is building a national highway system that expanded to 53,000 km in 2007, with construction taking place at a pace of about 2,000 km per year.

Road transportation has become the dominant land transport system today. Automobiles, buses and trucks require a road bed. Such infrastructures are moderately expensive to provide, but there is a wide divergence of costs, from a gravel road to a multi-lane urban expressway (Plate 4.1). Because vehicles have the means to climb moderate slopes, physical obstacles are less important than for some other land modes. Most roads are provided as a public good by governments, while the vast majority of vehicles are owned privately. The capital costs, therefore, are shared, and do not fall as heavily on one source as is the case for other modes.

All road transport modes have limited abilities to achieve scale economies. This is due to the size constraints imposed by governments and also by the technical and economic limits of the power sources. In most jurisdictions, trucks and buses have specific weight and length restrictions which are imposed for safety reasons. In addition, there are serious limits on the traction capacities of cars, buses and trucks because of the considerable increases in energy consumption that accompany increases in the weight of the unit. For these reasons the carrying capacities of individual road vehicles are limited.

Road transport, however, possesses significant advantages over other modes:

Plate 4.1 Access ramp to the Nanpu Bridge, Shanghai. *Credit*: Claude Comtois

- The **capital cost of vehicles** is relatively small. This produces several key characteristics of road transport. Low vehicle costs make it comparatively easy for new
  users to gain entry, which helps ensure that the trucking industry, for example,
  is highly competitive. Low capital costs also ensure that innovations and new
  technologies can diffuse quickly through the industry.
- Another advantage of road transport is the **high relative speed of vehicles**, the major
  constraint being government-imposed speed limits.
- One of its most important attributes is the **flexibility of route choice**, once a network of roads is provided. Road transport has the unique opportunity of providing
  door-to-door service for both passengers and freight.

These multiple advantages have made cars and trucks the modes of choice for a
great number of trip purposes, and have led to their market dominance for short-
distance trips. The success of cars and trucks has given rise to a number of serious
problems. Road congestion has become a feature of most urban areas around the world
(see Chapters 7 and 10). In addition, the mode is behind many of the major environmental externalities linked to transportation (see Chapter 8). Addressing these issues
is becoming an important policy challenge at all levels of jurisdiction, from the local
to the global (see Chapter 9).

Rail transportation

Railways require tracks along which locomotives and rail cars move. The initial capital costs are high because the construction of rail tracks and the provision of rolling
stock are expensive. Historically, the investments have been made by the same source
(either governments or the private sector). These expenditures have to be made before

any revenues are realized and thus represent important entry barriers that tend to limit the number of operators. It also serves to delay innovation, compared with road transport, since rail rolling stock has a service life of at least 20 years. Railway routing is affected by topography because locomotives have limited capacities to mount gradients. As a result railways either avoid important natural barriers or overcome them by expensive engineering solutions.

An important feature of rail systems is the width of the rails or the gauge. The standard gauge of 1.4351 meters has been adopted in many parts of the world, across North America and most of Western Europe for example. But other gauges have been adopted in other areas. This makes integration of rail service very difficult, since both freight and passengers are required to change from one railway system to the other. As attempts are being made to extend rail services across continents and regions, this is an important obstacle, as for example between France and Spain, Eastern and Western Europe, and between Russia and China. The potential of the Eurasian land bridge is limited in part by these gauge differences. Other factors that inhibit the movement of trains between different countries include signaling and electrification standards. These are particular problems for the European Union where the lack of "interoperability" of the rail systems between the member states is a factor limiting the wider use of the rail mode.

The ability of trains to haul large quantities of goods and significant numbers of people over long distances is the mode's primary asset. Once the cars have been assembled or the passengers have boarded, trains can offer a high-speed–high-capacity service. It was this feature that led to the train's pre-eminence in opening the interior of the continents in the nineteenth century, and is still its major asset. Passenger service is effective where population densities are high. Freight traffic is dominated by bulk cargo shipments, agricultural and industrial raw materials in particular. Rail transport is a "green" system, in that its consumption of energy per unit load per km is lower than road modes.

Although sometimes identified as a mode that enjoyed its heyday during the nineteenth century, rail transport is enjoying a resurgence because of technological advances in the latter part of the twentieth century. In passenger transport this has come about through significant breakthroughs in speed. For instance, in Europe and Japan high-speed rail systems reach speeds up to 515 km/hr. This gives rail a competitive advantage over road transport and even with air transport over short and medium distances. Japan saw the first comprehensive development of a high-speed train system, notably used along the Tokyo–Osaka corridor in 1964. By the 1990s the usage of the system peaked, in part because of competition from air transport. Europe has been the region where the adoption of the high-speed train has been the most significant since the 1990s. Close to a half of all the world's high-speed passengers-km now occur in Europe. South Korea is one of the latest countries to build a high-speed rail system along the Seoul–Pusan corridor, which was inaugurated in 2004. In 2007 the Taipei-Kaohsiung corridor began to be serviced by a high-speed service.

Unit trains, where trains are made up of wagons carrying one commodity-type only, allow scale economies and efficiencies in bulk shipments, and double stacking has greatly promoted the advantages of rail for container shipments. Rail transport is also enjoying a resurgence as a mode for commuters in many large cities.

Pipelines

Pipelines are an extremely important and extensive mode of land transport, although very rarely appreciated or recognized by the general public, mainly because they are

Plate 4.2   High-speed train, Gare de Lyon, Paris. *Credit*: Jean-Paul Rodrigue

buried underground (or under the sea as in the case of gas pipelines from North Africa to Europe). In the USA, for example, there are 409,000 miles of pipelines that carry 17 percent of all ton/miles of freight. The longest oil pipeline is the Trans-Siberian, extending over 9,344 km to Western Europe from the Russian arctic oilfields in Eastern Siberia. Two main products dominate pipeline traffic: oil and gas, although locally pipelines are significant for the transport of water, and in some rare cases for the shipment of dry bulk commodities, such as coal in the form of slurry.

Pipelines are almost everywhere designed for a specific purpose only, to carry one commodity from one location to another. They are built largely with private capital and because the system has to be in place before any revenues are generated, represent a significant capital commitment. They are effective in transporting large quantities of products where no other feasible means of transport (usually water) is available. Pipeline routes tend to link isolated areas of production with major centers of refining and manufacture in the case of oil, or major populated areas, as in the case of natural gas.

The routing of pipelines is largely indifferent to terrain, although environmental concerns frequently delay approval for construction. In sensitive areas, particularly in arctic/sub-arctic areas where the pipes cannot be buried because of permafrost, the impacts on migratory wildlife may be severe, and be sufficient to deny approval, as was the case of the proposed McKenzie Valley pipeline in Canada in the 1970s. The 1,300-km long Trans-Alaskan pipeline was built under difficult conditions and is above the ground for most of its path. Geopolitical factors play a very important role in the routing of pipelines that cross international boundaries. Pipelines from the Middle East to the Mediterranean have been routed to avoid Israel, and new pipelines linking Central Asia with the Mediterranean are being routed in response to the ethnic and religious mosaic of the republics in the Caucasus.

Pipeline construction costs vary according to the diameter of the pipe and increase proportionally with the distance and with the viscosity of fluids (need for pumping stations). Operating costs are very low, however, and as mentioned above, pipelines represent a very important mode for the transport of liquid and gaseous products. One major disadvantage of pipelines is the inherent inflexibility of the mode. Once built (usually at great expense), expansion of demand is not easily adjusted to. There exist specific limits to the carrying capacity. Conversely a lessening of supply or demand will produce a lowering of revenues that may affect the viability of the system. A further limit arises out of geographical shifts in production or consumption, in which a pipeline having been built from one location to another may not be able to easily adjust to changes. For example, the refineries in Montreal, Canada, were served by a pipeline from Portland, Maine, in order to receive shipments year-round because of ice on the St. Lawrence River. In the 1980s a pipeline from western Canada was built to provide domestic crude oil at a time when the price of the international supply was escalating. Since then the Portland pipeline has been lying idle.

Water transportation

Shipping exploits the water routes that cross oceans as well as rivers and lakes. Many of the oceanic routes are in international waters and are provided at no cost to the users. In many coastal and inland waters too, shipping lanes are "free", although national regulations may exclude foreign vessels from cabotage trade. Physical barriers represent a particular problem for shipping in two areas. First is the sections of inland waterways where water depths and/or rapids preclude navigation. The second is where land barriers separate seas. In both cases canals can provide access for shipping, but they may be tolled. An example of the first type is the St. Lawrence Seaway, while the Suez and Panama Canals are examples of the latter. Thus, except for canals, shipping enjoys rights of way that are at no cost to the users. The relatively low operating costs of ships are a further advantage. Ships have the ability to carry large volumes with small energy consumption and limited manpower requirements. Shipping, therefore, is a mode that can offer very low rates compared to other modes.

Even if maritime transportation has experienced remarkable improvements in its safety and reliability, maritime routes are still hindered by environmental factors such as dominant winds, currents and general weather patterns. The North Atlantic and the North Pacific (50 to 60 degrees north) are subject to heavy wave activity during the winter that sometimes impairs navigation, and may cause ships to follow routes at lower latitudes, thereby increasing the route lengths (see Figure 4.2). During the summer monsoon season (April to October), navigation may become more hazardous on the Indian Ocean and the South China Sea.

Rivers may not be useful for commercial navigation if their orientations do not correspond to the directions of transport demand. Thus, many of the major rivers of Russia flow north–south, while the main trade and passenger flows are east–west. Shallow draught and extensive obstacles, such as rapids, may also limit navigation. However, many rivers, such as the Rhine or the Yangtze, are significant arteries for water transport because they provide access from the oceans to inland markets.

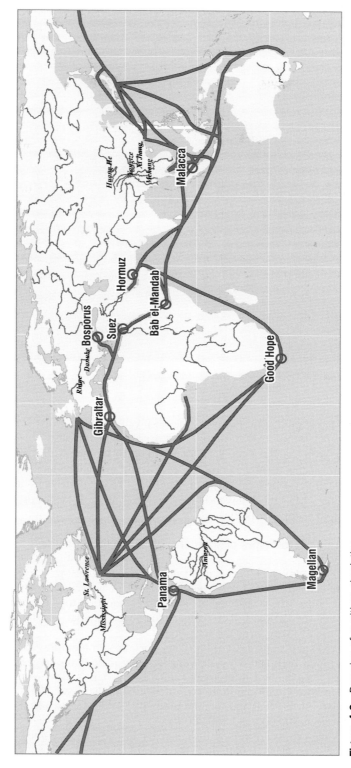

Figure 4.2 Domains of maritime circulation

Plate 4.3 Barges on the Upper Rhine. *Credit*: Brian Slack

Shipping has traditionally faced two drawbacks:

- It is **slow**, with speeds at sea averaging 15 knots (26 km/hr).
- **Delays** are encountered in ports where loading and unloading takes place. The latter may involve several days of handling. These drawbacks are particularly constraining where goods have to be moved over short distances or where shippers require rapid service deliveries.

There are four broad types of ships employed around the world:

- **Passenger vessels** can be further divided into two categories: passenger ferries, where people are carried across relatively short bodies of water in a shuttle-type service, and cruise ships, where passengers are taken on vacation trips of various durations, usually over several days. The former tend to be smaller and faster vessels, the latter are usually very large capacity ships.
- **Bulk carriers** are ships designed to carry specific commodities, and are differentiated into liquid bulk and dry bulk vessels. They include the largest vessels afloat. The largest tankers, the Ultra Large Crude Carriers (ULCC) are up to 500,000 deadweight tons (dwt), with the more typical size being between 250,000 and 350,000 dwt; the largest dry bulk carriers are around 350,000 dwt, while the more typical size is between 100,000 and 150,000 dwt.
- **General cargo ships** are vessels designed to carry non-bulk cargoes. The traditional ships were less than 10,000 dwt, because of extremely slow loading and off-

loading. More recently these vessels have been replaced by container ships because they can be loaded more efficiently and are becoming much larger, with 80,000 dwt being the largest today.

- **Roll-on/Roll-off** (RORO) vessels, which are designed to allow cars, trucks and trains to be loaded directly on board. Originally appearing as ferries, these vessels are used on deep-sea trades and are much larger than the typical ferry. The largest are the car carriers that transport vehicles from assembly plants to the main markets.

The distinctions in vessel types are further differentiated by the kinds of services on which they are deployed. Bulk ships tend to operate either on a regular schedule between two ports or on a voyage basis. In the latter case the ship may haul cargoes between different ports based on demand. General cargo vessels operate on liner services, in which the vessels are employed on a regular scheduled service between fixed ports of call, or as tramp ships, where the vessels have no schedule and move between ports based on cargo availability.

An important feature of the economics of shipping is the capital costs. Because of their size, ships represent a significant capital outlay. Cruise ships represent the most expensive class of vessels, with the *Queen Mary 2* costing $800 million, but even container ships represent initial capital outlays of $75 million and more. The annual cost of servicing the purchase of these vessels represents a large component of operating expenditures. Container shipping requires the deployment of many vessels to maintain a regular service (14 ships in the case of a typical Far East–Europe service), which is a severe constraint on the entry of new players. On the other hand, older second-hand vessels may be purchased for much smaller amounts, and sometimes the purchase price can be easily covered by a few successful voyages. In some regards, therefore, the shipping industry is quite open and historically has provided opportunities for entrepreneurs to accumulate large fortunes. Many of the largest fleets are in private hands, owned by individuals or by family groups (see Case Study at the end of this chapter).

The shipping industry has a very international character. This is reflected particularly in terms of ownership and flagging. The ownership of ships is very broad. While a ship may be owned by a Greek family or a US corporation, it may be flagged under another nationality. Flags of convenience are means by which ship owners can obtain lower registration fees, lower operating costs and fewer restrictions.

The share of open registry ships operated under a flag of convenience grew substantially after World War II. They accounted for 5 percent of world shipping tonnage in 1950, 25 percent in 1980, 50 percent in 1995 and 67 percent in 2007. The usage of a flag of convenience refers to a national owner choosing to register one or more vessels in another nation in order to avoid higher regulatory and manning costs. This enables three types of advantages for the ship owners:

- **Regulation**. Under maritime law, the owner is bound to the rules and regulations of the country of registration, which also involves requisition in emergency situations (war, humanitarian crisis, etc.). Being subject to less stringent regulations commonly confers considerable savings in operating costs.
- **Registry costs**. The state offering a flag of convenience is compensated according to the ship's tonnage. Registry costs are on average between 30 and 50 percent lower than those of North America and Western Europe.
- **Operating costs**. Operating costs for open registry ships are from 12 to 27 percent lower than traditional registry fleets. Most of the savings come from lower manning expenses. Flags of convenience have much lower standards in terms of salary and benefits.

The countries with the largest registered fleets offer flags of convenience (Panama, Liberia, Greece, Malta, Cyprus and the Bahamas) and have very lax regulations (see Figure 4.3). Ship registry is a source of additional income for these governments. Even the landlocked country of Mongolia offers ship registry services.

Because the costs of providing ship capacity to more and more markets are escalating beyond the means of many carriers, especially in container shipping, many of the largest shipping lines have come together by forming strategic alliances with erstwhile competitors. The alliances offer joint services by pooling vessels contributed by alliance members on the main commercial routes. In this way the individual companies are each able to commit fewer ships to a particular service route, and deploy their surplus ships on other routes that are maintained outside the alliance. The alliance services are marketed separately, but operationally involve close cooperation in selecting ports of call and in establishing schedules. The alliance structure has led to significant developments in route alignments and the economies of scale of container shipping.

Air transportation[1]

Speed is the major advantage of air transport compared to other modes. This feature has served to offset many of its limitations, among which operating costs, fuel consumption and limited carrying capacities are the most significant. Technology has worked to overcome some of the constraints, most notably the growth of capacity, in which aircraft are now capable of transporting 500 passengers or 100 tons of freight. Technology has also significantly extended the range of aircraft, so that while 40 years ago aircraft were just beginning to be capable of crossing the Atlantic without stopping at intermediate places such as Newfoundland, they are now capable of making trips of up to 18 hours duration. Surprisingly, the speed of commercial aircraft has not progressed since the 1960s, when the prospect of supersonic speed was being anticipated with the development of the Anglo-French Concorde. This plane was removed from service in 2003. There are three major categories of passenger jet planes:

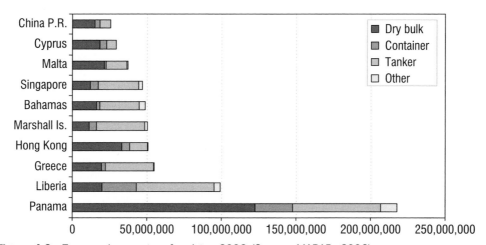

Figure 4.3 Tonnage by country of registry, 2006 (Source: MARAD, 2006)

---

1 Dr. John Bowen (Central Washington University) contributed to this section.

- **Short-range aircraft**. Bombardier's CRJ series and Embraer's ERJs are examples of planes with relatively small capacities (30–100 passengers) that travel over relatively short distances. They are usually referred to as regional jets that serve smaller markets and feed hub airports.
- **Medium-range aircraft**. The airbus A320, with a range of 3,700 km, and its Boeing equivalent, the B-737, are designed to service destinations within a continent. From New York, most of North America can be reached. This range can be applied to the European continent, South America, East Asia and Africa. This type of aircraft is also used for high demand regional services.
- **Long-range aircraft**. There are a variety of aircraft capable of crossing the oceans and linking the continents. Early variants such as the B-707 have evolved into planes offering high capacity, such as the B-747 series, or long-range abilities, such as B-777 or the A350 series which have ranges of up to 17,400 km (Figure 4.4).

Air transport makes use of air space that theoretically gives it great freedom of route choice. While the mode is less restricted than land transport to specific rights of way, it is nevertheless much more constrained than what might be supposed. In part this is due to physical conditions, in which aircraft seek to exploit (or avoid) upper atmospheric winds, in particular the jet stream, to enhance speed and reduce fuel consumption. In addition, specific corridors have been established in order to facilitate navigation and safety. Strategic and political factors have also influenced route choice. For example, the flights of South African Airways were not allowed to over-fly many African nations during the apartheid period, and Cubana Airlines has been routinely prohibited

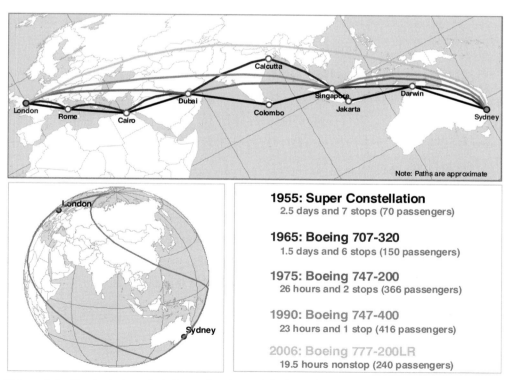

Figure 4.4 Shortest air route between London and Sydney, 1955–2006

from over-flying the USA. Even more significant was the opening up of Siberian airspace to Western airlines after the Cold War (Figure 4.5).

The older, more circuitous routings between Asia and North America, typified by R222, came at the cost of either a reduced payload or a refueling stop enroute, commonly at Anchorage. New polar routes, like Polar 2, permit fuller payloads and nonstop flights. The comparative difference between R222 and Polar 2 is about 2,000 km in favor of Polar 2, resulting in a reduced flight time of more than two hours. The new freedom permitted more direct routes not only between cities like London and Tokyo or New York and Hong Kong but also between transpacific city pairs like Vancouver–Beijing.

Like maritime transport, the airline industry is highly capital intensive. For instance, a new Boeing 747-400, used for high volume and long-distance travel, costs approximately $200 million, depending on the configuration, and a new Boeing 737-800, used for regional flights, costs about $60 million. However, unlike the maritime sector, air transportation is labor intensive, with limited room to lower labor requirements, although many airlines are now trying to reduce labor costs by cutting salaries and benefits. The industry has become a powerful factor of development, generating globally more than $700 billion in added value and creating more than 21 million jobs.

The initial development of air transportation took place in the 1920s and 1930s, not always for commercial reasons. It was seen as a means of providing a national air mail service (US) and of establishing long-haul air services to colonies and dependencies (UK, Netherlands and France). Airline companies were set up to provide these national goals, a trend that continued in the post-colonial period of the 1950s to the 1970s, as many African, Asian and Caribbean nations created their own airline companies while reserving them for specific markets and for specific routes. By convention, an air space exclusively belongs to the country under it, and this has led to significant government control over the industry.

Traditionally, an airline needs the approval of the governments of the various countries involved before it can fly in or out of a country, or even across another country without landing. Prior to World War II, this did not present too many difficulties since the range of commercial planes was limited and air transport networks were in their infancy and nationally oriented. In 1944, an International Convention was held in Chicago to establish the framework for all future bilateral and multilateral agreements for the use of international air spaces. Five freedom rights were designed, but a multilateral agreement went only as far as the first two freedoms (right to overfly and right to make a technical stop).

Freedoms are not automatically granted to an airline as a right, they are privileges that have to be negotiated. All other freedoms have to be negotiated by bilateral agreements, such as the 1946 agreement between the United States and the UK, which permitted limited "fifth freedom" rights. The 1944 Convention has been extended since then; as shown in Figure 4.6, there are currently nine different freedoms:

- **First Freedom**. The right to fly from a home country over another country (A) en route to another (B) without landing. Also called the transit freedom.
- **Second Freedom**. The right for a flight from a home country to land in another country (A) for purposes other than carrying passengers, such as refueling, maintenance or emergencies. The final destination is country B.
- **Third Freedom**. The right to carry passengers from a home country to another country (A) for purpose of commercial services.
- **Fourth Freedom**. The right to fly from another country (A) to a home country for purpose of commercial services.

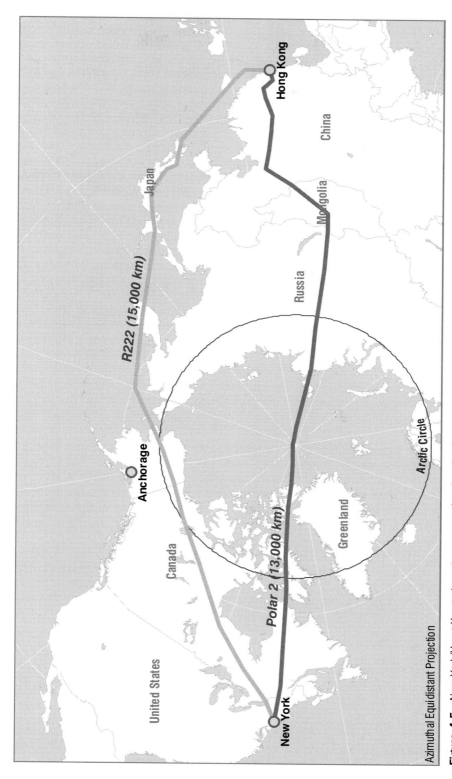

Figure 4.5 New York/Hong Kong air routes: conventional and polar

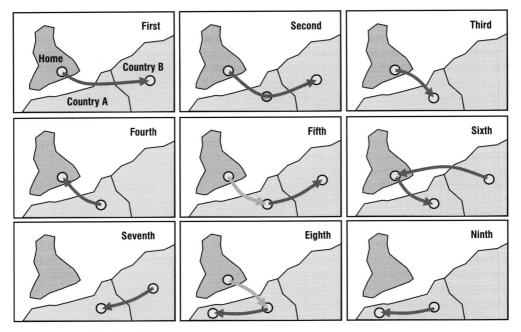

Figure 4.6 Air freedom rights

The Third and Fourth freedoms are the basis for direct commercial services, providing the rights to load and unload passengers, mail and freight in another country.

- **Fifth Freedom.** This freedom enables airlines to carry passengers from a home country to another intermediate country (A), and then fly on to a third country (B) with the right to pick up passengers in the intermediate country. Also referred to as "beyond right". This freedom divided into two categories: Intermediate Fifth Freedom Type is the right to carry from the third country to the second country. Beyond Fifth Freedom Type is the right to carry from the second country to the third country.
- **Sixth Freedom.** Not formally part of the original 1944 Convention, it refers to the right to carry passengers between two countries (A and B) through an airport in the home country. With the hubbing function of most air transport networks, this freedom has become more common, notably in Europe (London, Amsterdam).
- **Seventh Freedom.** Covers the right to operate a passenger service between two countries (A and B) outside the home country.
- **Eighth Freedom.** Also referred to as "cabotage" privileges. It involves the right to move passengers on a route from a home country to a destination country (A) that uses more than one stop along which passengers may be loaded and unloaded.
- **Ninth Freedom.** Also referred to as "full cabotage" or "open-skies" privileges. It involves the right of a home country to move passengers within another country (A).

In the 1970s, the perspective changed and air transport was increasingly seen as just another transport service. Market forces were considered to be the mechanism for fixing prices and it became widely accepted that airline companies should be given freedom within national markets to decide the nature and extent of their services, while the role of governments should be limited to operational and safety regulations. In the United States, the Air Deregulation Act of 1978 put an end to fixed markets and opened the industry to competition. This liberalization process has spread to many other countries, although with important local distinctions. Many of the former private firms in the USA and many former state-owned airlines elsewhere that were heavily protected and subsidized went bankrupt or have been absorbed by larger ones. Many new carriers have emerged, with several low-cost carriers such as Ryanair and South-West Air achieving industry leadership. Internationally, air transport is still dominated by bilateral agreements between nations.

As in the case of ocean shipping, there has been a significant development of alliances in the international airline industry. The alliances are voluntary agreements to enhance the competitive positions of the partners. Members benefit from greater scale economies, a lowering of transaction costs and a sharing of risks, while remaining commercially independent. The first major alliance was established in 1989 between KLM and Northwest Airlines. The "Star" alliance was initiated in 1993 between Lufthansa and United Airlines. In 1996 British Airlines and American Airlines formed the "One World" alliance. Other national carriers have joined different alliance groupings. They cooperate on scheduling, code sharing, equipment maintenance and schedule integration. It permits airlines that may be constrained by bilateral regulations to offer a global coverage.

As in shipping, there have been a number of mergers and acquisitions. The largest is the merger between Air France and KLM, members of the "Sky Team" alliance. In the USA other mergers are likely. This consolidation is sure to attract the attention of regulatory agencies.

Prior to deregulation movements (end of 1970s–early 1980s), many airline services were taking place on a point-to-point basis. Figure 4.7 shows two airline companies servicing a network of major cities. A fair amount of direct connections exists, but mainly

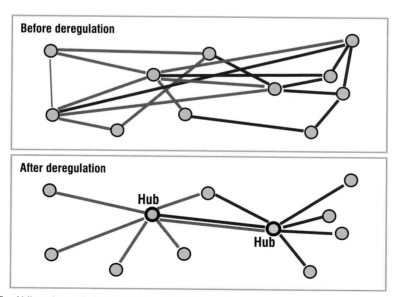

Figure 4.7 Airline deregulation and hub-and-spoke networks

at the expense of the frequency of services and high costs (if not subsidized). Also, many cities are serviced, although differently, by the two airlines, and connections are likely to be inconvenient. With deregulation, a system of hub-and-spoke networks emerges as airlines rationalize the efficiency of their services. A common consequence is that each airline assumes dominance over a hub and services are modified so the two hubs are connected to several spokes. Both airlines tend to compete for flights between their hubs and may do so for specific spokes, if demand warrants it. However, as this network matures, it becomes increasingly difficult to compete at hubs as well as at spokes, mainly because of economies of agglomeration. As an airline assumes dominance of a hub, it reaches oligopolistic (if not monopolistic) control and may increase airfares for specific segments. The advantage of such a system for airlines is the achievement of a regional market dominance and higher plane loads, while passengers benefit from better connectivity (although delays for connections and changing planes are more frequent) and lower costs.

Air transport is extremely important for both passenger and freight traffic. In 2000, 1.4 billion passengers traveled by air transport, representing the equivalent of 23 percent of the global population. Passenger traffic is made up of business travelers and the general public, many of whom are holiday-makers. Air transport is a very significant factor in the growth of international tourism. Figure 4.8 indicates the continued domination of US carriers in passenger transport.

Air transport accounts for a small proportion of all freight carried, but because it is made up of electronics, parcels and parts with a high value to weight ratio that are at the heart of contemporary just-in-time and flexible production systems, air freight plays an important role in the world economy. Historically, freight has been carried in the belly-hold of passenger airplanes, providing supplementary income for airline companies. However, with the growth of freight traffic, an increasing market share has been captured by integrators, firms that specialize in freight management and handling, which increasingly operate their own fleets of cargo planes. Two of these companies (FedEx and UPS) are the largest freight carriers (see Figure 4.8). A smaller market share has been captured by all-freight airlines, such as Cargolux. Nevertheless, the passenger airlines still account for the largest volume of air freight carried.

Modal competition

A general analysis of transport modes reveals that each has key operational and commercial advantages and properties. As a result, modal competition exists at various degrees and takes several dimensions. Modes can compete or complement one another in terms of cost, speed, accessibility, frequency, safety, comfort, etc. Cost is one of the most important considerations in modal choice. Because each mode has its own price/performance profile, the actual competition between the modes depends primarily upon the distance traveled, the quantities that have to be shipped and the value of the goods. Thus, while maritime transport might offer the lowest variable costs, over short distances and for small bundles of goods road transport tends to be most competitive. A critical factor is the terminal cost structure for each mode, where the costs (and delays) of loading and unloading the unit impose fixed costs that are incurred independent of the distance traveled. As shown in Figure 4.9, different transportation modes have different cost functions. Road, rail and maritime transport have respectively a C1, C2 and C3 cost functions. While road has a lower cost function for short distances, its cost function climbs faster than rail and maritime cost functions. At a distance D1, it becomes more profitable to use railway transport than road transport while from a distance D2, maritime transport becomes more advantageous. Point

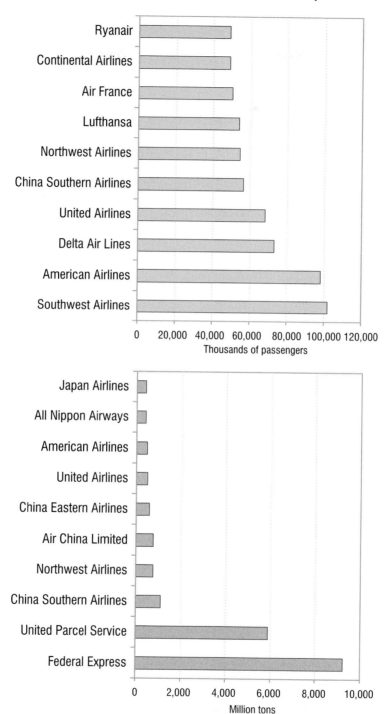

Figure 4.8 World's ten largest passenger and freight airlines, 2007

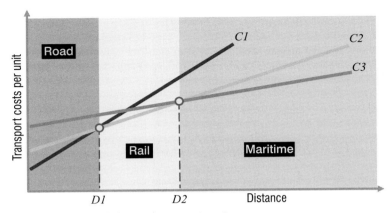

Figure 4.9 Distance, modal choice and transport costs

D1 is generally located between 500 and 750 km of the point of departure while D2 is near 1,500 km.

Modal competition can also been influenced by public policy where one mode could be advantaged over the others. This particularly takes place over government funding and regulation issues. For instance, in the United States the federal government would finance 80 percent of the costs of a highway project, leaving the state government to supply the remaining 20 percent. For public transit, this share is 50 percent, while for passenger rail the federal government will not provide any funding. Under such circumstances, public policy shapes modal preferences.

With increasing levels of income the propensity for people to travel rises. At the same time, international trade in manufactured goods and parts has increased. These trends in travel act differentially upon the modes. The modes that offer the faster and more reliable services gain over modes that might offer a lower cost, but slower, alternative. For passenger services, rail has difficulty in meeting the competition of road transport over short distances and aircraft for longer trips. For freight, rail and shipping have suffered from competition from road and air modes for high-value shipments. While shipping, pipelines and rail still perform well for bulkier shipments, intense competition over the last 30 years has seen road and air modes capture an important market share of the high revenue-generating goods.

A comparative modal split reflects different geographical conditions in which transport systems operate (Figure 4.10). While within the European Union and in Japan road and coastal shipping account for a significant share of ton-km, rail dominates in the United States. The continentality and the fragmentation of the American economy into specialized regions are prone to long-distance rail shipments as well as the reliance on pipelines to supply the large consumption of fossil fuels, namely petroleum and natural gas. The high densities and the relatively short distances involved in the case of the EU and Japan favor trucking, while the maritime exposure they have, in terms of coastline, favors a high usage of coastal shipping.

There are important geographical variations in modal competition. The availability of transport infrastructures and networks varies enormously. Some regions possess many different modes that in combination provide a range of transport services that ensure an efficient commercial environment. Thus, in contrast to the situation in the EU, rail transport occupies a more important market share in North America. In many parts of

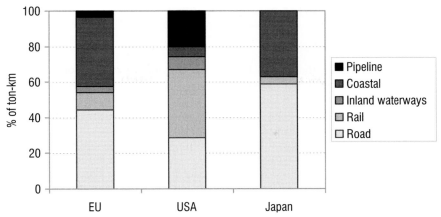

Figure 4.10 Modal split in the EU, United States and Japan, 2005 (Sources: Eurostat (EU); Bureau of Transport Statistics (USA); Ministry of Land, Infrastructure, Transport and Tourism (Japan))

the world, however, there are only limited services, and some important modes may be absent altogether. This limits the choices for people and shippers, and acts to limit accessibility. People and freight are forced to use the only available modes that may not be the most economic for the nature of the demand. Goods may not be able to find a market, and people's mobility may be impaired.

For these reasons, transport provision is seen as a major factor in economic development (see Chapter 3). Areas with limited modal choices tend to be among the least developed. The developed world, on the other hand, possesses a wide range of modes that can provide services to meet the needs of society and the economy.

Since 2000 the price of fuel has increased significantly as well as its volatility. All modes are affected, from the individual car owner to the corporation operating a fleet of hundreds of aircraft or ships. The higher costs are being passed on to the customer, either directly, as is the case of shipping where freight rates are climbing, or indirectly, as is the case of airlines where passengers are being charged additional fuel surcharges. These cost increases are likely to have significant impacts on mobility and trade, as well as on the modal split.

- Higher transport costs increase the **friction of distance** and constrain mobility. As a major consumer of petroleum the transport industry has to increase rates. Across the board increases cause people to rethink their patterns of movement and companies to adjust their supply and distribution chains. One of the expected effects of these cost increases is a decline in freight shipments and passenger carriers, such as airlines are anticipating a reduction in trips. Even school districts are anticipating reducing the number of buses and making children walk further to school.
- Because the impact of higher fuel costs hits the modes differentially, a **modal shift** is anticipated. Road and air transport are more fuel intensive than the other modes, and so fuel price increases are likely to impact upon them more severely than other modes. This could lead to a shift towards water and rail transport in particular.
- A further impact of fuel price increases is **greater fuel economy** across the modes. One of the best ways for all modes to reduce consumption is to lower speeds. A future of high energy prices is likely to have a major impact on just-in-time deliveries, and lead to a restructuring of supply chains.

Concept 2 – Intermodal transportation

The nature of intermodalism

Competition between the modes has tended to produce a transport system that is segmented and un-integrated. Each mode has sought to exploit its own advantages in terms of cost, service, reliability and safety. Carriers try to retain business by maximizing the line-haul under their control. All the modes saw the other modes as competitors, and viewed them with suspicion and mistrust. The lack of integration between the modes was also accentuated by public policy that has frequently barred companies from owning firms in other modes (as in the United States before deregulation), or has placed a mode under direct state monopoly control (as in Europe). Modalism was also favored because of the difficulties of transferring goods from one mode to another, thereby incurring additional terminal costs and delays.

The use of several modes of transport has frequently occurred as goods are shipped from the producer to the consumer. When several modes are used this is referred to as multimodal transport. Within the last 40 years efforts have been made to integrate separate transport systems through intermodalism. What distinguishes intermodal from multimodal transport is that the former involves the use of at least two different modes in a trip from origin to destination under a single transport rate. Intermodality enhances the economic performance of a transport chain by using the modes in the most productive manner. Thus, the line-haul economies of rail may be exploited for long distances, with the efficiencies of trucks providing local pickup and delivery. The key is that the entire trip is seen as a whole, rather than as a series of legs, each marked by an individual operation with separate sets of documentation and rates.

Figure 4.11 illustrates two alternatives to freight distribution. The first is a conventional point-to-point multimodal network where origins (A, B and C) are independently linked to destinations (D, E and F). In this case, two modes (road and rail) are used.

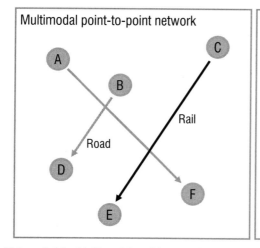
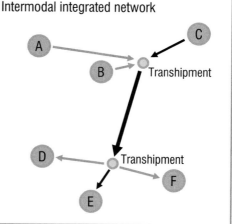

Figure 4.11 Multimodal and intermodal transportation

The second alternative involves the development of an integrated intermodal transport network. Traffic converges at two transhipment points, rail terminals, where loads are consolidated. This can result in higher load factors and/or higher transport frequency, especially between terminals. Under such circumstances, the efficiency of such a network mainly resides in the transhipment capabilities of transport terminals.

The emergence of intermodalism has been brought about in part by technology (Muller, 1995). Techniques for transferring freight from one mode to another have facilitated intermodal transfers. Early examples include piggyback (TOFC: trailers on flat cars), where truck trailers are placed on rail cars, and LASH (lighter aboard ship), where river barges are placed directly on board sea-going ships. The major development undoubtedly has been the container, which permits easy handling between modal systems. Containers have become the most important component for rail and maritime intermodal transportation.

While handling technology has influenced the development of intermodalism, another important factor has been the changes in public policy. Deregulation in the United States in the early 1980s liberated firms from government control. Companies were no longer prohibited from owning across modal types, and there developed a strong impetus towards intermodal cooperation. Shipping lines, in particular, began to offer integrated rail and road service to customers. The advantages of each mode could be exploited in a seamless system. Customers could purchase the service to ship their products from door to door, without having to concern themselves of modal barriers. With one bill of lading clients can obtain one through rate, despite the transfer of goods from one mode to another (Hayuth, 1987).

The provision of through bills of lading in turn necessitated a revolution in organization and information control. At the heart of modern intermodalism are data handling, processing and distribution systems that are essential to ensure the safe, reliable and cost-effective control of freight movements across several modes. Electronic data interchange (EDI) is an evolving technology that is helping companies and government agencies (customs documentation) cope with an increasingly complex global transport system.

Intermodalism, the container and maritime transport

Intermodalism originated in maritime space, with the development of the container in the late 1960s and has since spread to integrate other modes. It is not surprising that the maritime sector should have been the first mode to pursue containerization. It was the mode most constrained by the time taken to load and unload vessels. Containerization permits the mechanized handling of cargoes of diverse types and dimensions that are placed into boxes of standard sizes. In this way goods that might have taken days to be loaded or unloaded from a ship can now be handled in a matter of minutes.

One of the keys to the success of the container is that the International Organization for Standardization (ISO) very early on established base dimensions. The reference size is the 20-foot box, 20 feet long, 8 feet high and 8 feet wide, or 1 20-foot equivalent unit (TEU). The other major size is the 40-foot box, which can carry about 22 tons of cargo. Containers are either made of steel or aluminum and their structure confers flexibility and hardiness. Each year, about 1.5 million TEUs' worth of containers are manufactured. The global inventory of containers was estimated to be around 25 million TEUs in 2007, which approximately corresponds to 3 TEUs of containers for every TEU of maritime containership capacity. Among the numerous advantages related to the success of containers in international transport, are several elements:

Plate 4.4 Forty-foot container being handled, Trimodal Container Terminal, Willebroek, Belgium.
Credit: Jean-Paul Rodrigue

- **Standard transport product**. A container can be manipulated anywhere in the world as its dimensions are an ISO standard. Indeed, transfer infrastructures allow all elements (vehicles) of a transport chain to handle it with relative ease. The rapid diffusion of containerization was facilitated by the fact that its initiator, Malcolm McLean, purposely did not patent his invention. Consequently all segments of the industry, competitors alike, had access to the standard. It necessitated the construction of specialized ships and of lifting equipment.
- **Flexibility of usage**. Containers can transport a wide variety of goods ranging from raw materials (coal, wheat), manufactured goods and cars to frozen products. There are specialized containers for transporting liquids (oil and chemical products) and perishable food items in refrigerated containers or reefers. About 1 million TEUs of reefers were being used by 2002.
- **Management**. The container, as an indivisible unit, carries a unique identification number and a size type code, enabling transport management not only in terms of loads, but in terms of unit. Computerized management reduces waiting times considerably and allows the position of containers to be traced at any time. It enables containers to be assigned according to the priority, the destination and the available transport capacities.
- **Costs**. Containerization of shipping has reduced costs significantly. Before containerization maritime transport costs could account for between 5 and 10 percent of the retail price of manufactured products; this share has been reduced to 1.5 percent.

The main factors behind costs reduction reside in the speed and flexibility of containerization. It has permitted shipping to achieve ever greater economies of scale through the introduction of larger ships. A 5,000 TEU containership has operating costs per container 50 percent lower than a 2,500 TEU vessel.

- **Speed**. Transhipment operations are minimal and rapid. A modern container ship has a monthly capacity of three to six times more than a conventional cargo ship. This is notably attributable to gains in transhipment time as a crane can handle roughly 30 movements (loading or unloading) per hour. Port turnaround times have thus been reduced from 3 weeks to about 24 hours. It takes on average between 10 and 20 hours to unload 1,000 TEUs compared to between 70 and 100 hours for a similar quantity of general cargo. A regular freighter can spend between a half and two-thirds of its useful life in port. With less time in port, containerships can spend more time at sea, and thus be more profitable to operators. Further, containerships are on average 35 percent (19 knots versus 14 knots) faster than regular freighter ships. System-wide the outcome has been a reduction of costs by about 30 percent because of containerization.
- **Warehousing**. The container limits the risks for goods it transports because it is resistant to shocks and weather conditions. The packaging of goods it contains is therefore simpler and less expensive. Besides, containers fit together permitting stacking on ships and on the ground. The container is consequently its own warehouse.
- **Security**. The contents of the container are anonymous to outsiders as it can only be opened at the origin, at customs and at the destination. Thefts, especially those of valuable commodities, are therefore considerably reduced.

In spite of numerous advantages in the usage of containers, some drawbacks are evident:

- **Site constraints**. Containerization implies a large consumption of terminal space. A containership of 5,000 TEU requires a minimum of 12 hectares of unloading space, while unloading entirely its containers would require the equivalent of about 7 double-stack trains of 400 containers each. Conventional port areas are often not adequate for the location of container transhipment infrastructures, particularly because of draft issues as well as required space for terminal operations. Many container vessels require a draft of at least 14 meters (45 feet). A similar challenge applies to container rail terminals, many being relocated at the periphery of metropolitan areas. Consequently, major container handling facilities have modified the local geography.
- **Infrastructure costs**. Container handling infrastructures, such as gantry cranes, yard equipment, road and rail access, represent important investments for port authorities and load centers. Several developing countries cannot afford these infrastructures with local capital and so cannot participate effectively in international trade as efficient load centers unless concession agreements are reached with terminal operators.
- **Management logistics**. The management logistics of containers is very complex. This requires high levels of information technology for the recording, positioning and ordering of containers handled.
- **Empty travel**. Many containers are moved empty, which is not generating any income but conveys a cost that must be assumed somehow. Either full or empty, a container takes the same amount of space on a ship or in a storage yard and takes the same amount of time to be transhipped. Due to a divergence between production and consumption, it is uncommon to see an equilibrium in the distribution of containers. About 2.5 million TEUs of empty containers are stored in yards and depots

around the world, underlining the issue of the movement and accumulation of empty containers. They represent about 20 percent of the global container port throughput and of the volume carried by maritime shipping lines. Most container trade is imbalanced, and thus containers "accumulate" in some places and must be shipped back to locations where there have deficits (mostly locations having a strong export function). This is particularly the case for American container shipping. As a result, shipping lines waste substantial amounts of time and money in repositioning empty containers.

- **Illicit trade.** By its confidential character, the container is a common instrument used in the illicit trade of drugs and weapons, as well as for illegal immigrants. Concerns have also been raised about containers being used for terrorism. These fears have given rise to an increasing number of regulations aimed at counteracting illegal use of containers. In 2003, following US inspection requirements, the International Maritime Organization (IMO) introduced regulations regarding the security of port sites and the vetting of workers in the shipping industry (see Chapter 5). The USA itself established a 24-hour rule, requiring all shipments destined for the USA to receive clearance from US authorities 24 hours prior to the departure of the vessel. In 2008, the US Congress passed a regulation requiring all US-bound containers to be electronically scanned at the foreign port of loading, prior to departure. Needless to say, these measures incur additional costs and delays that many in the industry oppose.

Intermodalism and other modes

With the deregulation and privatization trends begun in the 1980s, containerization, which was already well established in the maritime sector, could spread inland. The shipping lines were among the first to exploit the intermodal opportunities that US deregulation permitted. They could offer door-to-door rates to customers by integrating rail services and local truck pickup and delivery in a seamless network. To achieve this they leased trains, managed rail terminals and in some cases purchased trucking firms. In this way they could serve customers across the country by offering door-to-door service from suppliers located around the world. The move inland also led to some significant developments, most notably the double-stacking of containers on rail cars. This produced important competitive advantages for intermodal rail transport.

Other parts of the world have not developed the same degree of synergies between rail and shipping as in North America. However, a trend towards closer integration in many regions is emerging. In Europe rail intermodal services are becoming well established between the major ports, such as Rotterdam and southern Germany, and between Hamburg and Eastern Europe. Rail shuttles are also making their appearance in China, although their market share remains modest.

While rail intermodal transport has been relatively slow to develop in Europe, there are extensive interconnections between barge services and ocean shipping, particularly on the Rhine. Barge shipping offers a low-cost solution to inland distribution where navigable waterways penetrate to interior markets. This solution is being tested in North America, although with limited success so far. While it is true that the maritime container has become the work horse of international trade, other types of containers are found in certain modes, most notably in the airline industry. High labor costs and the slowness of loading planes, that require a very rapid turnaround, made the industry very receptive to the concept of a loading unit of standard dimensions. The maritime container was too heavy and did not fit the rounded configuration of a plane's fuselage, and thus a box specific to the needs of the airlines was required. The major breakthrough came with the introduction of wide-bodied aircraft in the late 1970s. Lightweight aluminum boxes could be filled with passengers' baggage or parcels and freight, and loaded into the holds of planes using tracking that requires little human assistance.

A unique form of intermodal unit has been developed in the rail industry, particularly in the USA. Roadrailer is essentially a road trailer that can also roll on rail tracks. It is unlike the TOFC (piggyback) system that requires the trailer be lifted on to a rail flat car. Here the rail bogies may be part of the trailer unit, or be attached in the railway yard. The road unit becomes a rail car, and vice versa. It is used extensively by a major US rail company, Norfolk Southern, whose "Triple Crown" service provides just-in-time deliveries between the automobile parts manufacturers located in Michigan, and the assembly plants located in Georgia, Texas and Mexico and Canada.

Intermodalism and production systems

Transport chains are being integrated into production systems. As manufacturers are spreading their production facilities and assembly plants around the globe to take advantage of local factors of production, transportation becomes an ever more important issue. The integrated transport chain is itself being integrated into the production and distribution processes. Transport can no longer be considered as a separate service that is required only as a response to supply and demand conditions. It has to be built into the entire supply chain system, from multi-source procurement, to processing, assembly and final distribution. Supply Chain Management (SCM) has become an important facet of international transportation. As such, the container has become a transport, production and distribution unit.

While many manufacturing corporations may have in-house transportation departments, increasingly the complex needs of the supply chain are being contracted out to third parties. Third-party logistics providers (3PL) have emerged from traditional intermediaries such as the forwarders, or from the transport providers such as FedEx or Maersk-SeaLand. Because the latter are transporters themselves, they are referred to as fourth-party logistics providers (4PL). Both groups have been at the forefront of the intermodal revolution that is now assuming more complex organizational forms and importance. In offering door-to-door services, the customer is no longer aware of or necessarily concerned with how the shipment gets to its destination. The modes used and the routing selected are no longer of immediate concern. The preoccupation is with cost, reliability and level of service. This produces a paradox, that for the customer of intermodal services geographic space becomes meaningless; but for the intermodal providers routing, costs and service frequencies have significant geographical constraints. The effectiveness of intermodal transport systems is thus masking the importance of transportation to its users.

Concept 3 – Passengers or freight?

Advantages and disadvantages

With some exceptions, such as buses and pipelines, most transport modes have developed to handle both freight and passenger traffic. In some cases both are carried in the same vehicle, as for example in the airlines where freight is transported in the cargo holds of passenger aircraft. In others, different types of vehicle have been developed for freight and passenger traffic, but they both share the same road bed, as for example in rail and road traffic. In shipping, passengers and freight used to share the same vessel, but since the 1950s specialization has occurred, and the two are now quite distinct, except for ferries and some RORO services.

The sharing by freight and passengers of a mode is not without difficulties, and indeed some of the problems confronting transportation occur where the two seek to co-inhabit. For example, trucks in urban areas are seen as a nuisance and a cause of congestion by passenger transport users. The poor performance of some modes, such as rail, is seen as the outcome of freight and passengers having to share the same tracks. This raises the question as to whether freight and passengers are compatible. The main advantages of joint operations are:

- **High capital costs** can be justified more easily with a diverse revenue stream (rail, airlines, ferries).
- **Maintenance costs** can be spread over a wider base (rail, airlines).
- The **same traction sources** can be used for both freight and passengers, particularly for rail.

The main disadvantages of joint operations are:

- **Locations of demand** rarely match – origin/destination of freight is usually quite distinct spatially from passenger traffic.
- **Frequency of demand** is different – for passengers the need is for high frequency service, for freight it tends to be somewhat less critical.
- **Timing of service** – demand for passenger services has specific peaks during the day, for freight it tends to be more evenly spread throughout the day.
- **Traffic balance** – on a daily basis passenger flows tend to be in equilibrium, for freight, market imbalances produce empty flows.
- **Reliability** – although freight traffic increasingly demands quality service, for passengers delays are unacceptable.
- **Sharing routes** favors passenger traffic – passenger trains are given priority; trucks may be excluded from areas at certain times of the day.
- **Different operational speeds** – passengers demand faster service.
- **Security screening measures** for passengers and freight require totally different procedures.

A growing divergence

Passengers and freight are increasingly divergent activities as they reflect different transportation markets. In several modes and across many regions passenger and freight transport is being unbundled. For maritime shipping, mention has been made already of how in the maritime sector passenger services have become divorced from freight operations. The exception being some ferry services where the use of RORO ships on high frequency services adapt to the needs of both market segments. Deep sea passenger travel is now dominated by cruise shipping which has no freight-handling capabilities, and bulk and general cargo ships rarely have an interest or the ability to transport passengers.

Most rail systems still operate passenger and freight business. Where both segments are maintained the railways give priority to passengers, since rail persists as the dominant mode for inter-city transport in India, China and much of the developing world. In Europe the national rail systems and various levels of government have prioritized passenger service as a means of checking the growth of the automobile, with its resultant problems of congestion and environmental degradation (see Chapter 8). Significant investments have occurred in improving the comfort of trains and in passenger rail stations, but most notable have been the upgrading of track and equipment in order to achieve higher operational speeds. Freight transport has tended to lose out because of the emphasis on passengers. Because of their lower operational speeds, freight trains are frequently excluded from daytime slots, when passenger trains are most in demand. Overnight journeys may not meet the needs of freight customers. This incompatibility is a factor in the loss of freight business by most rail systems still trying to operate both freight and passenger operations.

In Europe, there are signs that the two markets are being separated. First, it is occurring at the management level. The liberalization of the railway system that is being forced by the European Commission is resulting in the separation of passenger and freight operations. This had already taken place in the UK when British Rail was privatized. Second, the move towards high-speed passenger rail service necessitated the construction of separate rights of way for the TGV trains. This has tended to move passenger train services from the existing tracks, thereby opening up more daytime slots for freight trains. Third, the Dutch are building a freight-only track, the Betuwe Line, from the port of Rotterdam to the German border, having already sold the freight business of the Netherlands railway (NS) to DB (Deutsche Bahn), and having opened up the freight business to other firms. It is in North America where the divorce between freight and passenger rail business is most complete. The private railway companies could not compete against the automobile and airline industry for passenger traffic, and consequently withdrew from the passenger business in the 1970s. They were left to operate a freight-only system, which has generally been successful, especially with the introduction of intermodality. The passenger business has been taken over by public agencies, Amtrak in the USA, and VIA Rail in Canada. Both are struggling to survive. A major problem is that they have to lease trackage from the freight railways, and thus slower freight trains have priority.

Freight and passenger vehicles still share the roads. The growth of freight traffic is helping increase road congestion, and in many cities concerns are being raised about the presence of trucks. Already restrictions are in place on truck dimensions and weights in certain parts of cities, and there are growing pressures to limiting truck access to non-daylight hours. Certain highways exclude truck traffic – the parkways in the USA for example. These are examples of what is likely to become a growing trend – the need to separate truck from passenger vehicle traffic. Facing chronic congestion around the access points to the port of Rotterdam and at the freight terminals at Schiphol Airport, Dutch engineers have worked on feasibility studies of developing separate underground road networks for freight vehicles.

Air transport is the mode where freight and passengers are most integrated. Yet even here a divergence is being noted. The growth of all-freight airlines and the freight-only planes operated by some of the major carriers, such as Singapore Airlines, are heralding a trend. The interests of the shippers, including the timing of the shipments and the destinations, are sometimes better served than in passenger aircraft. The divergence between passengers and freight is also being accentuated by the growing importance of charter and "low-cost" carriers. Their interest in freight is very limited, especially when their business is oriented towards tourism, since tourist destinations tend to be lean freight-generating locations.

Method 1 – Technical performance indicators

Macro indicators

Multimodal transportation networks rest upon the combinatory costs and performance of transport modes, or what is referred to as economies of scope. For instance, a single container shipped overseas at the lowest cost from its origin can go from road, to seaway, to railway and to road again before reaching its destination. Freight shippers and carriers therefore require quantitative tools for decision making in order to compare performances of various transport modes and transport networks. Time-efficiency becomes a set imperative for both freight and passenger transit in private as well as in public sector activities.

Performance indicators are widely used by geographers and economists to empirically assess the technical performance (not to be confused with economic performance, for there can exist a lag between the two) of differing transport modes, in other words their capacity to move goods or passengers around. Hence, basic technical performance calculations can be particularly useful for networks' global performance analysis as well as for modal comparison, analysis and evaluation by bridging both physical attributes (length, distance, configuration, etc.) and time-based attributes (punctuality, regularity, reliance, etc.) of networks. Some indicators are currently used to measure freight and passenger transport. Table 4.1 gives a few of the most common ones.

Passenger-km or ton-km are standard units for measuring travel that considers the number of people traveling or ton output and distance traveled. For example, 120 passenger-km represents 10 passengers traveling 12 kilometers or 2 passengers traveling 60 kilometers, and so on. More specifically, such indicators are of great utility by allowing cross-temporal analysis of a transport nexus or given transport modes.

Traffic performance indicators

There are two major operational types of traffic that influence the capacity of modern roads: continuous and discontinuous traffic. The capacity of a road is the maximal hourly flow of people or vehicles that can be supported by any link. This value is influenced by three major concepts:

Table 4.1 Commonly used macro performance indicators

Indicator	Passenger	Freight	Description
Passenger/freight density	passenger-km/km	ton-km/km	A standard measure of transport efficiency.
Mean distance traveled	passenger-km/ passenger	ton-km/ton	A measure of the ground-covering capacity of networks and different transport modes.
Mean per capita ton output (freight) Mean number of trips per capita (passenger)	passengers/population	tons/population	Used to measure the relative performance of transport modes.
Mean occupation coefficient	number of passengers aboard/total carrying capacity (%)	actual load (ton)/overall load capacity (ton) (%)	Especially useful with increasing complexity of logistics associated with containerization of freight (i.e. the problem of empty returns). Can also be used to measure transit ridership.

- **Road conditions**. Physical attributes of the road such as its type (paved, non-paved), number of lanes, width of lanes, design speed and the vertical and horizontal alignment.
- **Traffic conditions**. Attributes of the traffic using the road such as its temporal distribution and its direction.
- **Control conditions**. Attributes of the control structures and existing traffic laws such as speed limit, one-ways and priority.

Considering the above conditions, the capacity of a road is about 1,000 vehicles per lane per hour for continuous traffic roads and about 500 vehicles per lane per hour for discontinuous traffic roads. The operational goal of traffic planning is thus to make it so that road, traffic and control conditions ensure an adequate, if not optimal, service. Several guidelines will favor such a goal such as wide enough lanes for a safe maximum speed in both directions and limited grades to limit speed differentials. The capacity of a road is also linked to the level of service, which is a qualitative measure of operational conditions of roads and its perception by users. The spatial distribution of bottlenecks, notably within urban areas, also has a strong impact on capacity as they are the choking points of the whole road transport system. Traffic can be valued according to three primary measures, which are speed, volume or density:

- **Speed** is a rate of distance covered per unit of time. The average speed is the most commonly used measure to characterize traffic on a road.
- **Volume** is the number of vehicles observed at a point or a section over a period of time.
- **Density** is the number of vehicles that occupies a section at any point in time. For example, a road section having a volume of 1,000 vehicles per hour with an average speed of 50 km/hour will have a density of 20 vehicles/km.

The critical density is the density at which the volume is maximal and the critical speed is the speed at which the volume is maximal.

Economic impact indicators

Undoubtedly, transportation plays a considerable role in the economy with its omnipresence throughout the production chain, at all geographic scales. It is an integral constituent of the production–consumption cycle. Economic impact indicators help to appreciate the relationship between transport systems and the economy as well as to inform on the economic weight of this type of activity. Geographers should be familiar with basic econometric impact indexes (see Table 4.2).

Table 4.2 Measures of efficiency

| Factors of production | Scale-specific indicators | |
	Micro	*Macro*
Output/capital	Transport sector income/local income	
Output/labor	Output/local income	Output/GDP
Capital/labor		

Efficiency is usually defined as the ratio of input to output, or the output per each unit of input. Modal variations in efficiency will depend heavily on what is to be carried, the distance traveled, the degree and complexity of logistics required, as well as economies of scale. Freight transport chains rest upon the complementarity of cost-efficient and time-efficient modes, seeking most of the time a balanced compromise rather than an ideal or perfect equilibrium.

Maritime transport is still the most cost-efficient way to transport bulk merchandise over long distances. On the other hand, while air transport is recognized for its unsurpassed time-efficiency versus other modes over long distances, it remains an expensive option. Thus, vertical integration, or the absorption of transportation activities by producers, illustrates the search for these two efficiency attributes by gaining direct control over inputs.

Transportation and economic impacts

The relationship between transport systems and their larger economic frame becomes clear when looking at restructuring patterns which carriers and firms are currently undergoing. Structural mutations, best illustrated by the popularity of just-in-time practices, are fueled by two opposing yet effective forces: transporters seek to achieve economies of scale while having to conform to an increasingly "customized" demand.

Factor substitution is a commonly adopted path in order to reduce costs of production and attain greater efficiency. Containerization of freight by substituting labor for capital and technology is a good illustration of the phenomenon. Measures of capital productivity for such capital-intensive transport means are of central importance; an output/capital ratio is then commonly used. While the output/labor ratio performs the same productivity measurement but for the labor input (this form of indicator can be used for each factor of production in the system), a capital/labor ratio aims at measuring which factor predominates within the relationship between capital and labor productivity. The above set of indicators therefore provides insights on the relative weight of factors within the production process.

More scale-specific indicators can also be used to appreciate the role of transport within the economy. Knowing freight transport both contributes to and is fueled by a larger economic context, freight output can be confronted against macro-economic indicators: an output/GDP ratio measures the relationship between economic activity and traffic freight, in other words the traffic intensity. At the local level, the status of the transport industry within the local economy is given by a transport sector income/local income ratio. Still at a micro-scale, finally, a measure of the relative production value of freight output is provided by an output/local income ratio.

Underlying objectives of application of such indicators are as varied as they are numerous. Efficiency indicators constitute valuable tools to tackle project viability questions as well as to measure investment returns and cost/subsidy recovery of transport systems. Input–output analyses making use of some of the above indicators are also instrumental to the development of global economic impact indexes and productivity assessment concepts such as the Total Factor Productivity (TFP) and to identify sources of productivity gains.

Specialization index

In transport, to find out if a terminal is specialized in the transhipment and/or handling of a particular kind of merchandise or if, inversely, it transfers a wide variety of merchandise, we can calculate a specialization index. For example, the index can be used to know if a port is specialized in the handling of a certain type of product (e.g. containers) or if it handles a wide range of merchandise. As a consequence, such an index is quite versatile and has a variety of applications; it informs geographers on the activities of any type of terminal (port, train and airport). In the case of an airport terminal, one could ask if a given airport deals with only a single type of flights/passengers (local, national, international, etc.) or if it welcomes several. The specialization index (*SI*) is calculated using the following formula:

$$SI = \frac{\sum_i t_i^2}{\left(\sum_i t_i\right)^2}$$

which is the total of squares of tonnage (or monetary value) of each type of merchandise *i* (t_i) handled at a terminal over the square of the total volume tonnage (or monetary value) of merchandise handled at the terminal.

So, if the specialization index tends toward 1, such a result indicates that the terminal is highly diversified. If, inversely, the index tends toward 0, it means that the terminal's activity is specialized. Thus, the specialization index is called upon to appreciate the degree of specialization/diversification of a port, an airport, a train station or any type of terminal.

Location coefficient

Certain kinds of merchandise are often transhipped at particular terminals rather than at others. Thus, the degree of concentration of a certain type of traffic in a terminal (port, airport, train station) compared with the average for all the terminals, can be measured by using the location coefficient.

The **location coefficient** is the share of traffic occupied by a type of merchandise at a terminal over the share of traffic of the same type of merchandise among the total traffic of all terminals of the same type.

In the field of transportation, the location coefficient (*LC*) is calculated by using the following formula:

$$LC = \frac{\left(\dfrac{M_{ti}}{\Sigma_t M_{ti}}\right)}{\left(\dfrac{\Sigma_t M_t}{\Sigma M}\right)}$$

where M_{ti} is the traffic of a merchandise *t* at a terminal *i*, M_t is the total of all merchandises of type *t* for all terminals and *M* is the total of all types of merchandises for all terminals.

The greater the value of the index, the greater is the degree of traffic of a certain type of merchandise. Possible outcomes are of three types:

- A figure lower than 1 indicates that the traffic of the chosen merchandise in the terminal is under-represented compared to the same merchandise in all the terminals.
- A figure equal to 1 indicates that the quantity of traffic of the chosen merchandise in a terminal is proportional to its participation in total traffic.
- Finally, a coefficient above 1 indicates that the traffic of the chosen merchandise in a given terminal is preponderant in total traffic.

Beside using the location coefficient to evaluate the relative weight of a type of traffic in a terminal, the location coefficient can be used to appreciate the importance of an economic activity for a community compared with the importance of the same activity within a defined larger area (e.g. province, country, world, etc.). The larger geographic entity is also known as the benchmark and is critical in the calculation of the location coefficient.

Method 2 – Symbolization of transport features in a GIS

Cartography and symbolization

Cartography is a communication tool that conveys a message to a public through a medium: the map. The better the cartography, the more likely that this message will be conveyed effectively. Some forms of communication are better than others, so all maps are not equal, even if they are representing the same features. Since many transport projects have a high visibility and significant capital costs, it is surprising that the usage of visual resources, particularly of cartography, is often neglected or not used properly. The cartographic quality of many transport analyses is commonly poor. This stems from the fact that many transport practitioners are engineers or economists by training, disciplines in which cartographic expression is not emphasized or even considered. Among transport geographers using GIS-T (Geographic Information Systems for Transportation), the cartographic output is also commonly neglected, again an outcome of the priority placed on analytical methods. Even if cartography does not appear to be a feature which is analytically strong (in contradiction to the GIS packages that produce them), proper cartographic expression has become a crucial element of transportation research, particularly because of the following:

- Transportation systems, notably networks, are complex entities and the map offers a powerful **medium to visualize them**. Thus, cartography can be seen as a synthetic tool.
- Transportation is a field of application which is often planning driven. As such, many projects require the approval of various private (funding) and public (regulation) entities, and sometimes with the general public involved. Maps are thus a medium that can be used to **explain the nature of a project** and help persuade an audience.

Maps are using visual communication tools, thus implying that cartography is at the same time an art and a technique. It is an art since it is a visual expression; every map is to some extent a form of art that seeks to aesthetically please its audience. Considering maps as an artistic expression is often seen with a level of suspicion among practitioners. It is often perceived that the quality of the container is inversely proportional to the quality of the content. Cartography is also a technique since it abides to a set of rules and methods pertaining to the visual symbols it uses; their placement, the choice of colors and their size for instance. Cartography is a process of abstraction, also referred to as symbolization, which uses a set of defined graphical elements to communicate a message.

> **Symbolization** is the set of graphic methods used to convert cartographic information into a visual representation.

Symbolization implies that the features on a map are generalized and simplified since not all possible elements are relevant to the message a map conveys. It thus helps the message to be easier to understand. For instance, a map depicting a highway system often ignores all the roads of lesser importance, thus underlining the feature it seeks to emphasize.

With the maturation of GIS in recent years, the generation of maps has become a simpler and straightforward process. Graphic design capabilities, which were found lacking in earlier packages, are more extensive. GIS enable to produce maps at a very low cost and in large quantities. In addition, more information is available from a variety of sources, particularly in numerical format. Several databases and basemaps are made available at virtually no cost. The Internet has become a massive distribution medium of graphical images such as maps and enables access to a wide array of publicly available databases from international, national and local institutions. Many public or private agencies, from newspapers (e.g. the *New York Times*) to government offices employ professional cartographers and the quality of the cartographic output has considerably improved.

Visual resources

GIS automate several aspects of the cartographic process and assist cartographers for tasks that previously took a lot of training, time and manual expertise. The creation and revision process of maps is improved since previously created maps can be stored, retrieved and modified to suit new purposes. The layout, the composition and the symbolization can be modified at will. It is important to stress that GIS do not *per se* make good or bad maps, cartographers do. Consequently, the appropriate usage of visual resources is the first step in the efficient cartography of the transport phenomena.

The rapid diffusion of GIS and the improvement in computerized visualization techniques offers transport practitioners many opportunities to improve the visual quality of their work. This begins with the usage of visual resources, mainly two basic ones:

- **Color resources.** Considers the hue, texture and intensity of colors. A hue refers to the gradation of color within the optical spectrum (visible spectrum) of light. The texture is the variety of patterns that can be used to fill a shape, such as hatches, cross-hatches or dot density. The intensity is the relative saturation of a color, on a scale from bright to dull. Color resources are particularly useful for category ranges.
- **Shape resources.** Considers the wide variety of geometric figures available. In a vector-based GIS, shapes are mainly represented as points, lines and polygons. These shapes can be modified in terms of their nature, size and orientation.

Raster information, since it is grid-based, can only be modified through its color hue and intensity. For cartographic purposes, visual resources can be used to represent location, direction, distance, movement, function, process and correlation. On most maps, including those related to transportation, several elements, such as title, scale and legend, are almost always present. How all these elements are positioned on a map, also known as map composition, depend on the nature of the message as well as the potential audience. Each cartographer has his/her own visual style.

Symbolization strategies for transport attributes

Transportation deals with a set of issues that rely on a specific range of symbols. Most of the symbolization deals with networks, which are features that are commonly represented with lines and points (see graph theory). Other symbolization strategies, such as choropleth maps, are common with standard cartographic methods. The following are the most common symbolization strategies (Figure 4.12):

- **Nominal**. It includes only names, which are the result of a classification. These names are not ordered in a specific way; rather they describe different categories of the same rank. So, the only conclusion to be made is the inequality of each class. Transportation infrastructures are particularly suitable for nominal representations. Networks and terminals can be classified by function or ownership. On Figure 4.12 this involves a simple distinction between Interstate highways that are toll and those that are regular (non-toll). There is no order in this classification since each class is simply different from the other.
- **Ordinal**. Result of placing descriptive categories into a formal order which enables a comparison of rank without providing any information about the extent of the difference. There is thus an implicit qualitative order between classes. Networks and terminals can be classified by size, level of importance or congestion. On Figure 4.12 a distinction is made between major roads according to their capacity (low, average, high). In this case there is an implied order in the classification as some road segments have a higher level of importance than others.
- **Interval**. An interval scale is the result of arranging values on a scale with a point of reference and a unit of measure. These scales are quantitative, which means some computations are allowed, namely how the ranges between classes are set. The level of traffic on networks and terminals can be categorized. On Figure 4.12 a quantitative variable, the volume to capacity ratio (a good indicator of congestion), is classified in distinct and clearly bounded categories.
- **Proportional symbols**. The size of a symbol is a function of a quantitative variable. Thus, the radius of a circle or the thickness of a line can be set according to a variable. For transportation systems, proportional symbols are particularly important to express flows in a network or at terminals. On Figure 4.12, the size of a symbol, road segments in this case, is a direct function of a quantitative variable. The higher the traffic, the thicker the line segment.
- **Labeling**. Involves the positioning of descriptive text over specific geographical features. The labeling process which is particular to transportation mainly concerns assigning identification symbols (dominantly numbers) to road segments.

CASE STUDY Maersk shipping line

A global maritime shipper

Maersk is an old established company, founded in 1912 in Denmark by A.P. Moller. It has grown to become the world's largest container shipping line and is likely to be the most globalized transport company. The company has been an innovator,

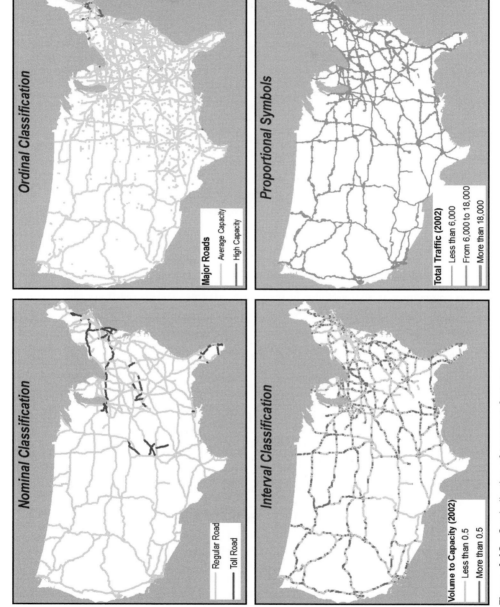

Figure 4.12 Symbolization of transport features

developing new approaches to shipping that have led to its present dominance. The company has from the very early days focused on general freight, tanker and shipbuilding. The son of the founder, Maersk-McInney Moller, whose mother was American, managed the company after the German invasion of Denmark, and operated the company from New York. After his father's death in 1965 he became CEO and Chairman, the latter position he occupied until his ninetieth birthday in 2003. Maersk is still largely under the ownership of the family trust. As a family business it has been able to respond quickly to commercial and technological changes that have contributed to its growth.

The company was relatively late in adopting containerization, its major activities remaining in general cargo trades on the Pacific, and oil transport. The company obtained its first cellular ship in 1973. Thereafter, however, it began a rapid conversion of its fleet, most of which were new purchases from German and Japanese shipyards. It was not until 1981 that the first container ships were built by its own yards, but thereafter it began producing ever larger ships, the owner recognizing the scale economies they provided. By the late 1980s Maersk shipyards were building the largest containerships afloat. The ability of Maersk to build ships itself enabled it to surprise the rest of the industry with ground-breaking developments. In 1996, for example, it introduced a class of ship with a capacity of 6,000 TEUs that was a breakthrough from the existing post-panamax ships. This was followed two years later by 8,000 TEU ships, and in 2006 it introduced a class of vessel with a capacity of 14,000 TEUs. By emphasizing capacities of ships, the shipping line has been the industry leader, forcing its competitors to follow.

Port and inland operations

The company was also a leader in reshaping services. The increasingly global nature of the container market along with the deployment of ever larger vessels caused Maersk to introduce a hub-and-spoke service network. In the late 1980s it established its own hub port at a little-used site in southern Spain, Algeciras, and quickly made this hub port the largest in the Mediterranean. Over the years it has progressively established other hubs, sometimes on greenfield sites, such as Tanjung Pelepas in Malaysia, and frequently in major ports such as New York. From these hubs local distribution and inter-service transfers are arranged.

In the 1990s, facing increasing costs of providing global coverage, Maersk, like most other major carriers, sought to share expenses by forming strategic alliances. Maersk joined forces with Sealand, the largest US carrier. Unlike the other alliances, however, this alliance developed into closer partnership, and culminated in the acquisition of Sealand by Maersk in 1999, establishing the company as the largest container shipping line. This marked a new phase in Maersk's growth, a growth maintained by mergers and acquisitions. In 1999 it also acquired the regional carrier Safmarine, and in 2006 it purchased P&O Nedlloyd, the world's fourth largest carrier. In 2007 Maersk accounted for 17 percent of the world's container carrying capacity.

Maersk is an excellent example of vertical integration. Not content to establish itself as the dominant ocean carrier, it sought to progressively bring more and more of the transport chain under its control. It is an important terminal operator, managing container berths for its own ships, as at Algeciras, but also for other carriers, such as at Kingston, Jamaica. It is actively involved in inland distribution in many markets, with the purchase of trucking firms in the USA, operating rail shuttles from Rotterdam and managing barge terminals in Germany. In addition, Maersk operates its own logistics company, providing its clients and others with supply chain management. In these ways the company is better able to manage its own traffic and realize profits from other parts of the supply chain. For a privately owned company from a small European country, Maersk's achievements are remarkable.

Bibliography

Agusdinata, B. and W. de Klein (2002) "The Dynamics of Airline Alliances", *Journal of Air Transport Management*, 8, 201–11.

Brooks, M. (2000) *Sea Change in Liner Shipping*, New York: Pergamon.

Fremont, A. (2007) "Global Maritime Networks: The Case of Maersk", *Journal of Transport Geography*, 15(6), 431–42.

Graham, B. (1995) *Geography and Air Transport*, Chichester: Wiley.

Hayuth, Y. (1987) *Intermodality*, Essex: Lloyds of London Press.

Levinson, M. (2006) *The Box: How the Shipping Container Made the World Smaller and the World Economy Bigger*, Princeton, NJ: Princeton University Press.

MARAD (2006) Office of Maritime Administration, US Department of Transportation, http://www.marad.dot.gov/library_landing_page/data_and_statistics/Data_and_Statistics.htm.

Muller, G. (1995) *Intermodal Freight Transportation*, 3rd edn, Westport, CT: Eno Transportation Foundation.

Notteboom, T. (1998) *Land Access to Sea Ports*, Round Table 113, Council of European Transport Ministers: Paris.

Notteboom, T. and J.-P. Rodrigue (2009) "The Future of Containerization: Perspectives from Maritime and Inland Freight Distribution", *Geojournal*, vol. 74, No. 1, pp. 7–22.

Notteboom, T. and Konings, R. (2004) "Network Dynamics in Container Transport by Barge", *Belgeo*, 5, 461–77.

Robinson, R. (2002) "Ports as Elements in Value-driven Chain Systems: The New Paradigm", *Maritime Policy and Management*, 29, 241–55.

Slack, B. (1998) "Intermodal Transportation", in B.S. Hoyle and R. Knowles (eds) *Modern Transport Geography*, Chichester: Wiley. 2nd edn, pp. 263–90.

Slack, B. (2004) "Corporate Realignment and the Global Imperatives of Container Shipping", in D. Pinder and B. Slack (eds) *Transport in the Twenty-First Century*, London: Routledge, pp. 25–39.

Stopford, M. (1997) *Maritime Economics*, 2nd edn, London: Routledge.

van Klink, A. and G.C. van den Berg (1998) "Gateways and Intermodalism", *Journal of Transport Geography*, 6, 1–9.

⑤ Transportation terminals

All spatial flows, with the exception of personal vehicular and pedestrian trips, involve movements between terminals. With these two exceptions, all the transport modes require assembly and distribution of their traffic, both passenger and freight. For example, passengers have to go to bus terminals and airports first in order to reach their final destinations, and freight has to be consolidated at a port or a rail yard before onward shipment. Terminals are, therefore, essential links in transportation chains. The goal of this chapter is to examine the strong spatial and functional character of transport terminals. They occupy specific locations and they exert a strong influence over their surroundings. At the same time they perform specific economic functions and serve as foci for clusters of specialized services.

Concept 1 – The function of transport terminals

The nature of transport terminals

A terminal may be defined as any facility where freight and passengers are assembled or dispersed. They may be points of interchange involving the same mode of transport. Thus, a passenger wishing to travel by train from Paris to Rotterdam may have to change trains in Brussels, or an air passenger wishing to fly between Montreal and Winnipeg may have to change planes in Toronto. They may also be points of interchange between different modes of transport, so that goods being shipped from the US Midwest to the Ruhr in Germany may travel by rail from Cincinnati to the port of New York, be put on a ship to Rotterdam, and then placed on a barge for delivery to Duisberg. Transport terminals, therefore, are central and intermediate locations in the movements of passengers and freight.

Differences in the nature, composition and timing of transfer activities give rise to significant differentiations in the form and function between terminals. A basic distinction is between passenger and freight transfers, because in order to carry out the transfer and bundling of each type, specific equipment and infrastructures are required.

Passenger terminals

With one exception, passenger terminals require relatively little specific equipment. This is because individual mobility is the means by which passengers access buses, ferries or trains. Certainly, services such as information, shelter, food and security are required, but the layouts and activities taking place in passenger terminals tend to be simple and require relatively little equipment. They may appear congested at certain times of the day, but the flows of people can be managed successfully with good design of platforms and access points, and with appropriate scheduling of arrivals and departures.

The amount of time passengers spend in such terminals tends to be brief. As a result bus termini and railway stations tend to be made up of simple components, from platforms, ticket offices and waiting areas to limited amounts of retailing.

Airports are of a different order. They are among the most complex of terminals functionally. Moving people through an airport has become a very significant problem, not least because of security concerns. Passengers may spend several hours in transiting, with check-in and security checks on departure, and baggage pickup and in many cases customs and immigration on arrival. Planes may be delayed for a multitude of reasons. The result is that a wide range of services not directly related to the transfer function have to be provided for passengers, including restaurants, bars, stores, hotels, in addition to the activities directly related to operations such as check-in halls, waiting areas, passenger loading ramps and baggage handling facilities (Plate 5.1). At the same time airports have to provide the very specific needs of the aircraft, from runways to maintenance facilities, from fire protection to air traffic control.

Measurement of activities in passenger terminals is generally straightforward. The most common indicator is the number of passengers handled, sometimes differentiated according to arrivals and departures. Transfer passengers are counted in the airport totals even though they do not originate there, and so airports that serve as major transfer facilities inevitably record high passenger totals. This is evident in Figure 5.1 where in-transit passengers at the two leading airports, Atlanta and Chicago, account for over 50 percent of the total passenger movements. High transfer passenger activity has been enhanced by the actions of many of the leading airlines adopting hub-and-spoke

Plate 5.1 Modern airport terminal, Madrid. *Credit*: Jean-Paul Rodrigue

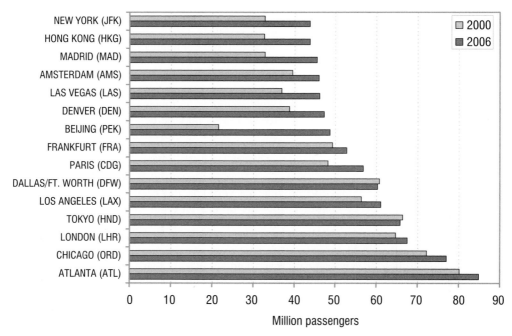

Figure 5.1 World's largest airports

networks (see Chapter 4). This results in many passengers being forced to change planes at the hub airports. By selecting certain airports as hubs, the carriers are able to dominate activity at those airports, thereby controlling most landing and departure slots and the best gates, thus fending off rival airlines. In this way they are able to extract monopoly profits.

A further measure of airport activity is the number of aircraft movements, a figure that must be used with some caution because it pays no regard to the capacity of planes. High numbers of aircraft movements may not be correlated with passenger traffic totals.

Freight terminals

Freight handling requires specific loading and unloading equipment. In addition to the facilities required to accommodate ships, trucks and trains (berths, loading bays and freight yards respectively) a very wide range of handling gear is required that is determined by the kinds of cargoes handled. The result is that terminals are differentiated functionally both by the mode involved and the commodities transferred. A basic distinction is that between bulk, general cargo and containers:

- **Bulk cargo** refers to goods that are handled in large quantities that are unpackaged and are available in uniform dimensions. Liquid bulk goods include crude oil and refined products that can be handled using pumps to move the product along hoses and pipes. Relatively limited handling equipment is needed, but significant storage facilities may be required. Dry bulk includes a wide range of products, such as ores, coal (Plate 5.2) and cereals. More equipment for dry bulk handling is required, because the material may have to utilize specialized grabs and cranes and conveyer-belt systems.

Plate 5.2 Bulk coal terminal, Shanghai. *Credit*: Claude Comtois

- **General cargo** refers to goods that are of many shapes, dimensions and weights such as machinery and parts. Because the goods are so uneven and irregular, handling is difficult to mechanize. General cargo handling usually requires a lot of labor.
- **Containers** are standard units that have been designed for simplicity and functionality. Container terminals have minimal labor requirements and perform a wide variety of intermodal functions. They however require a significant amount of storage space which are simple paved areas where containers can be stacked and retrieved with a set of cranes, straddlers and holsters. Depending on the intermodal function of the container terminal, specialized cranes are required, such as portainers (container cranes) (Plate 5.3). Intermodal terminals and their related activities are increasingly seen as agents of added value within supply chains.

A feature of most freight activity is the need for storage. Assembling the individual bundles of goods may be time-consuming and thus some storage may be required. This produces the need for terminals to be equipped with specialized infrastructures such as grain silos, storage tanks and refrigerated warehouses, or simply space to stockpile.

Measurement of freight traffic through terminals is more complicated than for passengers. Because freight is so diverse, standard measures of weight and value are difficult to compare and combine. Because bulk cargoes are inevitably weighty, terminals specialized in such cargoes will inevitably record higher throughputs measured in tons than others more specialized in general cargoes. This is evident from Figure 5.2, where the traffic of the two of the leading ports, Singapore and Rotterdam, is dominated by

Plate 5.3 Yantian Container Port, Shenzhen, China. *Credit*: Jean-Paul Rodrigue

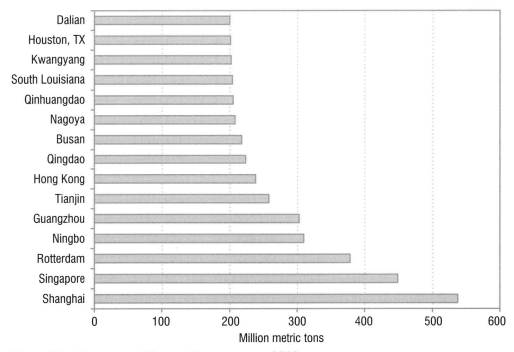

Figure 5.2 Throughput of the world's major ports, 2006

petroleum. The reverse may be true if value of commodities handled is the measure employed. The problem of measurement involving weight or volumes becomes very difficult when many types of freight are handled, because one is adding together goods that are inherently unequal. Care must be taken in interpreting the significance of freight traffic totals, therefore.

The difficulty of comparing traffic totals of different commodities has led to attempts to "weight" cargoes based upon some indication of the value-added they contribute to the terminal. The most famous is the so-called "Bremen" rule. This was developed in 1982 by the port of Bremen and was based on a survey of the labor cost incurred in the handling of one ton of different cargoes. The results found that handling one ton of general cargo equals three tons of dry bulk and 12 tons of liquid bulk. Although this is the most widely used method, other "rules" have been developed by individual ports, such as Rotterdam, and more recently by the port of Antwerp. The "Antwerp rule" indicates that the highest value-added is the handling of fruit. Using this as a benchmark, forest products handling requires 3.0 tons to provide the same value-added as fruit, cars 1.5 tons, containers 7 tons, cereals 12 tons and crude oil 47 tons.

Terminal costs

Because they jointly perform transfer and consolidation functions, terminals are important economically because of the costs incurred in carrying out these activities. The traffic they handle is a source of employment and benefits regional economic activities, notably by providing accessibility to suppliers and customers. Terminal costs represent an important component of total transport costs. They are fixed costs that are incurred regardless of the length of the eventual trip, and vary significantly between the modes. They can be considered as:

- **Infrastructure costs.** Include construction and maintenance costs of facilities such as piers, runways, cranes and structures (warehouses, offices, etc.).
- **Transhipment costs.** The costs of loading and unloading passengers or freight.
- **Administration costs.** Many terminal facilities are managed by institutions such as port or airport authorities or by private companies. In both cases administration costs are incurred.

Because ships have the largest carrying capacities, they incur the largest terminal costs, since it may take many days to load or unload a vessel. Conversely, a truck or a passenger bus can be loaded much more quickly, and hence the terminal costs for road transport are the lowest. Terminal costs play an important role in determining the competitive position between the modes. Because of their high freight terminal costs, ships and rail are unsuitable for short-haul trips.

Figure 5.3 represents a simplified assumption concerning transport costs for three modes. It should be noticed that the cost curves all begin at some point up the cost axis. This represents terminal costs, and as can be seen, shipping (*T3*) and rail (*T2*) start with a significant disadvantage compared to road (*T1*).

Competition between the modes is frequently measured by cost comparisons. Efforts to reduce transport costs can be achieved by using more fuel-efficient vehicles, increasing the size of ships and reducing the manpower employed on trains. However, unless terminal costs are reduced as well, the benefits would not be realized. For example, in water transportation, potential economies of scale realized by ever larger and more fuel-efficient vessels would be negated if it took longer to load and off-load the jumbo ships.

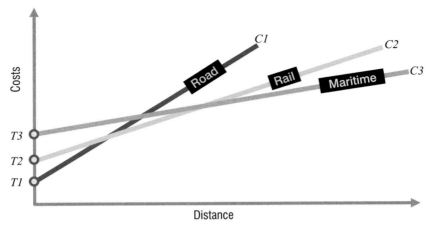

Figure 5.3 Terminal costs

Over the last 40 years, very significant steps to reduce terminal costs have been made. These have included introducing information management systems such as EDI (electronic data interchange) that have greatly speeded up the processing of information, removing delays typical of paper transactions. The most significant development has been the mechanization of loading and unloading activities. Mechanization has been facilitated by the use of units of standard dimensions such as the pallet and, most importantly, the container. The container, in particular, has revolutionized terminal operations (see Chapter 4). For the mode most affected by high terminal costs, ocean transport, ships used to spend as much as three weeks in a port undergoing loading and loading. The much larger ships of today spend less than a couple of days in port. A modern container ship requires approximately 750 man-hours to be loaded and unloaded. Prior to containerization it would have required 24,000 man-hours to handle the same volume of cargo. The rail industry too has benefited from the container, which permits trains to be assembled in freight yards in a matter of hours instead of days.

Reduced terminal costs have had a major impact on transportation and international trade. Not only have they reduced overall freight rates, and thereby reshaping competition between the modes, but they have had a profound effect on transport systems. Ships spend far less time in port, enabling ships to make many more revenue-generating trips per year. Efficiency in the airports, rail facilities and ports greatly improves the effectiveness of transportation as a whole. Terminals play a key role in transport systems.

Activities in transport terminals represent not just exchanges of goods and people, but also constitute an important economic activity. Employment of people in various terminal operations represents an advantage to the local economy. Dockers, baggage handlers and crane operators, air traffic controllers are examples of jobs generated directly by terminals. In addition there are a wide range of activities that are linked to transportation activity at the terminals. These include the actual carriers (airlines, shipping lines, etc.), intermediate agents (customs brokers, forwarders) required to carry out the transfers. It is no accident that centers that perform major airport, port and rail functions are also important economic locales.

Terminals favor the agglomeration of related activities in their proximity and often adjacent to them (see Figure 5.4). This terminal–client link mainly involves warehousing

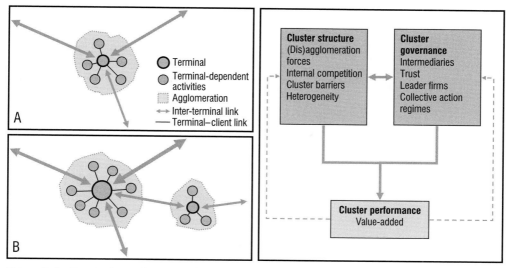

Figure 5.4 Terminals as clusters and growth poles

and distribution (A). The contribution of transport terminals to regional economic growth can often be substantial. As the regional demand grows, so does the traffic handled by the related terminal. This in turn can spur further investments to expand the capabilities of the terminal and the creation of a new terminal altogether (B).

Economists have identified clusters as a critical element in shaping competition between countries, regions and industries. Clusters are defined as a population of interdependent organizations that operate in the same value chain and are geographically concentrated. The seaport cluster is made up of firms engaged in the transfer of goods in the port and their onward distribution. It also includes logistics activities as well as processing firms and administrative bodies. The performance of the seaport cluster is defined as the value-added generated by the cluster, and is shaped by the interrelationships between the structure of the cluster and its governance. Cluster structure refers to the agglomeration effects and the degree of internal cohesion and competition. Cluster governance relates to the mix of, and relations between, organizations and institutions that foster coordination and pursue projects that improve the cluster as a whole. When applied to the Port of Rotterdam it was suggested that a key role was played by the intermediary firms, those that operated services and activities for core transport firms. High levels of trust between firms led to lower transaction costs, and leader firms were very significant because they helped strengthen the agglomeration.

Presented as a new approach, cluster theory is extending what others, including geographers, have recognized for some time, that port activity, historically at least, generates strong agglomeration economies that produce strong spatially distinct port communities. Despite similarities in results from economic impact studies, airports and rail terminals have not yet received the attention of cluster theorists.

Concept 2 – Terminals and location

Location and spatial relations play a significant role in the performance and development of transport terminals. As in all locational phenomena there are two dimensions involved. First is the issue of site, or absolute location. Terminals occupy very specific sites, usually with stringent requirements. Their site determinants may play an important role in shaping performance. The second component is relative location, or location relative to other terminals in the network. The spatial relations of terminals are an extremely important factor in shaping competition. Together, absolute and relative locations provide justification for the fundamental significance of geography in understanding transport terminals.

The nature of the function of the terminal is critical to understand its site features. Locations are determined according to the mode and by the types of activities carried on. As will be explained below, when site development took place is also a factor in site selection and elaboration.

Port sites

Ports are bound by the need to serve ships, and so access to navigable water has been historically the most important site consideration. Before the industrial revolution, ships were the most efficient means of transporting goods, and thus port sites were frequently chosen at the head of water navigation, the most upstream site. Many major cities owed their early pre-eminence to this fact: London on the Thames and Montreal on the St. Lawrence River. Sites on tidal waterways created a particular problem for shipping because of the twice-daily rise and fall of water levels at the berths, and there developed by the eighteenth century the technology of enclosed docks, with lock gates. Because ship transfers were slow, and vessels typically spent weeks in ports, a large number of berths were required. This frequently gave rise to the construction of piers and jetties to increase the number of berths per given length of shoreline, resulting in port expansion.

Over time changes in ships and handling gave rise to new site requirements. By the post-World War II period a growing specialization of vessels emerged, especially the development of bulk carriers. These ships were the first to achieve significant economies of scale, and their size grew very quickly. For example the world's largest oil tanker in 1947 was only 27,000 dwt, by the mid-1970s it was in excess of 500,000 dwt. There was thus a growing vessel specialization and increase in size which resulted in new site requirements, especially the need for dock space and greater depths of water. These site changes and developments in port infrastructure were captured in Bird's Anyport model (1963). The original five-stage model is condensed in Figure 5.5 to three phases.

Empirical research has confirmed the robustness of the Bird model and it has been amended to include more recent developments. In particular, it has been demonstrated that ports do undergo renovation of some old facilities, and that port physical development is cyclical.

Rotterdam, one of the largest ports of the world, provides a good example of morphological development, with a clear downstream progression of expansion (see Figure 5.6):

● The port originated adjacent to the old city center. With the growth of industrial activity in its hinterland, especially in the Ruhr (Germany) in the nineteenth century, the port began to expand downriver. The importance of the port resulted in its complete destruction during World War II.
● After 1945 there was some rebuilding of the larger older docks on the south bank of the river, but the major emphasis was the creation of new facilities further downriver at Botlek.

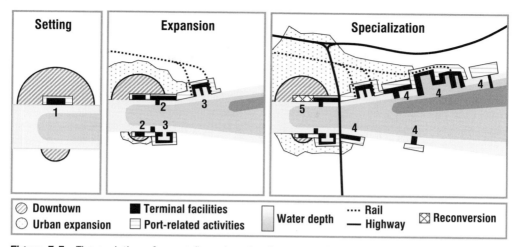

Figure 5.5 The evolution of a port (based on the Anyport model)

- By the 1950s the port authority realized that these were inadequate to meet the demands of ever larger oil tankers, and initially sought to build new terminals on the north bank of the river. The proposed sites were adjacent to urban development, and there was intense local opposition. The port authority then proposed development on reclaimed land south of the river, the Europoort complex. This was built in the 1960s, and became the heart of Europe's major oil refining and petrochemical industry.
- The advent of containers led to the conversion of several old sites in the Waalhaven and Botlek areas in the 1970s.
- The growth of container traffic along with continued expansion of bulk traffic caused the port to consider expansion out in to the North Sea. This led to the construction of an entirely new facility on reclaimed land at Maasvlatke in the 1980s.
- Subsequent traffic growth in the 1990s resulted in the port authority proposing a new facility further out in the North Sea: Maasvlatke II. After years of opposition by environmentalists, the project began construction in 2008 and should be open for traffic in 2010. By 2030 this phase is expected to be completed.

In form therefore the port has been squeezed between competing land uses, to the north by largely urban pressures and to the south by agricultural land.

One of the features that Anyport brings out is the changing relation between ports and their host cities. The model describes the growing repulsion by the rest of the urban milieu. This aspect has been worked on over the last two decades by a number of geographers investigating the redevelopment of harbor land. Hoyle (1988) proposed an Anyport-type model, which instead of stressing the port infrastructure development, emphasizes the changing linkages between the port and the city. One of these urban linkages is the redevelopment of old port sites for other urban uses, such as Docklands in London and Harborfront in Baltimore. More recently it has been shown that port–city relationships are quite varied and reflect both the size of the host city and the volume of the port traffic, which leads to a new typology of port–urban relations based on these two parameters (see Figure 5.7).

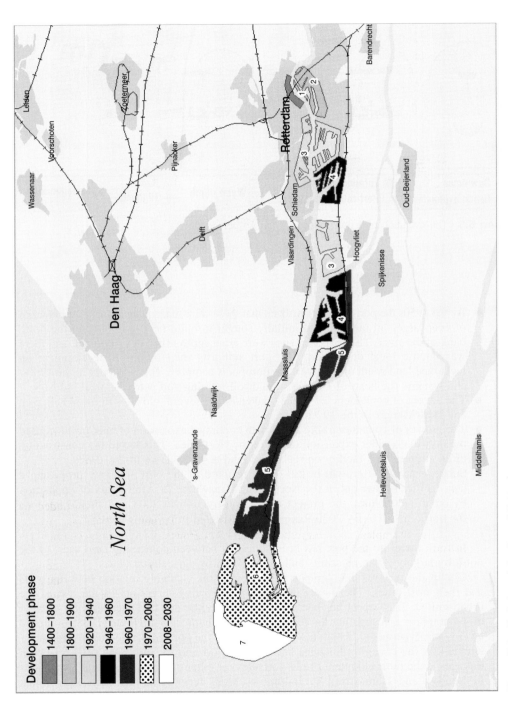

Figure 5.6 Evolution of the Port of Rotterdam

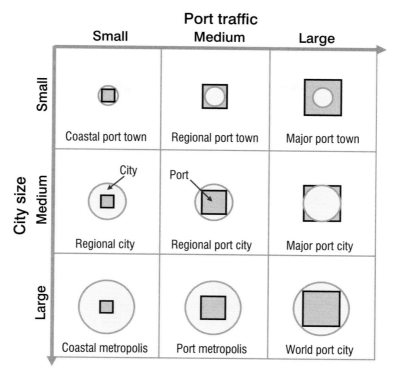

Figure 5.7 Typology of port cities (Source: adapted from Ducruet, 2007)

Airport sites

Airports require very large sites. They need space for runways, for terminal buildings, maintenance hangars and for parking. While there are considerable variations in the scale of different airports, minimum sizes in excess of 500 hectares represent enormous commitments of urban land. Thus, airports are sited at the periphery of urban areas, because it is only there that sufficient quantities of land are available. Many airports built in the 1940s and 1950s on the periphery now find themselves surrounded by subsequent metropolitan development. Pearson Airport (Toronto) and O'Hare Airport (Chicago) are examples. These airports have served as growth poles, drawing commercial, industrial as well as residential developments to those sectors of the city.

New site development today, in North America and Europe at least, is becoming very difficult because available sites are frequently so far from the urban core that even if planning permission could be obtained, it would lead to very significant diseconomies because of the distance from business and demographic cores. It is significant that there have been few new large-scale airport developments in North America over the last 30 years, and the examples of Denver and Montreal illustrate how difficult and contentious development has been. The result has been that most airports have to adjust to their existing sites, by reconfiguring runways and renovating existing terminal facilities, as for example at Chicago (see Figure 5.8).

In the growing economies of Asia, on the other hand, new airports are being built on greenfield sites to cope with passenger growth. Many of these airports are on sites far removed from the central cities. Examples include Pudong Airport that is 30 km

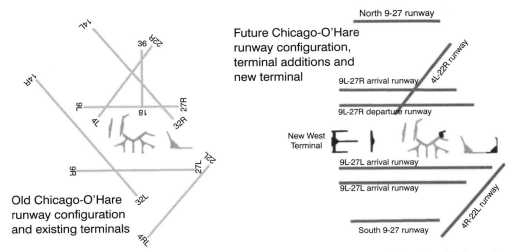

Figure 5.8 O'Hare Airport modernization program (Source: John Bowen, University of Wisconsin-Oshkosh 2007)

from Shanghai, and Chek Lap Kok Airport that has been built on reclaimed land on an island 35 km from the centre of Hong Kong. Their distance from the central cities is compensated by high-speed rail connections, and in the case of Shanghai using Maglev technology.

Rail terminal sites

Rail terminals, because they are not as space-extensive as airports and ports, suffer somewhat less from site constraints. Many rail terminals were established in the nineteenth century during the heyday of rail development, and while the sites may have been on the edge of urban areas at the time, they now find themselves surrounded by urban development. Individually rail terminals may not be as extensive as airports or ports, but cumulatively the area of all the rail sites in a city may exceed those of the other modes. For example, in Chicago the combined area of rail freight yards exceed that of the airports.

Passenger rail terminals are typically in the heart of downtown cores. At one time their sites may have been on the edge of the pre-industrial city, as is the case for London and Paris, but today they are very much part of the central business district. The stations are typically imposing buildings reflecting the power and importance represented by the railway in the nineteenth and early twentieth centuries. Grand Central Station in New York or St. Pancras Station in London are impressive architectural achievements unmatched in any other type of transportation terminal. As rail passenger traffic has declined the need for many of these stations has diminished, and a rationalization has resulted in the conversion of many stations to other uses, sometimes with striking effects, such as the Musée d'Orsay in Paris (Plate 5.4).

Rail freight yards did not have to be quite so centrally located, and because they required a great deal of space for multiple tracks for marshalling they were more likely to be located on entirely greenfield sites than passenger terminals. However, rail yards tended to attract manufacturing activities, and thus became important industrial zones.

Plate 5.4   Quai d'Orsay Museum, Paris. *Credit*: Brian Slack and Jean-Paul Rodrigue

By the end of the twentieth century many of the industries around rail freight yards had relocated or disappeared, and in many cities these former industrial parks have been targets of urban revitalization. This has been accompanied by closure of some of the rail yards, either because they were too small for contemporary operating activities, or because of shrinkage of traffic base. However, in North America many older rail freight yards have been converted into intermodal facilities because of the burgeoning traffic involving containers and road trailers. The ideal configuration for these terminals, however, is different from the typical general freight facility with their need for multiple spurs to permit the assembling of wagons to form train blocks. Intermodal trains tend to serve a more limited number of cities and are more likely to be dedicated to one destination. The need here is for long but fewer rail spurs. The configuration typically requires a site over three kilometers in length and over 100 hectares in area. In addition, good access to the highway system is a requisite as well as a degree of automation to handle the transhipment demands of modern intermodal rail operations (see Figure 5.9).

In some cases, the existing stock of terminals has been found to be wanting in terms of configuration or location with regards to expressways. Thus, new rail yards have been built on greenfield sites on the fringe of metropolitan areas, such as Canadian Pacific's Vaughan terminal or Canadian National's Brampton facilities in Toronto.

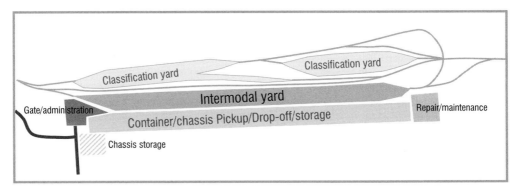

Figure 5.9   Configuration of an intermodal rail terminal

Relative location

Geographers have long recognized situation, or relative location, as an important component of location. It refers to the position of places with regards other places. Accessibility is relative, because the situation of places changes over time. For example ports in the Mediterranean used to be in the heart of the Western world during the Greek and Roman eras, and Genoa and Venice prospered during the Middle Ages. The exploitation of the Americas changed the location of these places, since the Mediterranean became a backwater. The opening of the Suez Canal in the nineteenth century refocused the relative location of the Mediterranean again.

Spatial relationships between terminals are a vital element in competition, particularly for ports and rail terminals, and geographers have developed a number of concepts to explore these locational features:

- **Centrality.** One of the most enduring concepts in urban geography is central place theory, with its emphasis on centrality as a feature of the urban hierarchy. Cities more centrally located to markets are larger with a wider range of functions. Transport accessibility is equated with size, and thus many large terminals arise out of centrality. Examples include Heathrow Airport, London, whose traffic pre-eminence is related to the city's location in the heart of the most developed part of Britain, as well as Britain's functional centrality to its former empire. The port of New York owes its pre-eminence in part to the fact that it is at the heart of the largest market area in the USA.
- **Intermediacy.** This term is applied to the frequent occurrence of places that gain advantage because they are between other places. The ability to exploit transhipment has been an important feature of many terminals. Anchorage, for example, was a convenient airport located on the great circle air routes between Asia and Europe and Continental USA and Asia. For many years passengers alighted here while planes refueled. The growth of long-haul jets has made this activity diminish considerably, and Anchorage now joins the list of once important airports, such as Gander, Newfoundland, that have seen their relative locations change because of technological improvements. It should be noted, however, that Anchorage continues to fulfill its intermediacy role for air freight traffic. Other examples include Chicago, the dominant US rail hub, that is not only a major market area in its own right (centrality) but also lies at the junction of the major eastern and western railroad networks. Ports too can exploit advantages of intermediate locations. The largest container port in the Mediterranean is Giao Tauro, located on the toe of Italy. A few years ago the port did not exist, but because of its location close to the main East–West shipping lanes through the Mediterranean it has been selected as a hub, where the large mother ships can transfer containers to smaller vessels for distribution to the established markets in the northern Mediterranean, a classic hub-and-spoke network.

Hinterland and foreland

One of the most enduring concepts in transport geography, especially applied to ports, is the hinterland. It refers to the market area of ports, the land areas from which the port draws and distributes traffic. Two types of hinterlands are sometimes noted. The term natural or primary hinterland refers to the market area for which the port is the closest terminal. It is assumed that this zone's traffic will normally pass through the port, because of proximity. The competitive hinterland is used to describe the market areas over which the port has to compete with other terminals for business (see Figure 5.10).

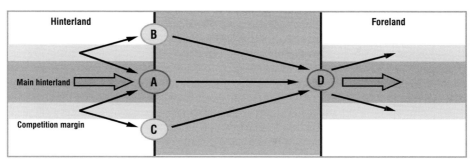

Figure 5.10 Port foreland and hinterland

The **hinterland** is a land space over which a transport terminal, such as a port, sells its services and interacts with its clients. It accounts for the regional market share that a terminal has relative to a set of other terminals servicing this region. It regroups all the customers directly bounded to the terminal. The terminal, depending on its nature, serves as a place of convergence for the traffic coming by roads, railways or by sea/fluvial feeders.

In recent years the validity of the hinterland concept has been questioned, especially in the context of contemporary containerization. The mobility provided by the container has greatly facilitated market penetration, so that many ports compete over the same market areas for business. The notion of discrete hinterlands with well-defined boundaries is questionable therefore. Nevertheless, the concept is still widely employed, and port authorities continue to emphasize their port's centrality to hinterland areas in their promotional literature.

The term foreland is the oceanward mirror of hinterland, referring to the ports and overseas markets linked by shipping services from the port. It is above all a maritime space with which a port performs commercial relationships. It includes overseas customers with which the port undertakes commercial exchanges. The provision of services to a wide range of markets around the world is considered to be an advantage.

In academic studies there have been far fewer assessments of foreland than hinterland, yet in port publicity documents the foreland is usually one of the elements stressed. Geographers have long criticized the distinction, arguing that foreland and hinterland should be seen as a continuum, rather than separate and distinct elements. This point has achieved greater weight recently, with the emergence of door-to-door services and networks, where the port is seen as one link in through transport chains (see Chapter 2 and Figure 5.11).

There is a clear trend involving the growing level of integration between maritime transport and inland freight transport systems. Until recently, these systems evolved separately but the development of intermodal transportation and deregulation provided new opportunities which in turn significantly impacted both maritime and inland logistics. One particular aspect concerns high inland transport costs, since they account for anywhere between 40 percent and 80 percent of the total costs of container shipping, depending on the transport chain. Under such circumstances, there is a greater involvement of maritime actors (e.g. port holdings) in inland transport systems.

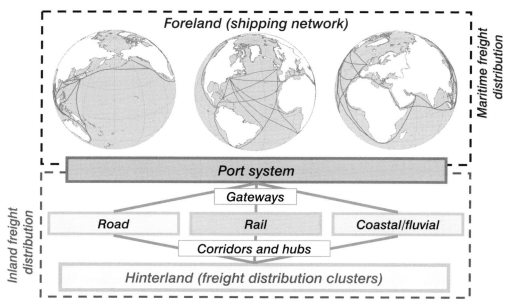

Figure 5.11 Elements in global transport networks

Concept 3 – Transport terminal governance

The nature of governance

Governance is the exercise of authority and institutional resources to manage activity in society and the economy. For transport terminals many different forms of governance are in place which shape modes of financing, operations, functioning and external relationships. There are two main components of terminal governance: ownership and operations. **Ownership** involves the owner of the terminal site and facilities:

- Because of their economic and strategic importance, many types of terminals are owned by **public authorities**. In many countries railroads are owned by the national government, and the passenger stations are thus under the control of the state-owned railway company, such as is the case in China, Europe and North America. Public ownership of airports is also prevalent, although in the USA this is at the state or municipal levels of government. Under public ownership, investment in infrastructure and planning future expansion is carried out by the public authority using public monies.
- **Private ownership** is less evident in transport terminals, except in the case of certain modes, such as road freight, rail freight transport in North America and where privatization has taken place, as for example in ports and airports in the UK and New Zealand. Here, private capital is used to provide infrastructure.

Operations involve the day-to-day management and carrying out of terminal activities.

- **Public** control of operations is typical in many ports, such as Singapore and Hampton Roads, Virginia, in many state-controlled railroads such as China, and at publicly owned airports such as in the USA. Here the public authority provides the handling equipment, contracts with the labor force and operates the rail, airport and port terminals.

- **Private** companies manage and carry out operations in privately owned terminals. They are also active as operators in many publicly owned facilities under a concession. The latter is a growing trend in ports and airports, where facilities are leased to terminal operators for fixed terms. The types of concession vary considerably, in terms of duration and conditions. Some are short term, a few years or so; more typically they are long-term concessions of 15 to 30 years. In some the owner provides some equipment, such as gantry cranes in ports, in others the concession holders are expected to invest in equipment. In some they are required to use public employees, while in others they may use their own workers. In Canada a halfway private/public system of governance of major airports and ports is in place. The airports and ports are leased to locally managed non-profit corporations that have to operate the facilities commercially, without access to public funds. Surpluses have to be reinvested.

Table 5.1 reveals the complexity in terminal governance. Public ownership and operations have been important in many modes because of the strategic importance of transport and the long-term investments required that the private sector may be incapable or unwilling to make. In this way the terminals can be owned and operated in the public good, and can be integrated with public regional and national economic policies. On the other hand, public facilities are seen by some as slow to respond to market conditions, with a propensity to over-invest in non-economic developments, and with high costs to the users.

Table 5.1 Forms of port terminal privatization

Type	Nature
Sale	Terminal is transferred on a freehold basis but with the requirement that it will be used only to provide terminal services.
Concession agreement	Long-term lease of terminal land and facilities and the requirement that the concessionaire undertakes specified capital investments to build, expand or maintain the cargo-handling facilities, equipment and infrastructure.
Capital lease	Similar to a concession except that the private sector is not explicitly required to invest in the facilities and equipment other than for normal maintenance and replacement over the life of the agreement.
Management contract	Private sector assumes responsibility for the allocation of terminal labor and equipment and provides services to the terminal users in the name of the public owner. The public sector retains control over all the assets.
Service contract	The private sector performs specific terminal activities. The arrangement differs from a management contract in that the private sector provides the management, labor and equipment required to accomplish these activities.
Equipment lease	Can be in various forms involving leaseback arrangements or supplier credits. These agreements are used to amortize the costs to the terminal for new equipment and to ensure a reliable supply of spare parts and, often, a guaranteed level of service/reliability from this equipment.

There is a growing tendency towards privatization in transport as a whole. In terminals this is manifested in the sale of ports and airports in some countries such as the UK, and in the break-up of state rail monopolies, as in the EU. Privatization is most evident, however, in the awarding of operational concessions to private companies. The trend towards concessions is warranted in part by the belief (not always justified) that the private sector is more efficient than the public in operating terminals, and that this form of governance keeps the ownership still under public control. It is also seen as a means of reducing public expenditures at a time when states are becoming less willing to make large investments.

Changing port governance

Even as late as the 1980s, ports around the world were amongst the types of terminal most dominated by public ownership and operations. While the form of governance differed greatly, from the municipally owned ports in Northern Europe and the USA, to the state-owned ports in France, Italy and much of the developing world, public ownership was dominant and publicly managed port operations were very prevalent. This contrasted with the shipping industry, where private ownership was almost universal. The changes, slow at first, came from two directions:

- First, there was the belief, promoted in the UK by Prime Minister Thatcher, that the transport industry as a whole should be **divested to the private sector to promote competition**. Ports were among the many sectors thus targeted. New Zealand actually carried out this policy before it was finally implemented in the UK. In both countries the state has relinquished control over the ports industry.
- Second, there was a policy recommendation from the World Bank that developing countries would do well to **free their highly controlled port industry**, by issuing **concessions** to companies capable of modernizing their port industries. To facilitate the changes required, the World Bank created a Port Reform Tool Kit to demonstrate to states how to go about affecting the reforms.

These developments helped create what has become a global snowball of port reform. It made governments around the world more open to considering reforming port governance. The growing demands for public investment in ports, precipitated by the growth in world trade, and the limited abilities of states to meet these needs because of competing investment priorities, were key factors. Thus while few were willing to go as far as the UK in the total privatization of ports, many countries were willing to consider awarding concessions. The result has been an almost global trend towards the award of port operational concessions, especially for container terminals.

If the opportunities to award operational concessions can be seen as an increase in demand, growth has also been greatly affected by an increase in the supply of companies seeking concessions. In Northern Europe and the USA many ports had already operated through concessions, awarded to local terminal handling companies. Because they were relatively small and locally based with only few exceptions they did not participate in the global growth of opportunities for concession awards. The exceptions were Stevedore Services of America (SSA), which was already active in several US West Coast ports, that obtained concessions to operate facilities in Panama and several other smaller ports in Central America, and Eurogate, a joint company formed by terminal handling companies from Bremen and Hamburg, that obtained concessions in Italy.

The major actors have come from Asia, with three large companies dominating:

- Hong Kong-based firm, **Hutchison Port Holdings** (HPH), part of a major conglomerate Hutchison Whampoa.
- **Port of Singapore Authority** (PSA), the government-owned operator of the port of Singapore.
- **Dubai Ports World** (DPW), which is mainly a component of a sovereign wealth fund.

HPH, which originated as a terminal operator in Hong Kong, first purchased Felixstowe, the largest UK container port, and today has a portfolio of 39 terminals around the world, including Rotterdam and Shanghai. PSA has been active securing concessions in China and Europe, including Antwerp. DPW has grown through purchases, such as P&O Ports and CSX World Terminals, and by securing concessions elsewhere (Table 5.2).

Shipping lines have also participated in terminal concessions. The most important is the in-house terminal operating company of Maersk, APM Terminals. In addition, Evergreen, MSC, NYK and CMA-CGM hold port terminal leases. Between the dedicated terminal operating companies and the shipping lines, a global pattern of concessions is evident.

Significance and consequences

The rapid expansion of terminal operating companies reflects two economic forces. First, the entry of former terminal operators into the global system represents a process of horizontal integration, in which the companies, constrained by the limits of their own ports, seek to apply their expertise in new circumstances and seek new sources of income. Second, the entry of shipping lines into terminal operations is an example of vertical integration, in which the companies seek to extend their control over other links in the transport chain (Figure 5.12).

Several other factors explain the growth of global terminal operating companies:

- **Profitability**. By modernizing port operating systems, HPH achieved a 35 percent per year return on investment in the early 2000s. Port management was very lucrative.
- **Commercial strategy**. By securing a concession at one port, the terminal operating company excluded competitors to give it monopoly power in that port.
- **Provide a global network**. Holding concessions from ports around the world permits the terminal operator more power to negotiate with global shipping lines and offer global solutions to terminal requirements in ports around the world.

Table 5.2 Dedicated maritime container terminals controlled by major port holdings, 2007

Holding	Australia	Europe	N. America	Pacific Asia	South Asia/ S&C America	Middle East	Total
APM		7	13	7	3	5	35
DPW	5	9	1	10	3	12	40
HPH		7		21	8	2	39
PSA		10		18		1	29
Total	5	33	14	56	14	20	143

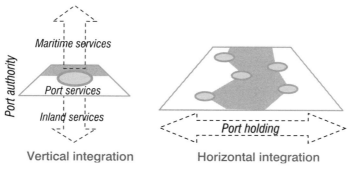

Figure 5.12 Vertical and horizontal integration in port development

The growth of multinational terminal operating companies has resulted in a concentration of power. In 2005 the top five global terminal operators accounted for 47 percent of global container port activity. What is perhaps most important is that they now dominate activity at the most important container ports in the world. They are able to wield monopoly power in many parts of the world. The consequences of this remain to be analyzed, but there is growing evidence of dissatisfaction in many ports about the actions of these companies that possess long-term leases. Thus, in Genoa there is concern about the lack of performance of the port since PSA took over the main container terminal. In Antwerp, there are concerns about the imposition of Singapore-based management systems on a European operation. In China there is opposition to HPH and how it is increasing terminal handling costs to enhance profitability. On the other hand, there is strong evidence to suggest that port performance has improved in most ports as a result of the award of concessions to international terminal companies. The question will be whether to regulate further concentration of power.

Method 1 – The Gini coefficient

Definition

The Gini coefficient was developed to measure the degree of concentration (inequality) of a variable in a distribution of its elements. It compares the Lorenz curve of a ranked empirical distribution with the line of perfect equality. This line assumes that each element has the same contribution to the total summation of the values of a variable. The Gini coefficient ranges between 0, where there is no concentration (perfect equality), and 1 where there is total concentration (perfect inequality).

Figure 5.13 is a graphical representation of the proportionality of a distribution (the cumulative percentage of the values). To build the Lorenz curve, all the elements of a distribution must be ordered from the most important to the least important. Then, each element is plotted according to their cumulative percentage of X and Y, X being the cumulative percentage of elements. For instance, out of a distribution of 10 elements (N), the first element would represent 10 percent of X and whatever percentage of Y it represents (this percentage must be the highest in the distribution). The second element would cumulatively represent 20 percent of X (its 10 percent plus the 10 percent of the first element) and its percentage of Y plus the percentage of Y of the first element.

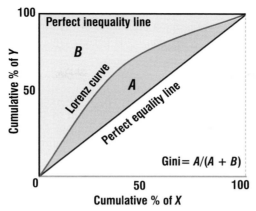

Figure 5.13 The Lorenz curve

The Lorenz curve is compared with the perfect equality line, which is a linear relationship that plots a distribution where each element has an equal value in its shares of X and Y. For instance, in a distribution of 10 elements, if there is perfect equality, the fifth element would have a cumulative percentage of 50 percent for X and Y. The perfect equality line forms an angle of 45 degrees with a slope of $100/N$. The perfect inequality line represents a distribution where one element has the total cumulative percentage of Y while the others have none.

The Gini coefficient is defined graphically as a ratio of two surfaces involving the summation of all vertical deviations between the Lorenz curve and the perfect equality line (A) divided by the difference between the perfect equality and perfect inequality lines ($A + B$).

Figure 5.14 shows a simple system of five ports along a coast. In case A, the traffic for each port is the same, so there is no concentration and thus no inequality. The Lorenz curve of this distribution is the same as the perfect equality line; they overlap. In case

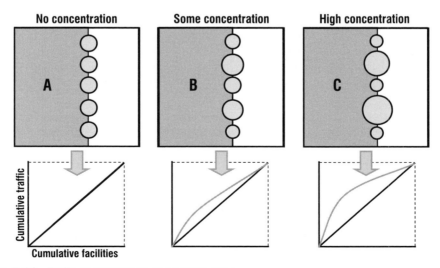

Figure 5.14 Traffic concentration and Lorenz curves

B, there is some concentration of the traffic in two ports and this concentration is reflected in the Lorenz curve. Case C represents a high level of concentration in two ports and the Lorenz curve is significantly different to the perfect equality line.

Calculating the Gini coefficient (G)

The coefficient represents the area of concentration between the Lorenz curve and the line of perfect equality as it expresses a proportion of the area enclosed by the triangle defined by the line of perfect equality and the line of perfect inequality. The closer the coefficient is to 1, the more unequal the distribution.

$$G = 1 - \sum_{i=0}^{N} (\sigma Y_{i-1} + \sigma Y_i)(\sigma X_{i-1} - \sigma X_i)$$

Table 5.3 shows a hypothetical set of terminals with varying amounts of traffic. X refers to the traffic proportion if the traffic was distributed evenly throughout all the terminals. Y refers to the actual proportion of traffic of each terminal. σX and σY are cumulative percentages of Xs and Ys (in fractions) and N is the number of elements (observations).

The Gini coefficient for this distribution is 0.427 ($|1-1.427|$).

Geographers have used the Gini coefficient in numerous instances, such as assessing income distribution among a set of contiguous regions (or countries) or to measure other spatial phenomena such as racial segregation and industrial location. Its major purpose as a method in transport geography has been related to measuring the concentration of traffic, mainly at terminals, such as assessing changes in port system concentration. Economies of scale in transportation favor the concentration of traffic at transport hubs, so the Gini coefficient of maritime traffic has tended to increase over the last decades, although perhaps not to the degree that has been expected.

Method 2 – Delphi forecasting

Introduction

Delphi forecasting is a non-quantitative technique for forecasting. Unlike many other methods that use so-called objective predictions involving quantitative analysis, the Delphi method is based on expert opinions. It has been demonstrated that predictions obtained this way can be at least as accurate as other procedures. The essence of the procedure

Table 5.3 Calculating the Gini coefficient

Terminal	Traffic	X	Y	σX	σY	$\sigma X_{i-1} - \sigma X_i$ (B)	$\sigma Y_{i-1} + \sigma Y_i$ (A)	A*B
A	25,000	0.2	0.438	0.2	0.438	0.2	0.438	0.088
B	18,000	0.2	0.316	0.4	0.754	0.2	1.192	0.238
C	9,000	0.2	0.158	0.6	0.912	0.2	1.666	0.333
D	3,000	0.2	0.053	0.8	0.965	0.2	1.877	0.375
E	2,000	0.2	0.035	1.0	1.000	0.2	1.965	0.393
Total	57,000	1.0	1.000					1.427

is to use the assessment of opinions and predictions by a number of experts over a number of rounds in carefully managed sequences.

One of the most important factors in Delphi forecasting is the selection of experts. The persons invited to participate must be knowledgeable about the issue, and represent a variety of backgrounds. The number must not be too small to make the assessment too narrowly based, nor too large to be difficult to coordinate. It is widely considered that 10 to 15 experts can provide a good base for the forecast.

Procedure

The procedure begins with the planner/researcher preparing a questionnaire about the issue at hand, its character, causes and future shape. These are distributed to the respondents separately who are asked to rate and respond. The results are then tabulated and the issues raised are identified.

The results are then returned to the experts in a second round. They are asked to rank or assess the factors, and justify why they made their choices. During a third or subsequent rounds their ratings along with the group averages and lists of comments are provided, and the experts are asked to re-evaluate the factors. The rounds continue until an agreed level of consensus is reached. The literature suggests that by the third round a sufficient consensus is usually obtained.

The procedure may take place in many ways. The first step is usually undertaken by mail. After the initial results are obtained the subsequent round could be undertaken at a meeting of experts, assuming it would be possible to bring them together physically. Or, the subsequent rounds could be conducted again by mail. E-mail has greatly facilitated the procedure. The basic steps are as follows:

- **Identification of the problem**. The researcher identifies the problem for which some predictions are required, e.g. what is the traffic of Port X likely to be in 10 years time? The researcher prepares documentation regarding past and present traffic activity. A questionnaire is formulated concerning future traffic estimates and factors that might influence such developments. A level of agreement between the responses is selected, i.e. if 80 percent of the experts can agree on a particular traffic prediction.
- **Selection of experts**. In the case of a port scenario this might include terminal managers, shipping line representatives, land transport company representatives, intermediaries such as freight forwarders, and academics. It is important to have a balance, so that no one group is overly represented.
- **Administration of questionnaire**. The experts are provided with background documentation and the questionnaire. Responses are submitted to the researcher within a narrow time frame.
- **Researcher summarizes responses**. Actual traffic predictions are tabulated and means and standard deviations calculated for each category of cargo as in the case of a port traffic prediction exercise. Key factors suggested by the experts are compiled and listed.
- **Feedback**. The tabulations are returned to the experts, either by mail or in a meeting convened to discuss first-round results. The advantage of a meeting is that participants can confront each other to debate areas of disagreement over actual traffic predictions or of key factors identified. The drawback is that a few individuals might exert personal influence over the discussion and thereby sway outcomes, a trend that the researcher must be alert to and seek to mitigate. The experts are invited to review their original estimates and choices of key factors in light of the results presented, and submit a new round of predictions.

- These new predictions are tabulated and returned to the experts either by mail or immediately to the meeting, if the level of agreement does not meet the predetermined level of acceptance. The **specific areas of disagreement are highlighted**, and the experts are again requested to consider their predictions in light of the panel's overall views.
- The process is continued until the level of **agreement has reached the predetemined value**. If agreement is not possible after several rounds, the researcher must terminate the process and try to pinpoint where the disagreements occur, and utilize the results to indicate specific problems in the traffic prediction process in this case.

This method could be applied in a classroom setting, with students serving as "experts" for a particular case study. The traffic at the local airport or port might be an appropriate example. On the basis of careful examination of traffic trends and factors influencing business activity, the class could be consulted to come up with predictions that could then be compared with those of some alternate method such as trend extrapolation.

CASE STUDY Chicago and intermodal rail terminals

Chicago is the undisputed rail hub of North America. The city's growth coincided with the expansion of railways in the USA in the latter part of the nineteenth century. It became the center in which the Eastern railroads met the Western rail companies, there being no transcontinental railroad in the USA (Figure 5.15).

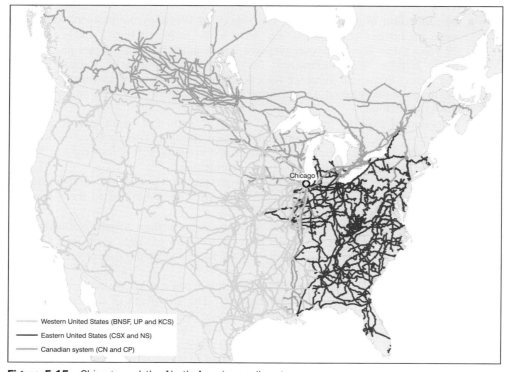

	Western United States (BNSF, UP and KCS)
	Eastern United States (CSX and NS)
	Canadian system (CN and CP)

Figure 5.15 Chicago and the North American rail system

Chicago's position was reinforced with the development of intermodal transportation and the establishment of rail land bridges for containers imported via US West Coast ports in the 1980s. Over 250 intermodal trains per day pass through the city and over 10 million containers are handled by its rail terminals every year, making it the largest US container complex, bigger than any port (Plate 5.5). Consolidation among the railroads since deregulation resulted in four major Western railroads and three Eastern companies (later reduced by merger to two Western and two Eastern railroads) handling the majority of the container traffic. All the railroads (as well as the two Canadian railroads) have terminals in Chicago, which thus became the leading city where containers could be transferred from one company to another. This is called **interlining**.

When containers from the port of Los Angeles are being transported to a market in the East on a block train the interline does not create any major problems, since the train block can be transferred at the terminal of the Western railroad to a locomotive and crew from an Eastern railroad for onward delivery. Difficulties arise when smaller container blocks and individual containers have to be interlined with other carriers for multiple destinations. Here containers have to be transferred from one rail yard to another, and since there are so many container yards in Chicago (see Figure 5.16), the interline can be complex. There is a railroad that links all the terminals, the Chicago Beltline, but it provides slow service between many of the yards, resulting in up to 48 hours to carry out an interline. As a result, the railroads prefer to use a rubber tire interchange, where containers are put on trucks for direct delivery to the onward terminal. Such interchanges are quicker, and until the increase in fuel prices in 2008, cheaper.

Plate 5.5 Double-stacked container train at the Corwith Yard, Chicago. *Credit*: Jean-Paul Rodrigue

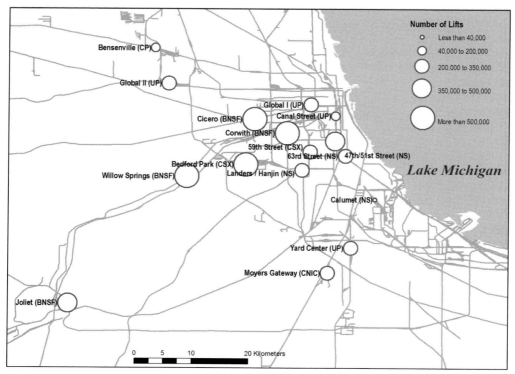

Figure 5.16 Intermodal rail terminals in Chicago

The result is that 20,000 trucks per day are added to the highways around Chicago. This source of congestion and pollution is opposed by citizens and local public officials. The question is what to do about it? The obvious solution would be to establish one major intermodal facility where all the railroads could manage interlines. The impossibility of this solution is the amount of land such a facility would require. The existing terminals in Chicago cover an area twice as large as the two airports, O'Hare and Midway. Where could such a site be found? At what cost?

A solution being pursued is a joint plan developed by the State of Illinois, the Chicago Department of Transportation and the American Association of Railroads. The CRE-ATE program is a $1.5 billion project to improve rail connections in Chicago and to remove obstacles at level crossings and rail crossing points. The railroads are contributing $212 million, since they will get improved connections. The remainder will come from public funds, since commuter rail service will be enhanced, and there will be fewer trucks on the roads engaged in rubber tire interchanges.

Bibliography

Bird, J.H. (1963) *The Major Seaports of the United Kingdom*, London: Hutchinson.

Caves, R.E. and G.D. Gosling (1999) *Strategic Airport Planning*, Oxford: Pergamon.

Charlier, J. (1992) "The Regeneration of Old Port Areas for New Port Uses", in B.S. Hoyle and D. Hilling (eds) *Seaport Systems and Spatial Change*, Chichester: Wiley, pp. 137–54.

De Langen, P.W. (2004) "Analysing Seaport Cluster Performance", in D. Pinder and B. Slack (eds) *Shipping and Ports in the Twenty-first Century*, London: Routledge, pp. 82–98.

Ducruet, C. (2007) "A Metageography of Port–city relationships", in J.J. Wang, D. Olivier, B. Slack and T. Notteboom (eds) *Ports, Cities, and Global Supply Chains*, Aldershot: Ashgate, pp. 157–72.

Fleming, D.K. and Y. Hayuth (1994) "Spatial Characteristics of Transportation Hubs: Centrality and Intermediacy", *Journal of Transport Geography*, 2, 3–18.

Goetz, A.R. and J.S. Szyliowicz (1997) "Revisiting Transport Planning and Decision Making: The Case of Denver International Airport", *Transport Research*, A 31, 263–80.

Graham, B. (1995) *Geography and Air Transport*, Chichester: Wiley.

Haezenonck, E. (2001) *Essays on Strategy Analysis for Seaports*, Leuven: Garant.

Hoyle, B.S. (1988) "Development Dynamics at the Port–City Interface", in B.S. Hoyle, D.A. Pinder and M.S. Husain (eds) *Revitalising the Waterfront*, Chichester: Wiley.

McCalla, R.J. (1999) "From St. John's to Miami: Containerisation at Eastern Seaboard Ports", *GeoJournal*, 48, 15–28.

McCalla, R.J. (2004) "From 'Anyport' to 'Superterminal' in D. Pinder and B. Slack (eds) *Shipping and Ports in the Twenty-first Century*, London: Routledge, pp. 123–42.

McCalla, R.J., B. Slack and C. Comtois (2001) "Intermodal Freight Terminals: Locality and Industrial Linkages", *Canadian Geographer*, 45, 404–13.

Notteboom, T.E. and W. Winkelmans (2001) "Structural Changes in Logistics: How Will Port Authorities Face the Challenge?" *Maritime Policy and Management*, 28, 71–89.

Porter, M.E. (1990) *The Competitive Advantage of Nations*, London: Macmillan.

Robinson, R. (2002) "Ports as Elements in Value-driven Chain Systems: The New Paradigm", *Maritime Policy and Management*, 29, 241–55.

Rodrigue, J.-P. (2008) "The Thruport Concept Transmodal Rail Freight Distribution in North America", *Journal of Transport Geography*, vol. 16, pp. 233–46.

Rodrigue, J.-P. and B. Slack (2002) "Logistics and National Security", in S.K. Majumdar *et al.* (eds) *Science, Technology and National Security*, Easton, PA: Pennsylvania Academy of Science, pp. 214–25.

Slack, B. (1989) "The Port Service Industry in an Environment of Change", *Geoforum*, 20, 447–57.

Slack, B. (1993) "Pawns in the Game: Ports in a Global Transportation System", *Growth and Change*, 24, 579–88.

6 International trade and freight distribution

International transportation takes place at the highest scales of mobility that involve intercontinental and inter-regional movements. Globalization processes have extended considerably the need for international transportation, notably because of economic integration, which grew on par with the fragmentation of production systems and the expansion of international trade. Both processes are interdependent and require an understanding of the transactional context in which multinational corporations are now evolving. There is thus a growing level of integration between production, distribution and consumption. The heightened integration and efficiency has been expanded by logistics.

Concept 1 – Transportation, globalization and international trade

Trade and the global economy

> Courtesy of ongoing trade liberalization, in conjunction with sharply declining communication and transportation costs, there has been a sharp increase in the tradable goods portion of world output over the past 15 years. At the same time, a veritable explosion in e-based connectivity since 1995, together with the emergence of an entirely new global IT outsourcing industry, has led to the networking of service providers around the world. As a result, rapidly expanding trade in both goods and services has become an increasingly powerful engine in driving the global growth dynamic.
>
> (Stephen Roach, Morgan Stanley, July 18, 2005)

In a global economy, no nation is self-sufficient. Each is involved at different levels in trade to sell what it produces, to acquire what it lacks and also to produce more efficiently in some economic sectors than its trade partners. As supported by conventional economic theory, trade promotes economic efficiency by providing a wider variety of goods, often at lower costs. The globalization of production is concomitant to the globalization of trade as one cannot function without the other. Even though international trade took place centuries before the modern era, as ancient trade routes such as the Silk Road can testify, trade occurred at an ever increasing scale over the last 600 years to play an even more active part in the economic life of nations and regions. This process has been facilitated by significant technical changes in the transport sector. The scale, volume and efficiency of international trade all have continued to increase since the 1970s. As such, a point has been reached where larger distances can be traded for a decreased amount of time, and this at similar or lower costs. It has become increasingly possible to trade between parts of the world that previously had limited access to international transportation systems. Further, the division and the fragmentation of production that went along with these processes also expanded trade. Trade thus contributes to lower manufacturing costs (Figure 6.1).

The economic benefits of international or inter-regional trade are numerous and well known since the seminal work of Adam Smith (1776) and David Ricardo (1817). Without trade, each country must produce a set of basic goods to satisfy the requirements of the national economy. On Figure 6.1, four countries are each producing four different goods. National markets tend to be small, impairing the potential economies of scale, which results in higher prices. Product diversity also tends to be limited because of the market size and standards (such as safety or component size) may even be different. With trade, competition increases and a redistribution of production often takes place as comparative advantages are being exploited. The outcome of trade liberalization involves a specialization of production of one good in each country and the trade of other goods between them. Greater economies of scale that are achieved through specialization result in lower prices. A situation of interdependency is thus created.

Without international trade, few nations could maintain an adequate standard of living. With only domestic resources, each country could only produce a limited number of products and shortages would be prevalent. Global trade allows for an enormous variety of resources – from Persian Gulf oil to Chinese labor – to be made more widely accessible. It also facilitates the distribution of many different manufactured goods that are produced in different parts of the world. Wealth becomes increasingly derived through regional specialization of economic activities. This way, production costs are lowered, productivity rises and surpluses are generated, which can be transferred or traded for

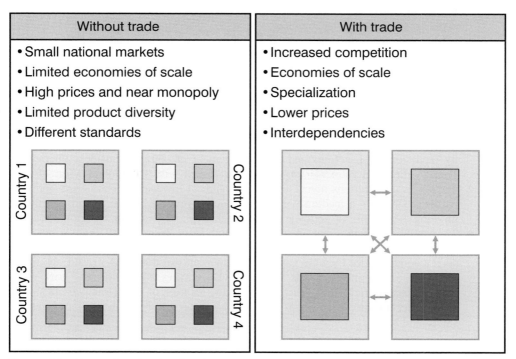

Figure 6.1 Economic rationale of trade

commodities that would be too expensive to produce domestically or would simply not be available. As a result, international trade decreases the overall costs of production worldwide. Consumers can buy more goods from the wages they earn, and standards of living should, in theory, increase. International trade consequently demonstrates the extent of globalization with increased spatial interdependencies between elements of the global economy and their level of integration. Interdependencies imply numerous relationships where exchanges of capital, goods, raw materials and services are established between regions of the world.

Trade facilitation

The volume of exchanged goods and services between nations is taking a growing share of the generation of wealth, mainly by offering economic growth opportunities in new regions and by reducing the costs of a wide array of manufacturing goods. By 2007, international trade surpassed for the first time 50 percent of the global GDP, a twofold increase since 1950. The facilitation of trade involves how the procedures regulating the international movements of goods can be improved, which concerns three main factors:

- **Integration processes**, such as the emergence of economic blocs and the decrease of tariffs at a global scale through agreements, promoted trade as regulatory regimes were harmonized. One straightforward measure of integration relates to custom delays, which can be a significant trade impediment. The higher the level of economic integration, the more likely the concerned elements are to trade. International trade has consequently been facilitated by a set of factors linked with growing levels of economic integration, the outcome of processes such as the European Union or the North American Free Trade Agreement. The transactional capacity is consequently facilitated with the development of transportation networks and the adjustment of trade flows that follow increased integration. Integration processes have also taken place at the local scale with the creation of free trade zones where an area is given a different governance structure in order to promote trade, particularly export-oriented activities. In this case, the integration process is not uniform as only a portion of a territory is involved, which can create dislocations.
- **Production systems** are more flexible and embedded, which encourages exchanges of commodities, parts and services. Information technologies have played a role by facilitating transactions and the management of complex business operations. Foreign direct investments are commonly linked with the globalization of production as corporations invest abroad in search of lower production costs and new markets. China is a leading example of such a process. There is consequently a growing availability of goods and services that can be traded on the global market.
- **Transport efficiency** has increased significantly because of innovations and improvements in the modes and infrastructures. Ports are particularly important in such a context since they are gateways to international trade. As a result, the transferability of commodities has improved. Decreasing transport costs does more than increasing trade, it also helps change the location of economic activities. Yet, transborder transportation issues remain to be better addressed in terms of capacity, safety and security.

Thus, the ability to compete in a global economy is dependent on the transport system as well as on a vast array of supporting service activities. These activities include:

- **Distribution-based**. A multimodal and intermodal freight transport system composed of modes, infrastructures and terminals that spans across the globe. It ensures a physical capacity to support trade.
- **Regulation-based**. Customs procedures, tariffs, regulations and handling of documentation. They ensure that trade flows abide to the rules and regulations of the jurisdictions they cross.
- **Transaction-based**. Banking, finance, legal and insurance activities where accounts can be settled. They ensure that the sellers of goods and services are receiving an agreed-upon compensation and that the purchasers are protected and have a legal recourse if the outcome of the transaction is judged unsatisfactory or are insured if a partial or full loss incurs.

The quality, cost and efficiency of these services influence the trading environment as well as the overall costs linked with the international trade of goods.

Global trade patterns

International trade, both in terms of value and tonnage, has been a growing trend in the global economy. The emergence of global trade patterns can mainly be articulated within three major phases (Figure 6.2).

- **First phase**. Concerns a conventional perspective on international trade that prevailed until the 1970s where factors of production were much less mobile. Particularly, there was a limited level of mobility of raw materials, parts and finished products in a setting which is fairly regulated with impediments such as tariffs, quotas and limitations to foreign ownership. Trade mainly concerned a range of specific products, namely commodities (and very few services), that were not readily available

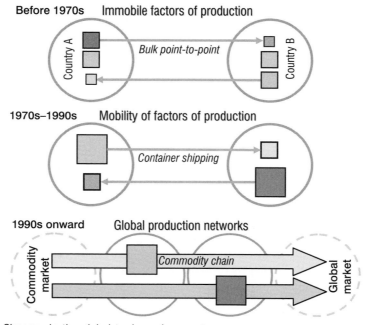

Figure 6.2 Changes in the global trade environment

in regional economies. Any product which could in theory be produced nationally was subject to a variety of protectionist policies. Due to regulations, protectionism and fairly high transportation costs, trade remained limited and delayed by inefficient freight distribution. In this context, trade was more an exercise to cope with scarcity than to promote economic efficiency.

- **Second phase.** From the 1980s, the mobility of factors of production, particularly capital, became possible. The legal and physical environment in which international trade was taking place led to a better realization of the comparative advantages of specific locations. Concomitantly, regional trade agreements emerged and the global trade framework was strengthened from a legal and transactional standpoint (the General Agreement on Tariffs and Trade (GATT) and World Trade Organization (WTO)). In addition, containerization provided the capabilities to support more complex and long-distance trade flows, as did increasing air traffic. Cheaper and more efficient containerized transportation supported such a process to locations that previously were mainly outside global economic trends, namely China. This process had for long been advocated by economic theory (e.g. Adam Smith and Ricardo) but had never taken place at a notable scale. Still, some integration existed before the 1970s, such as in North America (USA–Canada) and Western Europe (early stages of the EU). Due to high production (legacy) costs in old industrial regions, activities that were labor intensive were gradually relocated to lower cost locations. The process began as a national one, then went to nearby countries when possible and afterwards became a truly global phenomenon. Thus, foreign direct investments surged, particularly towards new manufacturing regions, as multinational corporations became increasingly flexible in the global positioning of their assets.
- **Third phase.** There is a growth in international trade, now including a wide variety of services that were previously fixed to regional markets, and a surge in the mobility of the factors of production. Since these trends are well established, the priority is now shifting to the geographical and functional integration of production, distribution and consumption with the emergence of global production networks. By setting or capturing a commodity chain, a corporation is able to generate added value and compete more effectively. Containerization has become embedded in freight distribution, from the global commodity market to distribution centers close to the final consumer. Complex networks involving flows of information, commodities, parts and finished goods have been set, which in turn demand a high level of command of logistics and freight distribution. In such an environment, powerful actors have emerged which are not directly involved in the function of production and retailing, but mainly take the responsibility for managing the web of flows.

The global economic system is thus one characterized by a growing level of integrated services, finance, retail, manufacturing and distribution, which in turn is mainly the outcome of improved transport and logistics, a more efficient exploitation of regional comparative advantages and a transactional environment supportive of the legal and financial complexities of global trade. The outcome has been a shift in global trade flows with many developing countries having a growing participation in international trade. The nature of what can be considered international trade has also changed, particularly with the emergence of global commodity chains. This trend obviously reflects the strategies of multinational corporations positioning their manufacturing assets in order to lower costs and maximize new market opportunities while maintaining the cohesion of their freight distribution systems. In addition, another important trade has been growing imports of resources from developing countries, namely energy, commodities and agricultural products.

The dominant factor behind the growth in international trade has been an increasing share of manufacturing activities taking place in developing countries as manufacturers are seeking low-cost locations for many stages of the supply chain. The evolution of international trade thus has a concordance with the evolution of production. There are however significant fluctuations in international trade that are linked with economic cycles of growth and recession, fluctuations in the price of raw materials, as well as disruptive geopolitical and financial events. The international division of production has been accompanied by increasing flows of manufactured goods, which take a growing share of international trade. There is relatively less bulk liquids (such as oil) and more dry bulk and general cargo being traded.

The geography of international trade still reveals a dominance of a small number of countries, mainly in North America and Europe. Alone, the United States, Germany and Japan account for about a third of all global trade, but this supremacy is being seriously challenged. Further, G7 countries account for half of the global trade, a dominance which has endured for over a 100 years. A growing share is being accounted for by the developing countries of Asia, with China accounting for the most significant growth both in absolute and relative terms. Those geographical and economic changes are also reflected over trans-oceanic trade with trans-Pacific trade growing faster than trans-Atlantic trade.

Regionalization has been one of the dominant features of global trade. The bulk of international trade has a regional connotation, promoted by proximity and the establishment of economic blocs such as NAFTA and the EU. The closer economic entities are, the more likely they are to trade, which explains why the most intense trade relations are within Western Europe and North America. The growth of the amount of freight being traded as well as a great variety of origins and destinations promotes the importance of international transportation as a fundamental element supporting the global economy.

International transportation

With the growth of international trade and the globalization of production, international transportation systems have been under increasing pressures to support additional demands in volume and distance carried. This could not have occurred without considerable technical improvements permitting to transport larger quantities of passengers and freight, and this more quickly and more efficiently. Few other technical improvements than containerization have contributed to this environment of growing mobility of freight. Since containers and intermodal transportation improve the efficiency of global distribution, a growing share of general cargo moving globally is containerized. Consequently, transportation is often referred to as an enabling factor that is not necessarily the cause of international trade, but a means without which globalization could not have occurred. A common development problem is the inability of international transportation infrastructures to support flows, undermining access to the global market and the benefits that can be derived from international trade.

International trade requires distribution infrastructures that can support trade between several partners. Three components of international transportation facilitate trade:

- **Transportation infrastructure**. Concerns physical infrastructures such as terminals, vehicles and networks. Efficiencies or deficiencies in transport infrastructures will either promote or inhibit international trade.
- **Transportation services**. Concerns the complex set of services involved in the international circulation of passengers and freight. It includes activities such as distribution, logistics, finance, insurance and marketing.

- **Transactional environment.** Concerns the complex legal, political, financial and cultural setting in which international transport systems operate. It includes aspects such as exchange rates, regulations, quotas and tariffs, but also consumer preferences.

About half of the global trade takes place between locations more than 3,000 km apart. Because of this geographical scale, most international freight movements involve several modes, especially when origins and destinations are far apart. Transport chains must thus be established to service these flows which reinforce the importance of intermodal transportation modes and terminals at strategic locations (Figure 6.3).

International trade is based on the notion of exchange which involves the nature of merchandises being traded, the partners involved as well as the transactional environment in which trade takes place, namely tariff and non-tariff barriers. The physical realization of international trade requires a transport chain that can provide a succession of modes and terminals, such as railway, maritime and road transportation systems. The first stage in the transport chain is composition where merchandises are assembled at origin (A), often on pallets and/or containers. The cargo being traded then moves along the transport chain and is transhipped at terminals from one mode to another. Once it enters another country customs inspection takes place. This activity is dominantly located at major terminals, or points of entry, namely ports and airports. The final stage of the transport chain, decomposition, takes place at destination (B). In this context, the container is useful to facilitate transfers between modes where the most important function in international transportation becomes transhipment. Among the numerous transport modes, two are specifically concerned with international trade:

- **Ports and maritime shipping.** The importance of maritime transportation in global freight trade is unmistakable, particularly in terms of tonnage as it handles about 90 percent of the global total. Thus, globalization is the realm of maritime shipping, with containerized shipping at the forefront of the process. The global maritime transport system is composed of a series of major gateways granting access to major production and consumption regions. Between those gateways are major hubs acting as points of interconnection and transhipment between systems of maritime circulation.

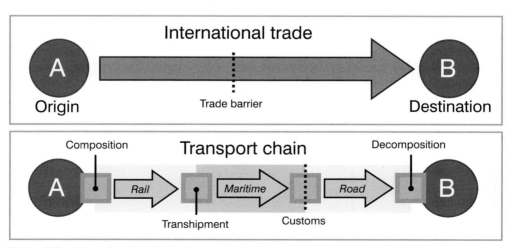

Figure 6.3 International trade and transportation chains

- **Airports and air transport**. Although in terms tonnage air transportation carries an insignificant amount of freight (0.2 percent of total tonnage) compared with maritime transportation, its importance in terms of the total value is much more significant: about 15 percent. International air freight is about 70 times more valuable than its maritime counterpart and about 30 times more valuable than freight carried overland, which is linked with the types of goods it transports (e.g. electronics). The location of freight airports corresponds to high-technology manufacturing clusters as well as to intermediary locations where freight planes are refueled and/or cargo is transhipped.

Road and railway modes tend to occupy a more marginal portion of international transportation since they are above all modes for national or regional transport services. Their importance is focused on their role in the "first and last miles" of global distribution. Freight is mainly brought to port and airport terminals by trucking or rail. There are however notable exceptions in the role of overland transportation in international trade. A substantial share of the NAFTA trade between Canada, the United States and Mexico is supported by trucking, as well as a large share of the Western European trade. In spite of this, these exchanges are a priori regional by definition, although intermodal transportation confers a more complex setting in the interpretation of these flows.

Economic development in Pacific Asia, and in China in particular, has been the dominant factor behind the growth of international transportation in recent years. Since the trading distances involved are often considerable, this has resulted in increasing demands on the maritime shipping industry and on port activities. As its industrial and manufacturing activities develop, China is importing growing quantities of raw materials and energy and exporting growing quantities of manufactured goods. The outcome has been a surge in demand for long-distance international transportation. The ports in the Pearl River delta in Guangdong province now handle almost as many containers as all the ports in the United States combined.

Concept 2 – Commodity chains and freight transportation

Contemporary production systems

Production and consumption are the two core components of economic systems and both are interrelated through the conventional supply/demand relationship. Basic economic theory underlines that what is being consumed has to be produced and what is being produced has to be consumed. Any disequilibrium between the quantity being produced and the quantity being consumed can be considered as a market failure. On one side, insufficient production involves shortages and price increases, while on the other overproduction involves waste, storage and price reductions. The realization of production and consumption cannot occur without flows of freight within a complex system of distribution that includes modes and terminals, but also facilities managing freight activities, namely distribution centers.

Contemporary production systems are the outcome of significant changes in production factors, distribution and industrial linkages:

- **Production factors**. In the past, the three dominant factors of production – land, labor and capital – could not be effectively used at the global level. For instance, a corporation located in one country had difficulties taking advantage of cheaper labor and land in another country, notably because regulations would not permit full (and often dominant) ownership of a manufacturing facility by foreign interests. This process has also been strengthened by economic integration and trade agreements. The European Union established a structure that facilitates the mobility of production factors, which in turn enabled a better use of the comparative productivity of the European territory. Similar processes are occurring in North America (NAFTA), South America (MERCOSUR) and in Pacific-Asia (ASEAN) with various degrees of success. Facing integration processes and massive movements of capital coordinated by global financial centers, factors of production have an extended mobility, which can be global in some instances. To reduce their production costs, especially labor costs, many firms have relocated segments (sometimes the entire process) of their industrial production systems to new locations. For instance, in 2003, American corporations were performing around 27 percent of their manufacturing activities abroad while this figure was about 15 percent for their Japanese counterparts.

- **Distribution**. In the past, the difficulties of overcoming distances were related to constraints in physical distribution as well as to telecommunications. Distribution systems had limited capabilities to ship merchandises between different parts of the world and it was difficult to manage fragmented production systems due to inefficient communication systems. In such a situation, freight alone could cross borders, while capital flows, especially investment capital, had more limited ranges. Trade could be international, but production systems were dominantly regionally focused. Production systems were thus mainly built through regional agglomeration economies with industrial complexes as an outcome. With improvements in transportation and logistics, the efficiency of distribution has reached a point where it is possible to manage large-scale production and consumption.
- **Industrial linkages**. In the past, the majority of relationships between elements of the production system took place between autonomous entities, which tended to be smaller in size. As such, those linkages tended to be rather uncoordinated. The emergence of multinational corporations underlines a higher level of linkages within production systems, as many activities that previously took place over several entities now occur within the same corporate entity. While in the 1950s, the share of the global economic output attributable to multinational corporations was in the 2 percent to 4 percent range, by the early twenty-first century this share has surged to 25 percent to 50 percent. About 30 percent of all global trade occurs within elements of the same corporation, with this share climbing to 50 percent for trade between developed countries.

The development of global transportation and telecommunications networks, ubiquitous information technologies, the liberalization of trade and multinational corporations are all factors that have substantially impacted production systems. In many cases, so called "platform companies" have become new paradigms where the function of manufacturing has been removed from the core of corporative activities (Figure 6.4). Corporations following this strategy, particularly mass retailers, have been active in taking advantage of the "China effect" in a number of manufacturing activities.

There has been a growing disconnection in many economic sectors between the manufacturing base and what can be called the "core base", which mainly includes research and development (R&D), distribution and marketing/retail. The term platform corporation has been used to describe a variety of multinational corporations that have removed the manufacturing component from their core activities (or never had

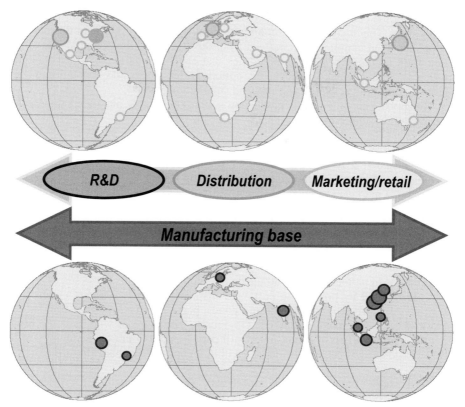

Figure 6.4 Disconnection of global production and distribution

manufacturing in the first place). They have done so by focusing on the activities that provide the most added value and subcontracted the manufacturing of the products they design. Their core activities include R&D, finance, marketing, retail and distribution. Many of them own globally recognized brand names and are actively involved in the development of new products. Their net worth is thus more a function of their brand names and capacity at innovation than from some tangible assets (like factories), outside those heavily involved in mass retailing where commercial real estate assets can be very significant. They outsource as much of the low margin work as possible and are very flexible in their choice of suppliers. Thus, the usage of the term "platform" to characterize a mobile core establishing temporary relationships with manufacturers: "Produce nowhere but to sell everywhere". Conceptually, they are reminiscent of the "cottage production system" that took place in the early phases of the industrial revolution where many labor-intensive activities (especially in garments) were subcontracted to households looking for additional income.

Commodity chains

Commodities are resources that can be consumed. They can be accumulated for a period of time (some are perishable while others can be virtually stored for centuries), exchanged as part of a transaction or purchased on specific markets (such as a futures market). Some commodities are fixed, implying that they cannot be transferred, except for the title. This includes land, mining, logging and fishing rights. In this context, the value of a fixed commodity is derived from the utility and the potential rate of extraction. Bulk commodities are commodities that can be transferred, which includes for instance grains, metals, livestock, oil, cotton, coffee, sugar and cocoa. Their value is derived from utility, supply and demand (market price).

The global economy and its production systems are highly integrated, interdependent and linked through commodity chains.

> **Commodity chain.** A functionally integrated network of production, trade and service activities that covers all the stages in a supply chain, from the transformation of raw materials, through intermediate manufacturing stages, to the delivery of a finished good to a market. The chain is conceptualized as a series of nodes, linked by various types of transactions, such as sales and intra-firm transfers. Each successive node within a commodity chain involves the acquisition or organization of inputs for the purpose of added value.

There are several stages through which a multinational corporation (or a group of corporations in partnership) can articulate its commodity chain. These stages are in large part conditioned by production costs and main markets. Commodity chains are also integrated by a transport chain routing goods, parts and raw materials from extraction and transformation sites to markets. Obviously, the nature of what is being produced and the markets where it is consumed will correspond to a unique geography of flows. Three major stages can be considered within a commodity chain (Figure 6.5):

- **First stage (parts and raw materials).** The cost structure for parts and raw materials often imposes sourcing at the international level, a process which has accelerated in recent years. It dominantly concerns the procurement of commodities. The flows occurring at this stage are mainly supported by international transportation systems in a wide variety of contexts, such as bulk cargo for raw materials and containers for parts. Distribution tends to involve high volumes and low frequency.
- **Second stage (manufacturing and assembly).** Mainly concerns intermediate goods. Some capital-intensive manufacturing and assembly activities will take place

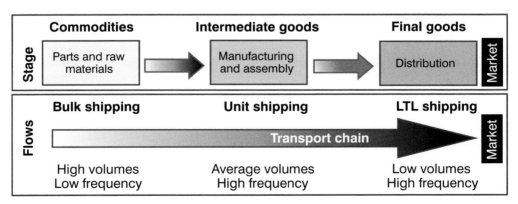

Figure 6.5 Commodity chain

inside of the national economy while labor-intensive activities will be outsourced. Flows are either containerized or on pallets, with average volumes and a tendency to have rather high frequencies, notably for commodity chains relying on timely deliveries.

- **Third stage (distribution)**. Distribution of final goods mainly takes place on the national market, although globally oriented distribution can take place, namely in the electronics sector. Depending on the scale of the distribution (international, national or regional), flows can be coordinated by distribution centers each having their own market areas. Flows are often in low volumes (less than truckload; LTL), but in frequency since they are related to retailing.

Commodity chains are thus a sequential process used by corporations within a production system to gather resources, transform them into parts and products and, finally, distribute manufactured goods to markets. Each sequence is unique and dependent on product types, the nature of production systems, where added-value activities are performed, market requirements as well as the current stage of the product life cycle. Commodity chains enable a sequencing of inputs and outputs between a range of suppliers and customers, mainly from a producer and buyer-driven standpoint. They also offer adaptability to changing conditions, namely an adjustment of production to adapt to changes in price, quantity and even product specification. The flexibility of production and distribution becomes particularly important, with a reduction of production, transaction and distribution costs as the logical outcome. The major types of commodity chains involve:

- **Raw materials**. The origin of these goods is linked with environmental (agricultural products) or geological (ores and fossil fuels) conditions. The flows of raw materials (particularly ores and crude oil) are dominated by a pattern where developing countries export towards developed countries. Transport terminals in developing countries are specialized in loading while those of developed countries unload raw materials and often include transformation activities next to port sites. Industrialization in several developing countries has modified this standard pattern with new flows of energy and raw materials.
- **Semi-finished products**. These goods already had some transformation performed conferring them an added value. They involve metals, textiles, construction materials and parts used to make other goods. The pattern of exchanges is varied in this domain, but dominated by regional transport systems integrated to regional production systems.
- **Manufactured goods**. These include goods that are shipped towards large consumption markets and require a high level of organization of flows to fulfill the demand. The majority of these flows concerns developed countries, but a significant share is related to developing countries, especially those specializing in export-based manufacturing. Containerization has been the dominant transport paradigm for manufactured goods with production systems organized around terminals and their distribution centers.

A significant trend has thus been a growing level of embeddedness between production, distribution and market demand. Since interdependencies have replaced relative autonomy and self-sufficiency as the foundation of the economic life of regions and firms, high levels of freight mobility have become a necessity. The presence of an efficient distribution system supporting global commodity chains (also known as global production networks) is sustained by:

- **Functional integration**. Its purpose is to link the elements of the supply chain in a cohesive system of suppliers and customers. A functional complementarity is then achieved through a set of supply/demand relationships, implying flows of freight, capital and information. Functional integration relies on distribution over vast territories where "just-in-time" and "door-to-door" strategies are relevant examples of interdependencies created by new freight management strategies. Intermodal activities tend to create heavily used transhipment points and corridors between them, where logistical management is more efficient.
- **Geographical integration**. Large resource consumption by the global economy underlines a reliance on supply sources that are often distant, as for example crude oil and mineral products. The need to overcome space is fundamental to economic development and the development of modern transport systems has increased the level of integration of geographically separated regions and with it better geographical complementarity. With improvements in transportation, geographical separation has become less relevant, as comparative advantages are exploited in terms of the distribution capacity of networks and production costs. Production and consumption can be more spatially separated without diminishing economies of scale, even if agglomeration economies are less evident.

Freight transport and commodity chains

As the range of production expanded, transport systems adapted to the new operational realities in local, regional and international freight distribution. Freight transportation has consequently taken an increasingly important role within commodity chains. Among the most important factors are:

- The improvement in transport efficiency facilitated an **expanded territorial range** to commodity chains.
- A reduction of telecommunication costs, enabling corporations to establish a **better level of control** over their commodity chains.
- Technological improvements, notably for intermodal transportation, enabled a **more efficient continuity** between different transport modes (especially land/maritime) and thus within commodity chains.

The results have been an improved velocity of freight, a decrease of the friction of distance and a spatial segregation of production. This process is strongly embedded with the capacity and efficiency of international and regional transportation systems, especially maritime and land routes. It is becoming rare for the production stages of a good to occur at the same location. Consequently, the geography of commodity chains is integrated to the geography of transport systems. The usage of resources, parts and semi-finished goods by commodity chains is an indication of the type of freight being transported. Consequently, transport systems must adapt to answer the needs of commodity chains, which force a level of diversification. Within a commodity chain, freight transport services can be categorized by:

- **Management of shipments**. Refers to cargo transported by the owner, the manufacturer or by a third party. The tendency has been for corporations to subcontract their freight operations to specialized providers who provide more efficient and cost-effective services.
- **Geographical coverage**. Implies a wide variety of scales ranging from intercontinental, within economic blocs, national, regional or local. Each of these scales often involves specific modes of transport services and the use of specific terminals.

- **Time constraint**. Freight services can have a time element ranging from express, where time is essential, to the lowest cost possible, where time is secondary. There is also a direct relationship between transport time and the level of inventory that has to be maintained in the supply chain. The shorter the time, the lower the inventory level, which can result in significant savings.
- **Consignment size**. Depending on the nature of production, consignments can be carried in full loads, partial loads (less than truck load; LTL), as general cargo, as container loads or as parcels.
- **Cargo type**. Unitized cargo (containers, boxes or pallets) or bulk cargo requires dedicated vehicles, vessels and transhipment and storage infrastructures.
- **Mode**. Cargo can be carried on a single mode (sea, rail, road or air) or in a combination of modes through intermodal transportation.
- **Cold chain**. A temperature-controlled supply chain linked to the material, equipment and procedures used to maintain specific cargo shipments within an appropriate temperature range. Often relates to the distribution of food and pharmaceutical products.

The globalization of the production is also concomitant – a by-product – of a post-Fordist environment where just-in-time and tense fluxes are becoming the norm in production and distribution systems. International transportation is shifting to meet the increasing needs of organizing and managing its flows through logistics. In spite of the diversity of transport services to support various commodity chains, containerization is adaptable enough to cope with a variety of cargo and time constraints.

Concept 3 – Logistics and freight distribution[1]

The nature of logistics

The growing flows of freight have been a fundamental component of contemporary changes in economic systems at the global, regional and local scales. These changes are not merely quantitative (more freight), but structural and operational. Structural changes mainly involve manufacturing systems with their geography of production, while operational changes mainly concern freight transportation with its geography of distribution. As such, the fundamental question does not necessarily reside in the nature, origins and destinations of freight movements, but how this freight is moving. New modes of production are concomitant with new modes of distribution, which brings forward the realm of logistics; the science of physical distribution.

> **Logistics** involves a wide set of activities dedicated to the transformation and distribution of goods, from raw material sourcing to final market distribution as well as the related information flows. Derived from Greek *logistikos* (to reason logically), the word is polysemic. In the nineteenth century the military referred to it as the art of combining all means of transport, revictualling and sheltering of troops. Today it refers to the set of operations required for goods to be made available on markets or to specific destinations.

Efficient logistics contributes to added value in four major interrelated ways (Figure 6.6):

---

1 Dr. Markus Hesse (University of Luxembourg) is the co-author of this section.

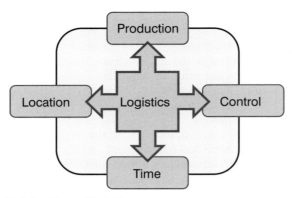

Figure 6.6 Value-added functions of logistics

- **Production**. Derived from the improved efficiency of manufacturing with appropriate shipment size, packaging and inventory levels. Thus logistics contributes to the reduction of production costs by streamlining the supply chain.
- **Location**. Derived from taking better advantage of locations implying expanded markets and lower distribution costs.
- **Time**. Derived from having goods and services available when required along the supply chain with better inventory and transportation management, and the strategic location of goods and services.
- **Control**. Derived from controlling most, if not all, the stages along the supply chain, from production to distribution. This enables better marketing and demand response, thus anticipating flows and allocating distribution resources accordingly.

The application of logistics enables a greater efficiency of movements with an appropriate choice of modes, terminals, routes and scheduling. Logistics is thus a multidimensional value-added activity including production, location, time and control of elements of the supply chain. It represents the material and organizational support of globalization. Activities comprising logistics include physical distribution – the derived transport segment – and materials management – the induced transport segment.

- **Physical distribution** is the collective term for the range of activities involved in the movement of goods from points of production to final points of sale and consumption. It must ensure that the mobility requirements of supply chains are entirely met. Physical distribution includes all the functions of movement and handling of goods, particularly transportation services (trucking, freight rail, air freight, inland waterways, marine shipping and pipelines), transhipment and warehousing services (e.g. consignment, storage, inventory management), trade, wholesale and, in principle, retail. Conventionally, all these activities are assumed to be derived from materials management demands.
- **Materials management** considers all the activities related in the manufacturing of commodities in all their stages of production along a supply chain. It includes production and marketing activities such as production planning, demand forecasting, purchasing and inventory management. Materials management must ensure that the requirements of supply chains are met by dealing with a wide array of parts for assembly and raw materials, including packaging (for transport and retailing) and,

ultimately, recycling discarded commodities. All these activities are assumed to be inducing physical distribution demands.

The close integration of physical distribution and materials management through logistics is blurring the reciprocal relationship between the induced transport demand function of physical distribution and the derived demand function of materials management. This implies that distribution, as always, is derived from materials management activities (namely production), but also, that these activities are coordinated within distribution capabilities. The functions of production, distribution and consumption are difficult to consider separately, thus recognizing the integrated transport demand role of logistics. Distribution centers are the main facilities from which logistics are coordinated.

> **Distribution center**. Facility or a group of facilities that perform consolidation, warehousing, packaging, decomposition and other functions linked with handling freight. Their main purpose is to provide value-added services to freight. DCs are often in proximity to major transport routes or terminals. They can also perform light manufacturing activities such as assembly and labeling.

Since it would be highly impractical to ship goods directly from producers to retailers, distribution centers essentially act as a buffer where products are assembled, sometimes from other distribution centers, and then shipped in batches. Distribution centers commonly have a market area in which they offer a service window defined by delivery frequency and response time to order. This structure looks much like a hub-and-spoke network.

The wide array of activities involved in logistics, from transportation to warehousing and management, have respective costs. Once compiled, they express the burden that logistics impose on distribution systems and the economies they support, which is known as the total logistics costs. The nature and efficiency of distribution systems is strongly related to the nature of the economy in which they operate. Worldwide logistics expenditures represent about 10–15 percent of the total world GDP. In economies dependent on the extraction of raw materials, logistical costs are comparatively higher than for service economies since transport costs account for a larger share of the total added value of goods. For the transport of commodities, logistics costs are commonly in the range of 20 to 50 percent of their total costs.

Infrastructure and technology

Contemporary logistics was originally dedicated to the automation of production processes, in order to organize manufacturing as efficiently as possible, with the least cost-intensive combination of production factors. A milestone that marked rapid changes in the entire distribution system was the invention of the concept of lean management, primarily in manufacturing. One of the main premises of lean management is eliminating inventories and organizing materials supply strictly on demand, replacing the former storage and stock keeping of inventory. The outcome is a specialization of production and a greater variety of products.

Modern distribution systems require a high level of control of their flows. Although this control is at start an organizational and managerial issue, its application requires a set of technical tools and expertise. If technology can be defined by the level of control over matter, technology applied to logistics can be defined as the level of control of its flows, both physical and information related. An important technological change relates to intermodal transportation, particularly containerization, which has shaped the logistics system in a fundamental way. Containerization is now embedded within production, distribution and transport.

Logistics and integrated transport systems are reciprocal endeavors. More recently, the application of information and communication technologies (ICT) for improving the overall management of flows, particularly their load units, has received attention. Thus, the physical as well as the ICT parts of technological change are being underlined. The ICT component is particularly relevant as it helps strengthen the level of control distributors have over the supply chain. The technological dimension of logistics can thus be considered from five perspectives:

- **Transportation modes**. Modes have been the object of very limited technological changes in recent decades. In some cases, modes have adapted to handle containerized operations such as road and rail (e.g. doublestacking). It is maritime shipping that has experienced the most significant technological change, which required the construction of an entirely new class of ships and the application of economies of scale to maritime container shipping. In this context, a global network of maritime shipping servicing large gateways has emerged.
- **Transportation terminals**. The technological changes have been very significant with the construction of new terminal facilities operating on a high turnover basis. Better handling equipment led to improvements in the velocity of freight at the terminals, which are among the most significant technological changes brought by logistics in materials movements. In such a context, the port has become one of the most significant terminals supporting global logistics. Port facilities are increasingly being supported by an array of inland terminals connected by high-capacity corridors.
- **Distribution centers and distribution clusters**. Technological changes impacted over the location, design and operation of distribution centers; the facilities handling the requirements of modern distribution. From a locational standpoint, distribution centers mainly rely on trucking, implying a preference for suburban locations with good road accessibility supporting a constant traffic. They service regional markets with a 48-hour service window on average, implying that replenishment orders from their customers are met within that time period. They have become one-storey facilities designed more for throughput than for warehousing with specialized loading and unloading bays and sorting equipment. Cross-docking distribution centers represent one of the foremost expressions of a facility that handles freight in a time-sensitive manner. Another tendency has been the setting of freight distribution clusters where an array of distribution activities agglomerate to take advantage of shared infrastructures and accessibility. This tends to expand the added value performed by logistics.
- **Load units**. Since logistics involves improving the efficiency of flows, load units have become particularly important. They are the basic physical management unit in freight distribution and take the form of pallets, swap bodies, semi-trailers and containers. Containers are the privileged load unit for long-distance trade, but the growing complexity of logistics required a more specific level of load management. The use of bar codes and increasingly of RFID (radio frequency identification device) enables a high level of control of the load units in circulation.
- **E-commerce**. Consider the vast array of information processing changes brought by logistics. The commodity chain is linked with physical flows as well as with information flows, notably through electronic data interchange. Producers, distributors and consumers are embedded in a web of reciprocal transactions. These transactions mostly take place virtually and their outcomes are physical flows. E-commerce

offers advantages for the whole commodity chain, from consumers being exposed to better product information to manufacturers and distributors being able to adapt quickly to changes in demand. The outcome is often more efficient production and distribution planning with the additional convenience of tracking shipments and inventories.

For logistics, ICT is particularly a time and embeddedness issue. Because of ICT, freight distribution is within a paradigm shift from inventory-based logistics to replenishment-based logistics. The shift from a push to pull logistics is particularly important in a market economy. Demand, particularly in the retailing sector, is very difficult to anticipate accurately. A closer integration (embeddedness) between supply and demand enables a more efficient production system with less waste in terms of unsold inventory. Logistics is thus a fundamental component of a market economy.

Distribution systems

In a broader sense distribution systems are embedded in a changing macro- and micro-economic framework, which can be roughly characterized by the terms of flexibilization and globalization:

- **Flexibilization** implies a highly differentiated, strongly market- and customer-driven mode of creating added value. Contemporary production and distribution is no longer subject to single-firm activity, but increasingly practiced in networks of suppliers and subcontractors. The supply chain bundles together all this by information, communication, cooperation and, last but not least, by physical distribution.
- **Globalization** means that the spatial frame for the entire economy has been expanded, implying the spatial expansion of the economy, more complex global economic integration and an intricate network of global flows and hubs.

The flow-oriented mode affects almost every single activity within the entire process of value creation. The core component of materials management is the supply chain, the time- and space-related arrangement of the whole goods flow between supply, manufacturing, distribution and consumption. Its major parts are the supplier, the producer, the distributor (e.g. a wholesaler, a freight forwarder, a carrier), the retailer and the end consumer, all of whom represent particular interests. Compared with traditional freight transport systems, the evolution of supply chain management and the emergence of the logistics industry are mainly characterized by three features:

- **Integration**. A fundamental restructuring of goods merchandising by establishing integrated supply chains with integrated freight transport demand. According to macro-economic changes, demand-side oriented activities are becoming predominant. While traditional delivery was primarily managed by the supply side, current supply chains are increasingly managed by the demand.
- **Time mitigation**. Whereas transport was traditionally regarded as a tool for overcoming space, logistics is concerned with mitigating time. Due to the requirements of modern distribution, the issue of time is becoming increasingly important in the management of commodity chains. Time is a major issue for freight shipping as it imposes inventory holding and depreciation costs, which becomes sensitive for tightly integrated supply chains.
- **Specialization**. This was achieved by shifts towards vertical integration, namely subcontracting and outsourcing, including the logistical function itself. Logistics

services are becoming complex and time-sensitive to the point that many firms are now subcontracting parts of their supply chain management to what can be called third-party logistics providers (3PL; asset based). More recently, a new category of providers, called fourth-party logistics providers (4PL; non asset based) have emerged.

Logistics is thus concomitantly concerned by distribution costs and time. In addition, many dimensions are added to the function of distribution. While in the past it was a simple matter of delivering an intact good at a specific destination within a reasonable time frame, several components have become linked with distribution:

- **Distribution time**, notably the possibility to set a very specific estimated time of arrival (ETA) for deliveries and a low tolerance for delays.
- The **reliability of distribution** measured in terms of the availability of the ordered goods and the frequency at which orders are correctly serviced in terms of quantity and time.
- The **flexibility of distribution** in terms of possible adjustments due to changes in the quantity, the location or the delivery time.
- The **quality of distribution** concerns the condition of delivered goods and whether the specified quantity was delivered.

Geography of freight distribution

Logistics has a distinct geographical dimension, which is expressed in terms of flows, nodes and networks within the supply chain. Space/time convergence, a well-known concept in transport geography where time was simply considered as the amount of space that could be traded with a specific amount of time, including travel and transhipment, is being transformed by logistics. Activities that were not previously considered fully in space/time relationships, such as distribution, are being integrated. This implies an organization and synchronization of flows through nodes and network strategies:

- **Flows**. The traditional arrangement of goods flow included the processing of raw materials to manufacturers, with a storage function usually acting as a buffer (Figure 6.7). The flow continued via wholesaler and/or shipper to retailer, ending at the final customer. Delays were very common on all segments of this chain and accumulated as inventories in warehouses. There was a limited flow of information from the consumer to the supply chain, implying the producers were not well informed (often involving a time lag) about the extent of consumption of their outputs. This procedure is now changing, mainly by eliminating one or more of the costly operations in the supply chain organization. Reverse flows are also part of the supply chain, namely for recycling and product returns. An important physical outcome of supply chain management is the concentration of storage or warehousing in one facility, instead of several. This facility is increasingly being designed as a flow- and throughput-oriented distribution center, instead of a warehouse holding cost-intensive large inventories.
- **Nodes and Locations**. Due to new corporate strategies, a concentration of logistics functions in certain facilities at strategic locations is prevalent. Many improvements in freight flows are achieved at terminals. Facilities are much larger than before, the locations being characterized by a particular connection of regional and long-distance relations. Traditionally, freight distribution has been located at major

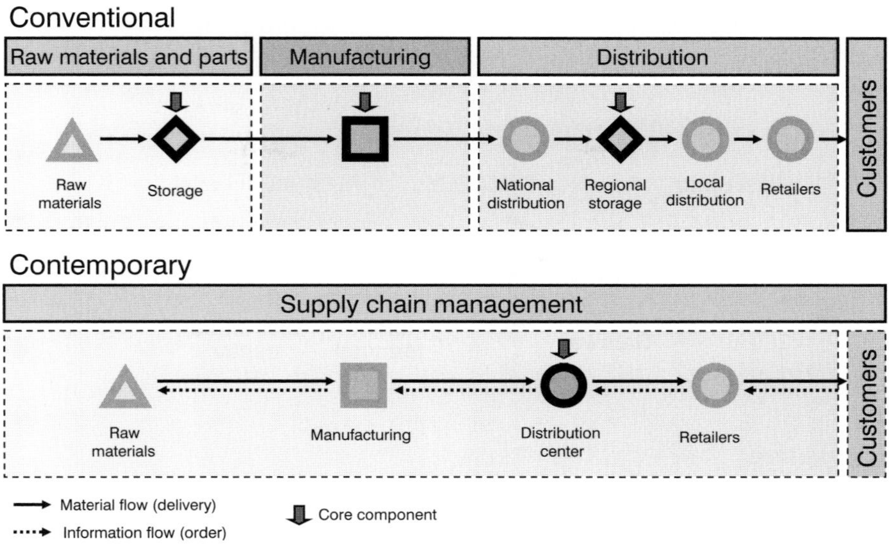

Conventional

Contemporary

Figure 6.7 Conventional and contemporary arrangement of goods flow

places of production, for instance in the manufacturing belt at the North American East Coast and in the Midwest, or in the old industrialized regions of England and continental Europe. Today, particularly the large-scale goods flows are directed through major gateways and hubs, mainly large ports and major airports but also highway intersections with access to a regional market. The changing geography of manufacturing and industrial production has been accompanied by a changing geography of freight distribution taking advantage of intermediary locations.

- **Networks**. The spatial structure of contemporary transportation networks is the expression of the spatial structure of distribution. The setting of networks leads to a shift towards larger distribution centers, often serving significant trans-national catchments. However, this does not mean the demise of national or regional distribution centers, with some goods still requiring a three-tier distribution system, with regional, national and international distribution centers. The structure of networks has also adapted to fulfill the requirements of an integrated freight transport demand, which can take many forms and operate at different scales. Most freight distribution networks, particularly in retailing, are facing the challenge of the "Last Mile" which is the final leg of a distribution sequence, commonly linking a distribution center and a customer (Figure 6.8).

The "Last Mile" (or "Last Kilometer") is a common distribution problem (Figure 6.8). Although it was initially conceived for the telecommunications sector (e.g. phone and cable services), it applies particularly well for freight distribution. Long-distance transportation tends to be well serviced by high capacity modes and terminals and is prone to economies of scale (massification). As we get closer to the final customer, economies of scale are increasingly difficult to apply as the size of batches tends to diminish (atomization). It would be rare, for instance, for a single customer to be the consignee of the cargo of a whole containership. For an international shipment, the global shipping network offers very high capacity levels and, depending on the routes, a reasonable frequency of services (for instance, one port call every two days). Hinterland

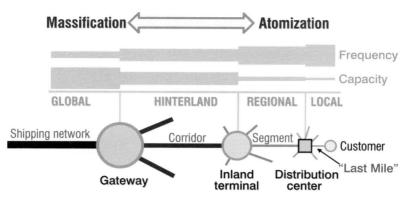

Figure 6.8 The "Last Mile" in freight distribution

transportation, which links gateways to inland terminals often using rail or barge services, is of lower capacity but of higher frequency. Once freight consignments arrive at an inland terminal they are collected and brought to distribution centers through regional segments, mostly by truck. The "Last Mile", notably for retailing, often consists of truck deliveries taking place over short distances, but likely in a congested urban setting. It is often one of the most complex elements of the commodity chain to organize as it reconciles many customers, a variety of shipments and reliability difficulties related to congestion. The "Last Mile" concept also applies to the "First Mile", albeit in reverse, which involves consolidation to a nearby transport terminal of the output of potentially several producers.

The containerization process is thus confronted with a growing tension between a massification at sea and an atomization on land. Growing vessel size has led to the massification of unit cargo at sea. On terminals and at the landside, massification makes place for an atomization process whereby each individual container has to find its way to its final destination. A major challenge consists in extending the massification concept as far inland as possible. Postponing the atomization of container batches shifts the container sorting function to inland and as such eases the pressure on port terminals. High-volume rail and barge corridors including inland terminals play a crucial role in this process.

Since cities are at the same time zones of production, distribution and consumption, the realm of city logistics is of growing importance. This issue is made even more complex by a growing dislocation between production, distribution and consumption, brought by globalization, global production networks and efficient freight transport systems (increasingly by logistics).

Method 1 – Spatial interactions

Overview

One methodology of particular importance to transport geography relates to how to estimate flows between locations, since these flows, known as spatial interactions, enable to evaluate the demand (existing or potential) for transport services.

A **spatial interaction** is a realized movement of people, freight or information between an origin and a destination. It is a transport demand/supply relationship expressed over a geographical space. Spatial interactions cover a wide variety of movements such as journeys to work, migrations, tourism, the usage of public facilities, the transmission of information or capital, the market areas of retailing activities, international trade and freight distribution.

Economic activities are generating (supply) and attracting (demand) flows. The simple fact that a movement occurs between an origin and a destination underlines that the costs incurred by a spatial interaction are lower than the benefits derived from such an interaction. As such, a commuter is willing to drive one hour because this interaction is linked to an income, while international trade concepts, such as comparative advantages, underline the benefits of specialization and the ensuing generation of trade flows between distant locations. Three interdependent conditions are necessary for a spatial interaction to occur (Figure 6.9):

- **Complementarity**. There must be a supply and a demand between the interacting locations. A residential zone is complementary to an industrial zone because the first is supplying workers while the second is supplying jobs. The same can be said concerning the complementarity between a store and its customers and between an industry and its suppliers (movements of freight). If location B produces/generates something that location A requires, then an interaction is possible because a supply/demand relationship has been established between those two locations; they have become complementary to one another. The same applies in the other direction (A to B), which creates a situation of reciprocity common in commuting or international trade.
- **Intervening opportunity**. There must not be another location that may offer a better alternative as a point of origin or as a point of destination. For instance, in order to have an interaction of a customer to a store, there must not be a closer store that offers a similar array of goods. If location C offers the same characteristics (namely complementarity) than location A and is also closer to location B, an interaction between B and A will not occur and will be replaced by an interaction between B and C.

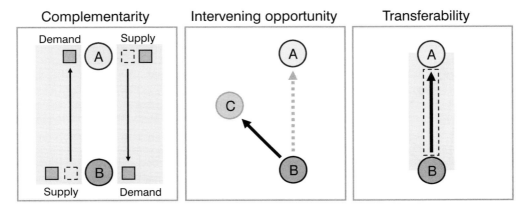

Figure 6.9 Conditions for the realization of a spatial interaction

- **Transferability**. Freight, persons or information being transferred must be supported by transport infrastructures, implying that the origin and the destination must be linked. Costs to overcome distance must not be higher than the benefits of related interaction, even if there is complementarity and no alternative opportunity. Transport infrastructures (modes and terminals) must be present to support an interaction between B and A. Also, these infrastructures must have a capacity and availability which are compatible with the requirements of such an interaction.

Spatial interaction models seek to explain spatial flows. As such it is possible to measure flows and predict the consequences of changes in the conditions generating them. When such attributes are known, it is possible for example to better allocate transport resources such as highways, buses, airplanes or ships since they would reflect the transport demand more closely.

Origin/destination matrices

Each spatial interaction, as an analogy for a set of movements, is composed of an origin/destination pair. Each pair can itself be represented as a cell in a matrix where rows are related to the locations (centroids) of origin, while columns are related to locations (centroids) of destination. Such a matrix is commonly known as an origin/destination matrix, or a spatial interaction matrix.

Figure 6.10 represents movements (O/D pairs) between five locations (A, B, C, D and E). From this graph, an O/D matrix can be built where each O/D pair becomes a cell. A value of 0 is assigned for each O/D pair that does not have an observed flow. In the O/D matrix the sum of a row (T_i) represents the total outputs of a location (flows originating from), while the sum of a column (T_j) represents the total inputs of a location (flows bound to). The summation of inputs is always equal to the summation of outputs. Otherwise, there are movements that are coming from or going to outside the considered system. The sum of inputs or outputs gives the total flows taking place within the system (T). It is also possible to have O/D matrices according to age group, income, gender, etc. Under such circumstances they are labeled sub-matrices since they account for only a share of the total flows.

In many cases where spatial interactions information is relied on for planning and allocation purposes, origin/destination matrices are not available or are incomplete. Palliating this lack of data commonly requires surveys. With economic development,

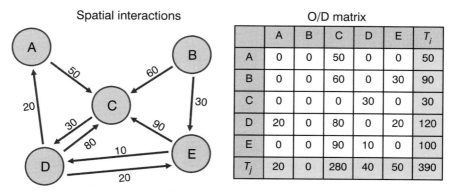

Figure 6.10 Constructing an origin/destination matrix

the addition of new activities and transport infrastructures, spatial interactions have a tendency to change very rapidly as flows adapt to a new spatial structure. The problem is that an origin/destination survey is very expensive in terms of effort, time and costs. In a complex spatial system such as a region, O/D matrices tend to be quite large. For instance, the consideration of 100 origins and 100 destinations would imply 10,000 separate O/D pairs for which information has to be provided. In addition, the data gathered by spatial interaction surveys are likely to rapidly become obsolete as economic and spatial conditions change. It is therefore important to find a way to estimate as precisely as possible spatial interactions, particularly when empirical data are lacking or are incomplete. A possible solution relies on using a spatial interaction model to complement and even replace empirical observations.

Spatial interaction models

The basic assumption concerning many spatial interaction models is that flows are a function of the attributes of the locations of origin, the attributes of the locations of destination and the friction of distance between the concerned origins and destinations. Figure 6.11 resumes the general formulation of the spatial interaction model.

- T_{ij}: interaction between location i (origin) and location j (destination). Its units of measurement are varied and can involve people, tons of freight, traffic volume, etc. It also relates to a time period such as interactions by the hour, day, month or year.
- V_i: attributes of the location of origin i. Variables often used to express these attributes are socio-economic in nature, such as population, number of jobs available, industrial output or gross domestic product.
- W_j: attributes of the location of destination j. It uses similar socio-economic variables to the previous attribute.
- S_{ij}: attributes of separation between the location of origin i and the location of destination j. Also known as transport friction. Variables often used to express these attributes are distance, transport costs or travel time.

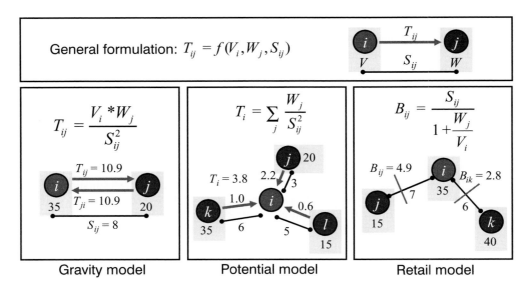

Gravity model   **Potential model**   **Retail model**

Figure 6.11 Three basic interaction models

The attributes of V and W tend to be paired to express complementarity in the best possible way. For instance, measuring commuting flows (work-related movements) between different locations would likely consider a variable such as working age population as V and total employment as W. From this general formulation, three basic types of interaction models can be constructed:

- **Gravity model**. Measures interactions between all the possible location pairs by multiplying their attributes, which is then divided by their level of separation. Separation is often squared to reflect the growing friction of distance. On Figure 6.11, two locations (i and j) have a respective "weight" (importance) of 35 and 20 and are at a distance (degree of separation) of 8. The resulting interaction is 10.9, which is reciprocal.
- **Potential model**. Measures interactions between one location and every other location by the summation of the attributes of each other location divided by their level of separation (again squared to reflect the friction of distance). On Figure 6.11, the potential interaction of location i (T_i) is measured by adding the ratio "weight"/ squared distance for each other location (j, k and l). The potential interaction is 3.8, which is not reciprocal.
- **Retail model**. Measures the boundary of the market areas between two locations competing over the same market. It assumes that the market boundary between two locations is a function of their separation divided by the ratio of their respective weights. If two locations have the same importance, their market boundary would be halfway between. On Figure 6.11, the market boundary between locations i and j (B_{ij}) is at a distance of 4.9 from i (and consequently at a distance of 2.1 from j).

Method 2 – The gravity model

Formulation

The gravity model is the most common formulation of the spatial interaction method. It is named as such because it uses a similar formulation than Newton's formulation of gravity. Accordingly, the attraction between two objects is proportional to their mass and inversely proportional to their respective distance. Consequently, the general formulation of spatial interactions can be adapted to reflect this basic assumption to form the elementary formulation of the gravity model:

$$T_{ij} = k \frac{P_i P_j}{d_{ij}}$$

- P_i and P_j: importance of the location of origin and the location of destination.
- d_{ij}: distance between the location of origin and then location of destination.
- k is a proportionality constant, related to the rate of the event. For instance, if the same system of spatial interactions is considered, the value of k will be higher if interactions were considered for a year comparatively to the value of k for one week.

Thus, spatial interactions between locations i and j are proportional to their respective importance divided by their distance.

Extension

The gravity model can be extended to include several parameters:

$$T_{ij} = k \frac{P_i^\lambda P_j^\alpha}{d_{ij}^\beta}$$

- P, d and k refer to the variables previously discussed.
- β (beta): a parameter of transport friction related to the efficiency of the transport system between two locations. This friction is rarely linear as the further the movement the greater the friction of distance. For instance, two locations serviced by a highway will have a lower beta index than if they were serviced by a road.
- λ (lambda): potential to generate movements (emissiveness). For movements of people, lambda is often related to an overall level of welfare. For instance, it is logical to infer that for retailing flows, a location having higher income levels will generate more movements.
- α (alpha): potential to attract movements (attractiveness). Related to the nature of economic activities at the destination. For instance, a center having important commercial activities will attract more movements.

Calibration

A significant challenge related to the usage of spatial interaction models, notably the gravity model, is related to their calibration. Calibration consists in finding the value of each parameter of the model (constants and exponents) to insure that the estimated results are similar to the observed flows. If it is not the case, the model is almost useless as it predicts or explains little. It is impossible to know if the process of calibration is accurate without comparing estimated results with empirical evidence.

In the two formulations of the gravity model that have been introduced, the simple formulation offers a good flexibility for calibration since four parameters can be modified. Altering the value of beta, alpha and lambda will influence the estimated spatial interactions. Furthermore, the value of the parameters can change in time due to factors such as technological innovations and economic development. For instance, improvements in transport efficiency generally have the consequence of reducing the value of the beta exponent (friction of distance). Economic development is likely to influence the values of alpha and lambda, reflecting a growth in the mobility.

Variations of the beta, alpha and lambda exponents have different impacts on the level of spatial interactions (Figure 6.12). For instance, the relationship between distance and spatial interactions will change according to the beta exponent. If the value of beta is high (higher than 0.5), the friction of distance will be much more important (steep decline of spatial interactions) than with a low value of beta (e.g. 0.25). A beta of 0 means that distance has no effects and that interactions remain the same whatever the concerned distance. Alpha and lambda exponents have the same effect on the interaction level. For a value of 1, there is a linear relationship between population (or any attribute of weight) and the level of interactions. Any value higher than 1 implies an exponential growth of the interaction level as population grows.

Often, a value of 1 is given to the parameters, and then they are progressively altered until the estimated results are similar to observed results. Calibration can also be considered for different O/D matrices according to age, income, gender, type of merchandise and modal choice. A great part of the scientific research in transport and regional

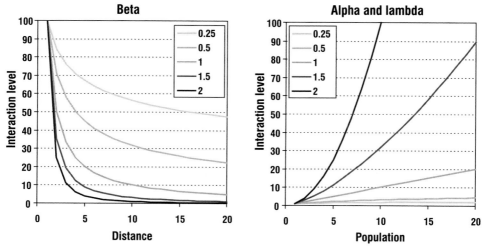

Figure 6.12 Effects of beta, alpha and lambda on spatial interactions

planning aims at finding accurate parameters for spatial interaction models. This is generally a costly and time-consuming process, but a very useful one. Once a spatial interaction model has been validated for a city or a region, it can then be used for simulation and prediction purposes, such as how many additional flows would be generated if the population increased or if better transport infrastructures (lower friction of distance) were provided.

CASE STUDY Commodity chain analysis

The structure of commodity chains

Commodity chains reveal a lot about the global structure of production, the global economy and thus represent a notable field of investigation that has yet to be fully considered by transport geographers. Understanding the significance of commodity chains requires a comprehensive approach since they include much more than a simple transport consideration; a multitude of activities are involved.

> **Commodity chain analysis.** The identification of the actors and processes that contribute to the origination of a product that is consumed by a market, such as raw materials, products or consumption goods. Thus, a commodity chain includes a sequence of operations ranging from the extraction of raw materials, the assembly of intermediate goods, to distribution to consumption markets. Commodity chain analysis can also consider only a specific segment related to a single product (or group of products).

The analysis of such a complex chain of agents and processes considers several perspectives:

● **Transactional perspective.** Identification of the flows and of the transactions that create them. This particularly concerns the decision-making process in the establishment and management of commodity chains.

- **Comparative perspective**. Assesses the relative competitiveness of the elements of the commodity chains in terms of added value.
- **Functional perspective**. Identifies the physical processes involved in the circulation of goods, including the capacity constraints in distribution, namely modal, intermodal and terminal effectiveness.

The analysis of commodity chains, depending on the perspective, can consider several factors:

- **Origin and destination**. A basic issue of supply and demand which reveals comparative advantages, locational preferences and market size. Intermediary locations where activities such as warehousing need also to be considered.
- **Cost function**. Evaluates the costs incurred to the set of activities taking place along the commodity chain such as procurement costs, manufacturing costs, distribution costs and retailing costs.
- **Load unit**. Considers how the material flows in the commodity chain are circulating, often related to how fragile, perishable or valuable a product is. It is more than simply an issue of containerization, but also in which way the containerized load unit is used.
- **Modal and intermodal use**. A matter of the nature of the transport chains used to accommodate the commodity chains in terms of modes, terminals and freight forwarders.
- **Regulation and ownership**. The set of rules and regulations related to the circulation of goods within the commodity chain, including compliance. Also considers the nature and the level of control shipping companies have over the commodity chains they use through agreements, mergers and alliances.
- **Distribution channel**. Relationships with logistical service providers, particularly with manufacturers and retailers. In many cases, distribution activities are subcontracted.
- **Added value**. The consideration of which parts of the commodity chain contributes the most to added value. This is an important strategic goal as added value is linked with profit margins. The organization of the commodity chain thus seeks to increase added value through locational and organizational strategies.

The China connection

China has become a crucial element in the emergence of global commodity chains. After more than 20 years of export-oriented industrialization, China has captured a whole range of manufacturing activities, from the most simple and labor-intensive to those with a growing level of sophistication. The footwear commodity chain is a notable example of a mature industry heavily dependent on low production costs and efficient distribution channels. Products tend to be relatively simple and success is commonly based on design, brand name and costs. It is thus a manufacturing sector that has achieved a high level of fragmentation due to globalization. From modest beginnings in the 1980s, footwear manufacturing has boomed in China, which now accounts for about 50 percent the world's shoe production. A brief commodity chain analysis for this sector reveals the following:

- **Origin and destination**. The Pearl River Delta has become one of the most intensive manufacturing clusters in the world, the outcome of more than two decades of foreign investments (from the mid-1980s), initially in special economic zones like Shenzhen. Mainly due to poor inland transportation, most of the manufacturing activities are clustered in the delta along main road corridors and close to port facilities. The production is exported to the rest of the world. In particular, 95 percent of the shoes sold in the United States are manufactured in China, which in itself represents a significant commodity chain. The nature of production commonly reveals a "platform" structure where large fashion companies (American, European and Japanese) controlling brand names subcontract their production. In many cases, a brand name designer is directly interacting with a retailer.

- **Cost function**. A typical cost structure in shoe manufacturing reveals that because of the low Chinese labor costs, labor became a marginal component of the production costs. Transport costs are low because of a significant value (at retail)-to-volume ratio. The most important costs are actually related to retailing and marketing, underlining the level of maturity this industry has achieved.

- **Load unit**. The typical factory output is a completed product including the wrapping and packaging (often including the price tag), ready to be put on a store shelf. Orders are placed on pallets, which are then assembled in container loads (Plate 6.1). A growing trend to maximize the usage of the container unit is to forgo the usage of pallets at the expense of additional loading and unloading costs. The products are put on pallets close to their destination. The load units are containerized but the assembly can include a variety of goods bound for the same distribution center, particularly if the retailer is diversified. At the distribution center, these loads will be broken down, often in LTL bound to specific retailing stores.

Plate 6.1 Palletized goods waiting to be loaded into containers, Shenzhen, China. *Credit*: Jean-Paul Rodrigue

- **Modal and intermodal use.** Since the export market is global, the commodity chain involves a variety of modes. The first step is commonly truck deliveries at a distribution center where loads are assembled in containers. Those containers are then delivered to a port facility. Since the shoe commodity chain is globally oriented with a multitude of markets being serviced by a fairly centralized production structure, a set of complex activities are performed at the port. A particular problem is linked with containerized trade imbalances and the loading of container ships, considering that each services different markets and has a set of port calls. In 2004, about 160,000 TEUs of containers carrying shoes were imported in the United States through West Coast ports. Then, the inland freight distribution system carries these containers to their destinations.
- **Regulation and ownership.** This commodity chain takes place in a context where the global apparel industry operates in a free trade environment. Since shoes are simple and labor-intensive products, few countries maintain duties for this type of product, which can circulate with relative ease from a regulatory perspective. The ownership of global commodity chains is increasingly concentrated since many international logistics providers have vested interests in physical distribution activities, notably in distribution centers.
- **Distribution channel.** In the case of shoe manufacturing in China, like many manufacturing activities, locational issues are simple as manufacturers choose sites close to port facilities. The challenge resides in the distribution of shoe production to a multitude of customers in a multitude of countries. Many logistics and distribution firms (3PL) have started offering comprehensive freight services in China, particularly around its export-oriented zones. There is thus a setting of more efficient distribution channels within China, which helps cope with the surge of exports.
- **Added value.** It is typical in this commodity chain that the designers and retailers capture the great majority of the added value (25 percent and 50 percent respectively).

The sequence provided here has focused more specifically on the transportation and distribution aspects of the commodity chain. It reveals a globalized and fragmented industry seeking to extract as much added value as possible from a mature product that is the object of intense competition for its production and retailing.

Bibliography

Bernstein, W.J. (2008) *A Splendid Exchange: How Trade Shaped the World*, New York: Atlantic Monthly Press.

Bowersox, D.J., E. Smykay and B. LaLonde (1968) *Physical Distribution Management. Logistics Problems of the Firm*, New York/London: Macmillan.

Bowersox, D.J., D. Closs and T. Stank (2000) "Ten Mega-Trends that will Revolutionize Supply Chain Logistics", *Journal of Business Logistics*, 21, 1–16.

Braudel, F. (1982) *The Wheels of Commerce. Civilization and Capitalism 15th–18th Century*, Vol. II, New York: Harper & Row.

CEMT (2005) *Transport and International Trade*, Conclusions of Round Table 131, European Conference of Ministers of Transport, http://www.cemt.org/online/conclus/rt131e.pdf.

Coyle, J.J., E.J. Bardi and R.A. Novack (1994) *Transportation*, 4th edn, New York: West Publishing.

Daniels, J.D. and L.H. Radebaugh (2000) *International Business: Environments and Operations*, 9th edn, New York: Prentice Hall.

Dicken, P. (1992) *Global Shift: The Internationalization of Economic Activity*, 2nd edn, New York: Guilford Press.

Gave, C., A. Kaletsky and L.V. Gave (2005) *Our Brave New World*, New York: GaveKal Research.

Hesse, M. and J.-P. Rodrigue (2004) "The Transport Geography of Logistics and Freight Distribution", *Journal of Transport Geography*, 12(3), 171–84.

Klare, M.T. (2001) *Resource Wars: The New Landscape of Global Conflict*, New York: Henry Holt.

Lakshmanan, T.J., U. Subramanian, W. Anderson and F. Leautier (2001) *Integration of Transport and Trade Facilitation: Selected Regional Case Studies*, Washington: World Bank.

McKinnon, A. (1988) "Physical Distribution", in J.N. Marshall (ed.) *Services and Uneven Development*, Oxford: Oxford University Press, pp. 133–59.

Tallec, F. and L. Bockel (2005) *Commodity Chain Analysis: Constructing the Commodity Chain Functional Analysis and Flow Charts*, Food and Agriculture Organization of the United Nations, http://www.fao.org/docs/up/easypol//330/cca_043EN.pdf.

Ullman, E.L. (1956) "The Role of Transportation and the Bases for Interaction", in W.L. Thomas Jr. (ed.) *Man's Role in Changing the Face of the Earth*, Chicago, IL: University of Chicago Press.

US Department of Transport (US-DOT), Federal Highway Administration (FHA) (1998) *US Freight: Economy in Motion*, http://ntl.bts.gov/data/freightus98.pdf.

7 Urban transportation

Urbanization has been one of the dominant contemporary processes as a growing share of the global population lives in cities. Considering this trend, urban transportation issues are of foremost importance to support the passengers and freight mobility requirements of large urban agglomerations. Transportation in urban areas is highly complex because of the modes involved, the multitude of origins and destinations, and the amount and variety of traffic. Traditionally, the focus of urban transportation has been on passengers as cities were viewed as locations of utmost human interactions with intricate traffic patterns linked to commuting, commercial transactions and leisure/cultural activities. However, cities are also locations of production, consumption and distribution, activities linked to movements of freight. Conceptually, the urban transport system is intricately linked with urban form and spatial structure. Urban transit is an important dimension of mobility, notably in high density areas.

Concept 1 – Transportation and urban form

Global urbanization

No discussion about the urban spatial structure can take place without an overview of urbanization, which has been one of the dominant trends of economic and social change of the twentieth century, especially in the developing world.

> **Urbanization**. The process of transition from a rural to a more urban society. Statistically, urbanization reflects an increasing proportion of the population living in settlements defined as urban, primarily through net rural to urban migration. The level of urbanization is the percentage of the total population living in towns and cities, while the rate of urbanization is the rate at which it grows.

This transition will go on well into the second half of the twenty-first century. Urban mobility problems have increased proportionally, and in some cases exponentially, with urbanization; a trend reflected in the growing size of cities and in the increasing proportion of the urbanized population. Since 1950, the world's urban population has more than doubled, to reach nearly 3.16 billion in 2005, about 48.7 percent of the global population. This is due to two main demographic trends:

- **Natural increase**. The outcome of more births than deaths in urban areas, a direct function of the fertility rate as well as the quality of healthcare systems.
- **Rural to urban migrations**. A strong factor of urbanization, particularly in the developing world where migration accounted for between 40 and 60 percent of the urban growth. Such a process has endured since the beginning of the industrial

revolution in the nineteenth century but has become prevalent in the developing world. The reasons for urban migration are numerous and may involve the expectation to find employment, improved agricultural productivity which frees rural labor or even political and environmental problems where populations are constrained to leave the countryside.

The outcome has been a fundamental change in the socio-economic environment of human activities as urbanization involves new forms of employment, economic activity and lifestyle. Thus, industrialization in the developing world is directly correlated with urbanization, the case of China being particularly eloquent. The industrialization of coastal China has led to the largest rural to urban migration in history. According to the United Nations Population Fund, about 18 million people migrate from rural areas to cities each year in China alone. Current global trends indicate a growth of about 50 million urbanites each year, roughly a million a week. More than 90 percent of that growth occurs in developing countries which places intense pressure on urban infrastructures, particularly transportation, to cope. By 2050, 6.2 billion people, about two-thirds of humanity, are likely to be urban residents.

The urban form

At the urban level, demographic and mobility growth have been shaped by the capacity and requirements of urban transport infrastructures, be they roads, transit systems or simply walkways. Consequently, there is a wide variety of urban forms, spatial structures and associated urban transportation systems.

> **Urban form**. The spatial imprint of an urban transport system as well as the adjacent physical infrastructures. Jointly, they confer a level of spatial arrangement to cities.

> **Urban (spatial) structure**. The set of relationships arising out of the urban form and its underlying interactions of people, freight and information.

Elements of the urban transport system – modes, infrastructures and users – have a spatial imprint which shapes the urban form (Figure 7.1). Considering that each city has different socio-economic and geographical characteristics, the spatial imprint of transportation varies accordingly. For instance, while North American cities tend to have

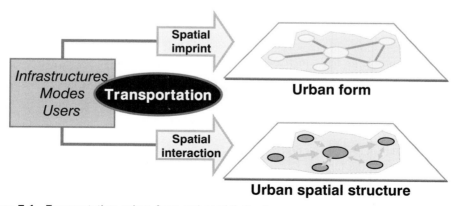

Figure 7.1 Transportation, urban form and spatial structure

an urban form that has been shaped by the automobile, cities in other parts of the world, because of different modal preferences and infrastructure developments, have different urban forms. The urban transport system is also composed of spatial interactions reflecting its spatial structure.

Even if the geographical setting of each city varies considerably, the urban form and its spatial structure are articulated by two structural elements:

- **Nodes**. These are reflected in the centrality of urban activities, which can be related to the spatial accumulation of economic activities or to the accessibility to the transport system. Terminals, such as ports, railyards and airports, are important nodes around which activities agglomerate at the local or regional level. Nodes have a hierarchy related to their importance and contribution to urban functions, such as production, management, retailing and distribution.
- **Linkages**. The infrastructures supporting flows from, to and between nodes. The lowest level of linkages includes streets, which are the defining elements of the urban spatial structure. There is a hierarchy of linkages moving up to regional roads and railways and international connections by air and maritime transport systems.

Urban transportation is organized in three broad categories of collective, individual and freight transportation. In several instances, they are complementary to one another, but sometimes they may be competing for the usage of available land and/or transport infrastructures:

- **Collective transportation (public transit)**. The purpose of collective transportation is to provide publicly accessible mobility over specific parts of a city. Its efficiency is based upon transporting large numbers of people and achieving economies of scale. It includes modes such as tramways, buses, trains, subways and ferryboats.
- **Individual transportation**. Includes any mode where mobility is the outcome of a personal choice and means, such as the automobile, walking, cycling and the motorcycle. The majority of people walk to satisfy their basic mobility, but this number varies according to the city considered. For instance, walking accounts for 88 percent of all movements inside Tokyo while this figure is only 3 percent for Los Angeles.
- **Freight transportation**. As cities are dominant centers of production and consumption, urban activities are accompanied by large movements of freight. These movements are mostly characterized by delivery trucks moving between industries, distribution centers, warehouses and retail activities as well as from major terminals such as ports, railyards, distribution centers and airports.

Historically, movements within cities tended to be restricted to walking, which made medium- and long-distance urban linkages rather inefficient and time-consuming. Thus, activity nodes tended to be agglomerated and urban forms compact. Many modern cities have inherited an urban form created under such circumstances, even though they are no longer prevailing. The dense urban cores of many European, Japanese and Chinese cities, for example, enable residents to make between one-third and two-thirds of all trips by walking and cycling. At the other end of the spectrum, the dispersed urban forms of most Australian, Canadian and American cities, which were built recently, encourages automobile dependency and are linked with high levels of mobility. Many major cities are also port cities with maritime accessibility playing an enduring role not only for the economic vitality but also in the urban spatial structure with the port district being an important node.

Urban transportation is thus associated with a spatial form which varies according to the modes being used. What has not changed much is that cities tend to opt for a grid street pattern. This was the case for many Roman cities as it is for American cities. The reasons behind this permanence are relatively simple: a grid pattern jointly optimizes accessibility and available real estate. In an age of motorization and personal mobility, an increasing number of cities are developing a spatial structure that increases reliance on motorized transportation, particularly the privately owned automobile. Dispersion, or urban sprawl, is taking place in many different types of cities, from dense, centralized European metropolises such as Madrid, Paris and London, to rapidly industrializing metropolises such as Seoul, Shanghai and Buenos Aires, to those experiencing recent, fast and uncontrolled urban growth, such as Bombay and Lagos.

For a commuter, the relationship between space and travel time changes dramatically with the transportation mode used (Figure 7.2):

- **Walking**. Assuming a willingness to commute for one hour, a pedestrian walking at 5 km per hour could cross about 5 km. The space/time relationship of such a commute would be a circle of 10 km diameter.
- **Streetcar**. A streetcar, like those operating in the first half of the twentieth century, could travel around 15 km per hour along fixed lines. In this case, the space/time relationship would be star-shaped to reflect walking to the streetcar line, and 15 km diameter along the lines.
- **Cycling**. The bicycle became a mode of mass transportation in the late nineteenth and early twentieth centuries. With approximately the same speed of a streetcar, but with no fixed line limitations, the space/time relationship of commuting by bicycle would be a circle of 15 km diameter.

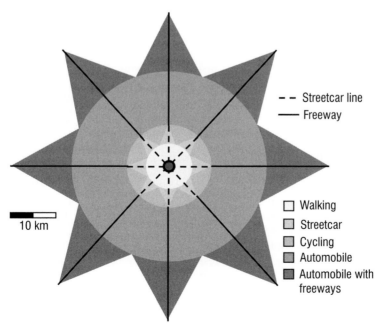

Figure 7.2 One-hour commuting according to different urban transportation modes (Source: adapted from Hugill, 1995)

- **Driving (no freeways).** With a driving speed of about 30 km per hour (taking into account of stops, lights, congestion and parking), an automobile creates a spherical space/time relationship of about 30 km in diameter.
- **Driving (with freeways).** Along a freeway, a fixed infrastructure, the driving speed is doubled to 60 km per hour. The space/time relationship is then star-shaped with a 60-km diameter along its axis.

Evolution of transportation and urban form

The evolution of transportation has generally led to changes in urban form. The more radical the changes in transport technology, the more the urban form has been altered. Among the most fundamental changes in the urban form is the emergence of new clusters expressing new urban activities and new relationships between elements of the urban system. In many cities, the central business district (CBD), once the primary destination of commuters and serviced by public transportation, has been changed by new manufacturing, retailing and management practices. Whereas traditional manufacturing depended on centralized workplaces and transportation, technological and transportation developments rendered modern industry more flexible. In many cases, manufacturing relocated in a suburban setting, if not altogether to entirely new low-cost locations. Retail and office activities are also suburbanizing, producing changes in the urban form. Concomitantly, many important transport terminals, namely port facilities and railyards, have emerged in suburban areas following new requirements in modern freight distribution brought in part by containerization. The urban spatial structure shifted from a nodal to a multi-nodal character (Figure 7.3).

The urban spatial structure basically considers the location of different activities as well as their relationships. Core activities are those of the highest order in the urban spatial structure, namely tertiary and quaternary activities involved in management (finance and insurance) and consumption (retailing). Central activities concern production and distribution with activities such as warehousing, manufacturing, wholesaling and

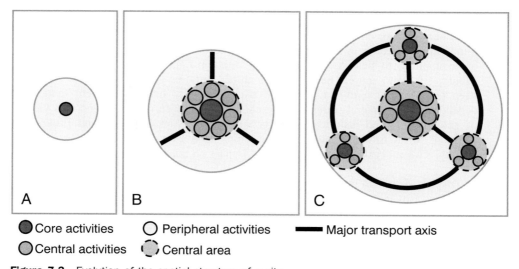

● Core activities   ○ Peripheral activities   ▬ Major transport axis

◉ Central activities   Central area

Figure 7.3 Evolution of the spatial structure of a city

transportation. Peripheral activities are dominantly residential or servicing local needs. A central area refers to an agglomeration of core and/or central activities within a specific location. The emergence of a CBD (central business district; the central area of a city) is the result of an historical process, often occurring over several centuries (depending on the age of a city), that has changed the urban form and the location of economic activities. Obviously, each city has its own history, but it is possible to establish a general common process:

- **Pre-industrial era (A).** For cities that existed before the industrial revolution, the CBD was limited to a small section of the city, generally near the waterfront, the market and/or a site of religious or political importance. These were locations where major transactions took place and thus required financial, insurance, warehousing and wholesale services.
- **Industrial revolution (B).** With the industrial revolution came mass production and mass consumption. This permitted the emergence of a distinct retailing and whole-saling part of the CBD, while manufacturing located outside the core. Managing these expanding activities also created an increasing need for office space that located near traditional places of financial interaction. As the industrial revolution matured, major transportation axes spurred from the central area towards the periphery.
- **Contemporary era (C).** After World War II, industries massively relocated away from central areas to suburban areas, leaving room for the expansion of adminis-trative and financial activities. The CBD was thus the object of an important accumulation of financial and administrative activities, particularly in the largest cities, as several corporations became multinational enterprises. These activities were even more willing to pay higher rents than retailing, thereby pushing some retail activities out of the CBD. New retailing sub-centers emerged in suburban areas because of road accessibility and because of the need to service these new areas. Warehousing and transportation, no longer core area activities, also relocated to new peripheral locations. The spatial structure of many cities became increasingly multi-nodal.

Initially, suburban growth mainly took place adjacent to major road corridors, leaving a lot of vacant or farm land in between. Later, intermediate spaces were gradually filled up, more or less coherently. Highways and ring roads, which circled and radiated from cities, favored the development of suburbs and the emergence of important sub-centers that compete with the central business district for the attraction of economic activities. As a result, many new job opportunities have shifted to the suburbs (if not to entirely new locations abroad) and the activity system of cities has been considerably modified. Different parts of a city have a different dynamism depending on its spatial pattern. These changes have occurred according to a variety of geographical and historical contexts, notably in North America and Europe. In addition, North American and European cities have seen different changes in urban density. Two processes have a substantial impact on contemporary urban forms:

- **Dispersed urban land development patterns** have been dominant in North America over the last 50 years, where land is abundant, transportation costs low and where the economy became dominated by service and technology industries. Under such circumstances, it is not surprising to find that there is a strong relationship between urban density and automobile use. For many cities their built-up areas have grown at a faster rate than their populations. In addition, commuting became relatively inexpensive compared with land costs, so households had an incentive to buy lower-priced housing at the urban periphery. Similar patterns can be found in many European cities, but this change is occurring at a lower pace and involving a smaller range.

- **The decentralization of activities** resulted in two opposite effects. First, commuting time has remained relatively stable in duration. Second, commuting increasingly tends to be longer and made by using the automobile rather than by public transit. Most transit and road systems were developed to facilitate suburb-to-city, rather than suburb-to-suburb, commuting. As a result, suburban highways are often as congested as urban highways.

Although transportation systems and travel patterns have changed considerably over time, one enduring feature remains that most people travel less than 30–40 minutes in one direction. Globally, people are spending about 1.2 hours per day commuting, wherever this takes place in a low or a high mobility setting. Different transport technologies, however, are associated with different travel speeds and capacity. As a result, cities that rely primarily on non-motorized transport tend to be different than auto-dependent cities. Transport technology thus plays a very important role in defining urban form and the spatial pattern of various activities.

The spatial imprint of urban transportation

The amount of urban land allocated to transportation is often correlated with the level of mobility. In the pre-automobile era, about 10 percent of urban land was devoted to transportation which were simply roads for a traffic that was dominantly pedestrian. As the mobility of people and freight increased, a growing share of urban areas is allocated to transport and the infrastructures supporting it. Large variations in the spatial imprint of urban transportation are observed between different cities as well as between different parts of a city, such as between central and peripheral areas. The major components of the spatial imprint of urban transportation are:

- **Pedestrian areas**. Refer to the amount of space devoted to walking. This space is often shared with roads as sidewalks may use between 10 percent and 20 percent of a road's right of way. In central areas, pedestrian areas tend to use a greater share of the right of way and in some instances whole areas are reserved for pedestrians. However, in a motorized context, most pedestrian areas are for servicing people's access to transport modes such as parked automobiles.
- **Roads and parking areas**. Refer to the amount of space devoted to road transportation, which has two states of activity: moving or parked. In a motorized city, on average 30 percent of the surface is devoted to roads while another 20 percent is required for off-street parking. This implies for each car about two off-street and two on-street parking spaces. In North American cities, roads and parking lots account for between 30 and 60 percent of the total surface.
- **Cycling areas**. In a disorganized form, cycling simply shares access to pedestrian and road space. However, many attempts have been made to create spaces specifically for bicycles in urban areas, with reserved lanes and parking facilities.
- **Transit systems**. Many transit systems, such as buses and tramways, share road space with automobiles, which often impairs their respective efficiency. Attempts to mitigate congestion have resulted in the creation of road lanes reserved to buses either on a permanent or temporary (during rush hour) basis. Other transport systems such as subways and rail have their own infrastructures and, consequently, their own rights of way.

- **Transport terminals**. Refer to the amount of space devoted to terminal facilities such as ports, airports, transit stations, railyards and distribution centers. Globalization has increased the mobility of people and freight, both in relative and absolute terms, and consequently the amount of urban space required to support those activities. Many major terminals are located in the peripheral areas of cities, which are the only locations where sufficient amounts of land are available.

The spatial importance of each transport mode varies according to a number of factors, density being the most important. If density is considered as a gradient, rings of mobility represent variations in the spatial importance of each mode at providing urban mobility. Further, each transport mode has unique performance and space consumption characteristics. The most relevant example is the automobile. It requires space to move around (roads) but it also spends 98 percent of its existence stationary in a parking space. Consequently, a significant amount of urban space must be allocated to accommodate the automobile, especially when it does not move and is thus economically and socially useless. At an aggregate level, measures reveal a significant spatial imprint of road transportation among developed countries. In the United States, more land is thus used by the automobile than for housing. In Western Europe, roads account for between 15 and 20 percent of the urban surface, while for developing countries, this figure is about 10 percent (6 percent on average for Chinese cities).

Transportation and urban structure

Rapid and expanded urbanization occurring around the world involves an increased number of trips in urban areas. Cities have traditionally responded to growth in mobility by expanding the transportation supply, by building new highways and/or transit lines. In the developed world, that has mainly meant building more roads to accommodate an ever-growing number of vehicles, therefore creating new urban structures. Several urban spatial structures have accordingly emerged, with the reliance on the automobile being the most important discriminatory factor. Four major types can be identified at the metropolitan scale (Figure 7.4):

- **Type I – completely motorized network**. Represents an automobile-dependent city with a limited centrality. Characterized by low to average land use densities, this automobile-oriented city assumes free movements between all locations. Public transit has a residual function while a significant share of the city is occupied by structures servicing the automobile, notably highways and large parking lots. Most activities are designed to be accessed with an automobile. This type of urban structure requires a massive network of high capacity highways to the point that urban efficiency is based on individual transportation. Secondary roads converge at highways, along which small centers are located, notably near interchanges. This system characterizes recent cities in a North American context where urban growth occurred in the second half of the twentieth century, such as Los Angeles, Phoenix, Denver and Dallas.
- **Type II – weak center**. Represents the spatial structure of many American cities where many activities are located in the periphery. These cities are characterized by average land use densities and a concentric pattern. The central business district is relatively accessible by the automobile and is the point of convergence of the transit system, which tends to be under-used and requires subsidies. The urban area cannot be cost-effectively serviced with the transit system, so services are often oriented along major corridors. In many cases, ring roads favored the emergence

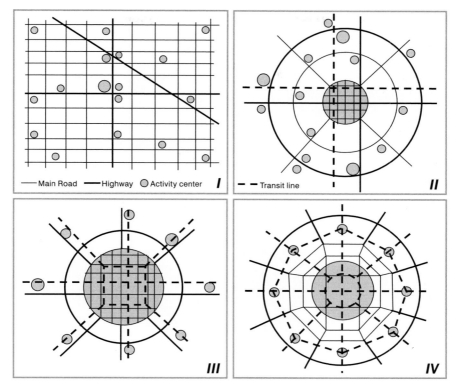

Figure 7.4 Four main types of urban spatial structures

of a set of small centers at the periphery, notably at the convergence of radial lines, some of them effectively competing with the central business district for the location of economic activities. This system is often related to older cities, which emerged in the first half of the twentieth century, such as Melbourne and San Francisco, and were afterwards substantially impacted by motorization.

- **Type III – strong center**. Represents high density urban centers with well-developed public transit systems, particularly in Europe and Asia. Characterizes cities having a high land use density and high levels of accessibility to urban transit. There are thus limited needs for highways and parking space in the central area, where a set of high capacity public transit lines service most mobility needs. The productivity of this urban area is thus mainly related to the efficiency of the public transport system. The convergence of radial roads and ring roads favors the location of secondary centers, where activities that were no longer able to afford a central location located. This system characterizes cities with important commercial and financial functions that grew in the nineteenth century, such as Paris (Plate 7.1), New York, Toronto, Sydney and Hamburg.
- **Type IV – traffic limitation**. Represents urban areas that have efficiently implemented traffic control and modal preference in their spatial structure. Commonly, the central area is dominated by public transit. They have a high land use density and were planned to limit the usage of the automobile in central areas for a variety of reasons, such as to preserve its historical character or to avoid congestion.

Plate 7.1 High density structured urban form, Paris. *Credit*: Claude Comtois

Through a "funnel" effect, the capacity of the road transport system is reduced the closer one gets to the central area. Public transit is used in central areas, while individual transportation takes a greater importance in the periphery. Between suburbs and the central city are places of interface between individual (automobile) and collective transportation or between low capacity collective transportation (bus) and high capacity collective transportation (metro, rail). Several cities are implementing this strategy, namely through congestion pricing, as it keeps cars from the central areas while supporting mobility in the suburbs. This system typifies cities with a long planning history favoring public transit, particularly in formally socialist economies. London, Singapore, Hong Kong, Vienna and Stockholm are good examples of this urban transport structure.

There are different scales on which transportation systems influence the structure of communities, districts and the whole metropolitan area. For instance, one of the most significant impacts of transportation on the urban structure has been the clustering of activities near areas of high accessibility. The impact of transport on the spatial structure is particularly evident in the emergence of suburbia. Although many other factors are important in the development of suburbia, including low land costs, available land (large lots), the environment (clean and quiet), safety and car-oriented services (shopping malls), the spatial imprint of the automobile is dominant. Although it can be argued that roads and the automobile have limited impacts on the extent of urban sprawl itself, they are a required condition for sprawl to take place. Initially an American invention, suburban developments have occurred in many cities worldwide, although no other places have achieved such a low density and high automobile dependency than in the United States.

Facing the expansion of urban areas, congestion problems and the increasing importance of inter-urban movements, several ring roads have been built around major cities. They became an important attribute of the spatial structures of cities, notably in North America. Highway interchanges in suburban areas are notable examples of new clusters of urban development. The extension (and the over-extension) of urban areas has created what may be called peri-urban areas. They are located well outside the urban core and the suburbs, but are within reasonable commuting distances.

Concept 2 – Urban land use and transportation

Land use – transport system

Urban land use comprises two elements: the nature of land use which relates to what activities are taking place where, and the level of spatial accumulation, which indicates their intensity and concentration. Central areas have a high level of spatial accumulation and corresponding land uses, such as retail, while peripheral areas have lower levels of accumulation. Most economic, social or cultural activities imply a multitude of functions, such as production, consumption and distribution. These functions take place at specific locations and are part of an activity system. Activities have a spatial imprint, therefore. Some are routine activities, because they occur regularly and are thus predictable, such as commuting and shopping. Others are institutional activities that tend to be irregular, and are shaped by lifestyle (e.g. sports and leisure) or by special needs (e.g. healthcare). Still others are production activities that are related to manufacturing and distribution, whose linkages may be local, regional or global. The behavioral patterns of individuals, institutions and firms have an imprint on land use. The representation of this imprint requires a typology of land use, which can be formal or functional:

- **Formal land use** representations are concerned with qualitative attributes of space such as its form, pattern and aspect and are descriptive in nature.
- **Functional land use** representations are concerned with the economic nature of activities such as production, consumption, residence and transport, and are mainly a socio-economic description of space.

Land use, both in formal and functional representations, implies a set of relationships with other land uses. For instance, commercial land use involves relationships with its suppliers and customers. While relationships with suppliers will dominantly be related with movements of freight, relationships with customers would include movements of people. Thus, a level of accessibility to both systems of circulation must be present. Since each type of land use has its own specific mobility requirements, transportation is a factor of activity location, and is therefore associated intimately with land use. Within the urban system each activity occupies a suitable, but not necessarily optimal, location, from which it derives rent. Transportation and land use interactions mostly consider the retroactive relationships between activities, which are land-use related, and accessibility, which is transportation related. These relationships often have been described as a "chicken-and-egg" problem since it is difficult to identify the triggering cause of change: do transportation changes precede land use changes or vice versa?

Activities have spatial locations creating a land use pattern, which is influenced by the existing urban form and spatial structure (Figure 7.5). This form is strongly related to the types of activities that can roughly be divided in three major classes:

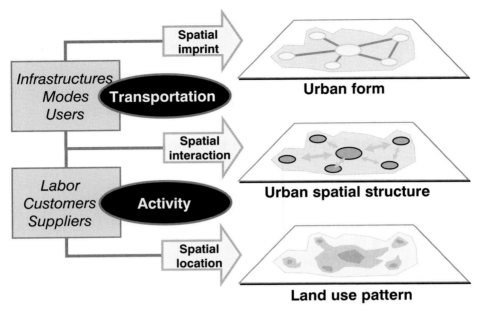

Figure 7.5 Transportation, activity systems and land use

- **Routine activities**. This class of activities is occurring regularly and is thus predictable. They involve journey to work (residential to industrial/commercial/administrative) and shopping (residential to retailing). The land use pattern generated is thus stable and coherent. Generally, these activities are zonal and links are from areas to areas.
- **Institutional activities**. Most institutions are located at specific points and generally have links with individuals. This activity system is linked to an urban environment where links occur irregularly and according to lifestyle (students, sports, leisure, etc.) or special needs (health).
- **Production activities**. This involves a complex network of relationships between firms, such as control, distribution, warehousing and subcontracting. This activity system can be linked to a specific urban environment, but also to a region, nation or even the world. Some activities are strongly linked to the local urban area, while others are far more linked to the global economy. The land use pattern of an activity may thus be linked to an external (international) process.

These activity systems underline the importance of linkages between land uses, which require movements of people, freight and information. The results of these linkages are land use patterns. Thus, understanding the set of relationships an industrial district has with its labor, suppliers and customers will provide an overview of the land use patterns in an urban area, but also with other urban areas.

Urban transportation aims at supporting transport demands generated by the diversity of urban activities in a diversity of urban contexts. A key for understanding urban entities thus lies in the analysis of patterns and processes of the transport/land use system.

Urban land use models

The relationships between transportation and land use are rich in theoretical representations that have contributed much to geographical sciences. Several descriptive and analytical models of urban land use have been developed over time, with increased levels of complexity. All involve some consideration of transport in the explanations of urban land use structures.

- **Von Thunen's regional land use model** is the oldest. It was initially developed in the early nineteenth century (1826) for the analysis of agricultural land use patterns in Germany. It used the concept of economic rent to explain a spatial organization where different agricultural activities are competing for the usage of land (Figure 7.6). The underlying principles of this model have been the foundation of many models where economic considerations, namely land rent and distance-decay, are incorporated. The core assumption of the model is that agricultural land use is patterned in the form of concentric circles around a market. Many concordances of this model with reality have been found, notably in North America.
- **The Burgess concentric model** was among the first attempts to investigate spatial patterns at the urban level (1925). Although the purpose of the model was to analyze social classes, it recognized that transportation and mobility were important factors behind the spatial organization of urban areas. The formal land use representation of this model is derived from commuting distance from the CBD, creating concentric circles (Figure 7.7). Each circle represents a specific socioeconomic urban landscape. This model is conceptually a direct adaptation of Von Thunen's model to urban land use since it deals with a concentric representation.

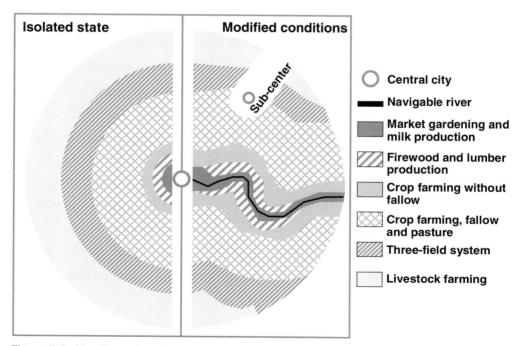

Figure 7.6 Von Thunen's regional land use model

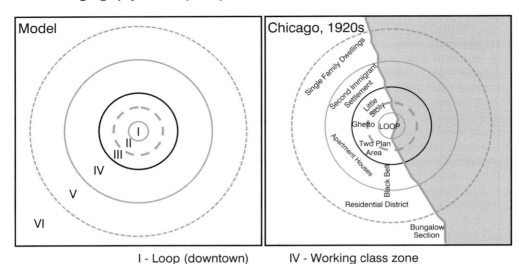

I - Loop (downtown)   IV - Working class zone

II - Factory zone   V - Residential zone

III - Zone of transition VI - Commuter zone

Figure 7.7 The Burgess urban land use model

- **Sector and multiple nuclei land use models** were developed to take into account numerous factors overlooked by concentric models, namely the influence of transport axes and multiple nuclei on land use and growth. Both representations consider the emerging impacts of motorization on the urban spatial structure.
- **Hybrid models** tried to include the concentric, sector and nuclei behavior of different processes in explaining urban land use. They are an attempt to integrate the strengths of each approach since none of these appear to provide a completely satisfactory explanation. Thus, hybrid models, such as that developed by Isard (1955), consider the concentric effect of nodes (CBDs and sub-centers) and the radial effect of transport axes, all overlain to form a land use pattern. Also, hybrid representations are suitable to explain the evolution of the urban spatial structure as they combine different spatial impacts of transportation on urban land use, let them be concentric, radial or nodal, and this at different points in time (Figure 7.8).
- **Land rent theory** was also developed to explain land use as a market where different urban activities are competing for land usage at a location. It is strongly based on the market principle of spatial competition. The more desirable the location, the higher its rent value. Transportation, through accessibility and distance-decay, is a strong explanatory factor on the land rent and its impacts on land use. However, conventional representations of land rent are being challenged by structural modifications of contemporary cities. Figure 7.9 illustrates the basic principles of the land rent theory. It assumes a center which represents a desirable location with a high level of accessibility. The closest area, within a radius of 1 km, has about 3.14 square kilometers of surface ($S = \pi D^2$). Under such circumstances, the rent is a function of the availability of land, which can be expressed in a simple fashion as $1/S$. As we move away from the center the rent drops substantially since the amount of available land increases exponentially.

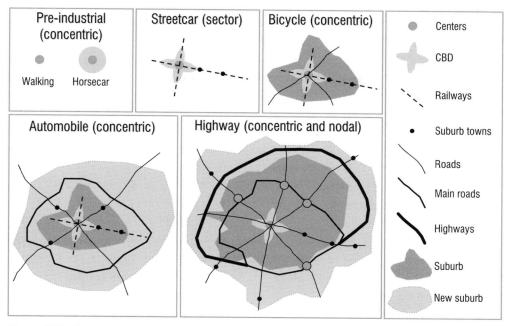

Figure 7.8 Transportation and the constitution of urban landscapes (Source: adapted from Taaffe *et al.*, 1996)

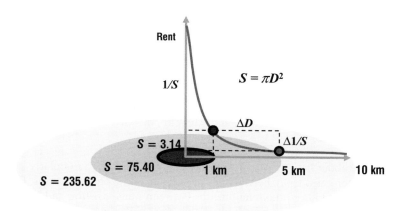

Figure 7.9 Land rent theory and rent curve

Most of these models are essentially static as they explain land use patterns. They do not explicitly consider the processes that are creating or changing them.

Transportation and urban dynamics

Both land use and transportation are part of a dynamic system that is subject to external influences. Each component of the system is constantly evolving due to changes in technology, policy, economics, demographics and even culture/values, among others. As a result, the interactions between land use and transportation are played out as the outcome of the many decisions made by residents, businesses and governments. The field of urban dynamics has extended the scope of conventional land use models, which tended to be descriptive, by trying to consider relationships behind the evolution of the urban spatial structure. This has led to a complex modeling framework including a wide variety of components. Among the concepts supporting urban dynamics representations are retroactions, where one component influences others. The changes then influence the initial component, either positively or negatively. The most significant components of urban dynamics are:

- **Land use**. This is the most stable component of urban dynamics, as changes are likely to modify the land use structure over a rather long period of time. This comes as little surprise since most real estate is built to last at least several decades. The main impact of land use on urban dynamics is its function of a generator and attractor of movements.
- **Transport network**. This is also considered to be a rather stable component of urban dynamics, as transport infrastructures are built for the long term. This is particularly the case for large transport terminals and subway systems that can operate for a very long time. For instance, many railway stations are more than one hundred years old. The main contribution of the transport network to urban dynamics is the provision of accessibility. Changes in the transport network will impact accessibility and movements.
- **Movements**. This is the most dynamic component of the system since movements of passengers or freight reflect changes almost immediately. Movements thus tend more to be an outcome of urban dynamics than a factor shaping them.
- **Employment and workplaces**. They account for significant inducement effects over urban dynamics since many models often consider employment as an exogenous factor. This is specifically the case for employment that is categorized as basic, or export-oriented, which is linked with specific economic sectors such as manufacturing. Commuting is a direct outcome of the number of jobs and the location of workplaces.
- **Population and housing**. They act as the generators of movements, because residential areas are the sources of commuting. Since there is a wide array of incomes, standards of living, preferences and ethnicity, this diversity is reflected in the urban spatial structure.

The issue about how to articulate these relations remains, particularly in the current context of interdependency between local, regional and global processes. Globalization has substantially blurred the relationships between transportation and land use as well as its dynamics. The main paradigm is that factors that used to be endogenous to a regional setting have become exogenous. Consequently, many economic activities that provide employment and multiplying effects, such as manufacturing, are driven by forces that are global in scope and may have little to do with regional dynamics. For instance, capital investment could come from external sources and the bulk of the output could be bound to international markets.

Concept 3 – Urban mobility

Evolution of urban mobility

Rapid urban development occurring across much of the globe implies increased quantities of passengers and freight moving within urban areas. Movements also tend to involve longer distances, but evidence suggests that commuting times have remained relatively similar through the last hundred years, approximately 1.2 hours per day. This means that commuting has gradually shifted to faster transport modes and consequently greater distances could be traveled using the same amount of time. Different transport technologies and infrastructures have been implemented, resulting in a wide variety of urban transport systems around the world. In developed countries, there have been three general eras of urban development, and each is associated with a different form of urban mobility:

- **The walking/horse-car era (1800–90).** Even during the onslaught of the industrial revolution, the dominant means of getting around was on foot. Cities were typically less than 5 kilometers in diameter, making it possible to walk from downtown to the city edge in about 30 minutes. Land use was mixed and density was high (e.g. 100 to 200 people per hectare). The city was compact and its shape was more-or-less circular. The development of the first public transit in the form of an omnibus service extended the diameter of the city but did not change the overall urban structure. The railroad facilitated the first real change in urban morphology. These new developments, often referred to as trackside suburbs, emerged as small nodes that were physically separated from the city itself and from one another. The nodes coincided with the location of rail stations and stretched out a considerable distance from the city center, usually up to a half-hour train ride. Within the city proper, rail lines were also laid down and horse-cars introduced mass transit.
- **The electric streetcar or transit era (1890–1920s).** The invention of the electric traction motor created a revolution in urban travel. The first electric trolley line opened in 1888 in Richmond, Virginia. The operating speed of the electric trolley was three times faster than that of horse-drawn vehicles. The city spread outward 20 to 30 kilometers along the streetcar lines, creating an irregular, star-shaped pattern. The urban fringes became areas of rapid residential development. Trolley corridors became commercial strips. The city core was further entrenched as a mixed-use, high density zone. Overall densities were reduced to between 50 and 100 people per hectare. Land use patterns reflected social stratification where suburban outer areas were typically middle class while the working class continued to concentrate in the central city. As street congestion increased in the first half of the twentieth century, the efficiency of streetcar systems deteriorated and fell out of favor; many were abandoned.
- **The automobile era (1930 onward).** The automobile was introduced in European and North American cities in the 1890s, but only the wealthy could afford this innovation. From the 1920s, ownership rates increased dramatically, with lower prices made possible by Henry Ford's revolutionary assembly-line production techniques. As automobiles became more common, land development patterns changed. Developers were attracted to greenfield areas located between the suburban rail axes, and the public was attracted to these single-use zones, thus avoiding many inconveniences associated with the city, mainly pollution, crowding and lack of space. Transit companies ran into financial difficulties and eventually transit services throughout North America and Europe became subsidized, publicly owned enterprises. As time went on, commercial activities also began to suburbanize. Within a short time, the automobile was the dominant mode of travel in all cities of North America. The automobile has reduced the friction of distance considerably which has led to urban sprawl.

In many areas of the world where urbanization is more recent, the above synthetic phases did not take place. In the majority of cases fast urban growth led to a scramble to provide transport infrastructure, often in an inadequate fashion. Each form of urban mobility, be it walking, the private car or urban transit, has a level of suitability to fill mobility needs. Motorization and the diffusion of personal mobility has been an ongoing trend linked with substantial declines in the share of public transit in urban mobility.

Types of urban movements

Movements are linked to specific urban activities and their land use. Each type of land use involves the generation and attraction of a particular array of movements. This relationship is complex, but is linked to factors such as recurrence, income, urban form, spatial accumulation, level of development and technology. Urban movements are either obligatory, when they are linked to scheduled activities (such as home-to-work movements) or voluntary, when those generating it are free to decide their own scheduling (such as leisure). The most common types of urban movements are (Table 7.1):

- **Pendular**. These are obligatory movements involving commuting between locations of residence and work. They are highly cyclical since they are predictable and recurring on a regular basis, most of the time a daily occurrence, thus the term pendulum.
- **Professional**. These are movements linked to professional, work-based activities such as meetings and customer services, dominantly taking place during work hours.
- **Personal**. These are voluntary movements linked to the location of commercial activities, which includes shopping and recreation.
- **Touristic**. These are important movements for cities having historical and recreational features. They involve interactions between landmarks and amenities such as hotels and restaurants. They tend to be seasonal in nature or occur at specific moments. Major sport events such as the World Cup or the Olympics are important generators of urban movements.
- **Distribution**. These are movements concerned with the distribution of freight to satisfy consumption and manufacturing requirements. They are linked to distribution centers and retail outlets.

The consideration of urban movements involves their generation, the modes and routes used and their destination:

- **Trip generation**. On average, an urban resident undertakes about three or four trips per day. Moving in an urban area is usually done to satisfy a purpose such as employment, leisure or access to goods and services. Each time a purpose is satisfied, a trip is generated. Important temporal variations of the number of trips by purpose are observed.

Table 7.1 Types of urban movements

Movement type	Pattern	Dominant time	Destination
Pendular	Structured	Morning and afternoon	Localized (employment)
Professional	Varied	Workdays	Localized
Personal	Structured	Evening	Varied with some foci
Touristic	Seasonal	Day	Highly localized
Distribution	Structured	Night-time	Localized

- **Modal split**. Implies which transportation mode is used for urban trips and is the outcome of a modal choice. Modal choice depends on a number of factors such as technology, availability, preference, travel time and income.
- **Trip assignment**. Involves which routes will be used for journeys within the city. For instance, a commuter driving a car most of the time has a fixed route. This route may be modified if there is congestion or if another activity (such as shopping) is linked with that trip; this is often known as trip chaining. Several factors influence trip assignment, the two most important being transport costs and availability.
- **Trip destination**. Changes in the spatial distribution of economic activities in urban areas have caused important modifications to the destination of movements, notably those related to work. The central city used to be a major destination for movements, but its share has substantially declined in most areas and suburbs now account for the bulk of urban movements.

The share of the automobile in urban trips varies in relation to location, social status, income, quality of public transit and parking availability. Mass transit is often affordable, but several social groups, such as students, the elderly and the poor, are a captive market. There are important variations in mobility according to age, income, gender and disability. The so-called gender gap in mobility is the outcome of socio-economic differences as access to individual transportation is dominantly a matter of income. Consequently, in some instances modal choice is more a modal constraint linked to economic opportunities.

In central locations, there are generally few transport availability problems because private and public transport facilities are present. However, in locations outside the central core that are accessible only by the automobile, a significant share of the population is isolated if they do not own an automobile. Limited public transit and high automobile ownership costs have created a class of spatially constrained (mobility deprived) people. They do not have access to the services in the suburb, but more importantly to the jobs that are increasingly concentrated in those areas.

Urban transit

Transit is dominantly an urban transportation mode, particularly in large urban agglomerations. The urban environment is particularly suitable for transit because it provides conditions fundamental to its efficiency, namely high density and significant short-distance mobility demands. Since transit is a shared public service, it potentially benefits from economies of agglomeration related to high densities and from economies of scale related to high mobility demands. The lower the density in which a transit system is operating, the lower the demand, with the greater likelihood that it will be run at a loss. In fact, the great majority of public transit systems are not financially sound and have to be subsidized. Transit systems are made up of many types of services, each suitable to a specific set of market and spatial contexts. Different modes are used to provide complementary services within the transit system and in some cases between the transit system and other transport systems.

Figure 7.10 represents an hypothetical urban transit system. Each of its components is designed to provide a specific array of services. Among the defining factors of urban transit are capacity, frequency, flexibility, costs and distance between stops:

- **Metro system**. A heavy rail system, often underground in central areas (parts above ground at more peripheral locations), with fixed routes, services and stations (Plate 7.2). Transfers between lines or to other components of the transit systems

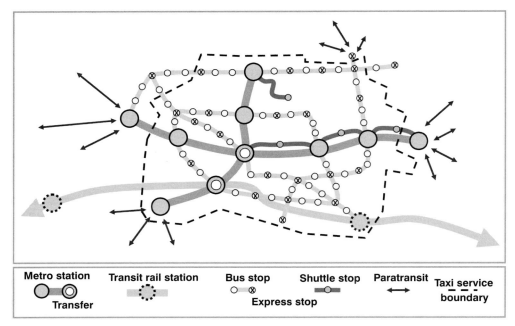

Figure 7.10 Components of an urban transit system

Plate 7.2 Metro station in Bangkok, Thailand. *Credit*: Jean-Paul Rodrigue

(mainly buses and light rail) are made at connected stations. The service frequency tends to be uniform throughout the day, but increases during peak hours. Fares are commonly access driven and constant, implying that once a user has entered the system the distance traveled has no impact on the fare. However, with the computerization of many transit fare systems, zonal/distance driven fares are becoming more common.

- **Bus system**. Characterized by scheduled fixed routes and stops serviced by motorized multiple-passenger vehicles (45–80 passengers). Services are often synchronized with other heavy systems, mainly metro and transit rail, where they act as feeders. Express services, using only a limited number of stops, can also be available, notably during peak hours. Since metro and bus systems are often managed by the same transit authority, the user's fare is valid for both systems.

- **Transit rail system**. Fixed rail comes in two major categories. The first is the tram rail system, which is mainly composed of streetcars (tramways) operating in central areas. They can be composed of up to four cars. The second is the commuter rail system, which are passenger trains mainly developed to service peripheral/suburban areas through a heavy (faster and longer distances between stations) or light rail system (slower and shorter distances between stations). Frequency of services is strongly linked with peak hours and traffic tends to be imbalanced. Fares tend to be separate from the transit system and proportional to distance or service zones.

- **Shuttle system**. Composed of a number of privately (dominantly) owned services using small buses or vans. Shuttle routes and frequencies tend to be fixed, but can be adapted to fit new situations. They service numerous specific functions such as expanding mobility along a corridor during peak hours, linking a specific activity center (shopping mall, university campus, industrial zone, hotel, etc.) or aimed at servicing the elderly or people with disabilities.

- **Paratransit system**. A flexible and privately owned collective demand–response system composed of minibuses, vans or shared taxis commonly servicing peripheral and low density zones. Their key advantage is the possibility of a door-to-door service, less loading and unloading time, less stops and more maneuverability in traffic. In many cities of developing countries this system is informal, dominant and often services central areas because of inadequacies or high costs of the formal transit system.

- **Taxi system**. Comprises privately owned cars or small vans offering an on-call, individual demand–response system. Fares are commonly a function of a metered distance/time, but sometimes can be negotiated. A taxi system has no fixed routes, but is rather servicing an area where a taxi company has the right (permit) to pick up customers. Commonly, rights are issued by a municipality and several companies may be allowed to compete on the same territory. When competition is not permitted, fares are set by regulations.

Contemporary transit systems tend to be publicly owned, implying that many decisions related to their development and operation are politically motivated. This is in sharp contrast to the past when most transit systems were private and profit-driven initiatives. With the fast diffusion of the automobile in the 1950s, many transit companies faced financial difficulties, and the quality of their service declined because in a declining market there were limited incentives to invest. Gradually, they were purchased by public interests and incorporated into large agencies, mainly for the sake of providing mobility. As such, public transit often serves more a social function of public service and a tool of social equity than having any sound economic role. Transit has become dependent on government subsidies, with little if any competition permitted as wages and fares are regulated. As a result, it tends to be disconnected from market forces and subsidies are constantly required to keep a level of service. In suburban areas transit systems tend to be even less related to economic activities.

Reliance on urban transit as a mode of urban transportation tends to be high in Asia, intermediate in Europe and low in North America. Since their inception in the early nineteenth century, comprehensive urban transit systems had significant impacts on the urban form and spatial structure, but this influence is receding. Three major classes of cities can be found in terms of the relationships they have with their transit systems:

- **Adaptive cities**. Represent true transit-oriented cities where urban form and urban land use developments are coordinated with transit developments. While central areas are adequately serviced by a metro system and are pedestrian friendly, peripheral areas are oriented along transit rail lines.
- **Adaptive transit**. Represent cities where transit plays a marginal and residual role and where the automobile accounts for the dominant share of movements. The urban form is decentralized and of low density.
- **Hybrids**. Represent cities that have sought a balance between transit development and automobile dependency. While central areas have an adequate level of service, peripheral areas are automobile-oriented.

Contemporary land development tends to precede the introduction of urban transit services, as opposed to concomitant developments in earlier phases of urban growth. Thus, new services are established once a demand is deemed to be sufficient, often the subject of public pressure. Transit authorities operate under a service warrant and often run a recurring deficit as services are becoming more expensive to provide. This has led to a set of considerations aimed at a higher integration of transit in the urban planning process, especially in North America, where such a tradition is not well established. Still, in spite of decades of investment, North American public transit ridership has roughly remained the same.

From a transportation perspective, the potential benefits of a better integration between transit and local land uses are reduced trip frequency and increased use of alternative modes of travel (i.e. walking, biking and transit). Evidence is often lacking to support such expectations as the relative share of public transit ridership is declining across the board. Community design can consequently have a significant influence on travel patterns. Local land use impacts can be categorized in three dimensions of relationships and are influenced by levels of use. Land use initiatives should be coordinated with other planning and policy initiatives to cope with automobile dependence. However, there is a strong bias against transit in the general population because of negative perceptions, especially in North America, but increasingly globally. As personal mobility is a symbol of status and economic success, the users of public transit are perceived as the least successful segment of the population. This bias may undermine the image of transit use within the general population.

Concept 4 – Urban transport problems

Challenges facing urban transportation

Cities are locations with a high level of accumulation and concentration of economic activities and are complex spatial structures that are supported by transport systems. The most important transport problems are often related to urban areas, when transport systems, for a variety of reasons, cannot satisfy the numerous requirements of urban mobility. Urban productivity is highly dependent on the efficiency of its transport system to move labor, consumers and freight between multiple origins and destinations. Additionally, important transport terminals such as ports, airports and railyards are located within urban areas, contributing to a specific array of problems. Some problems are ancient, like congestion (which plagued cities such as Rome), while others are new, like urban freight distribution or environmental impacts. Among the most notable urban transport problems are:

- **Traffic congestion and parking difficulties**. Congestion is one of the most prevalent transport problems in large urban agglomerations. It is particularly linked with motorization and the diffusion of the automobile, which has increased the demand for transport infrastructures. However, the supply of infrastructures has often not been able to keep up with the growth of mobility. Since vehicles spend the majority of the time parked, motorization has expanded the demand for parking space, which has created a space consumption problem particularly in central areas (Plate 7.3).
- **Public transport inadequacy**. Many public transit systems, or parts of them, are either over or under used. During peak hours, crowdedness creates discomfort for users. Low ridership makes many services financially unsustainable, particularly in suburban areas. In spite of strong subsidies almost every public transit system cannot generate sufficient income to cover its operating and capital costs.

Plate 7.3 Automobiles parked in a public park, Brussels. *Credit*: Jean-Paul Rodrigue

- **Difficulties for pedestrians**. These difficulties are the outcome of intense traffic, where the mobility of pedestrians and vehicles is impaired, and also a blatant lack of consideration for pedestrians in the physical design of facilities.
- **Loss of public space**. The majority of roads are publicly owned and free of access. Increased traffic has adverse impacts on public activities which once crowded the streets such as markets, agoras, parades and processions, games and community interactions. These have gradually disappeared to be replaced by automobiles. In many cases, these activities have shifted to shopping malls, while in other cases they have been abandoned altogether. Traffic flows influence the life and interactions of residents and their usage of street space. More traffic impedes social interactions and street activities. People tend to walk and cycle less when traffic is high.
- **Environmental impacts and energy consumption**. Pollution, including noise, generated by circulation has become a serious impediment to the quality of life and even the health of urban populations. Further, energy consumption by urban transportation has dramatically increased and so also the dependency on petroleum.
- **Accidents and safety**. Growing traffic in urban areas is linked with a growing number of accidents and fatalities, especially in developing countries. Accidents account for a significant share of recurring delays. As traffic increases, people feel less safe to use the streets.
- **Land consumption**. The territorial imprint of transportation is significant, particularly for the automobile. Between 30 and 60 percent of a metropolitan area may be devoted to transportation, an outcome of the over-reliance on some forms of urban transportation.
- **Freight distribution**. Globalization and the materialization of the economy has resulted in growing quantities of freight moving within cities. As freight traffic commonly shares infrastructures with the circulation of passengers, the mobility of freight in urban areas has become increasingly problematic. City logistics strategies can be established to mitigate the variety of challenges faced by urban freight distribution.

Many dimensions to the urban transport problem are linked with the dominance of the automobile.

Automobile dependency

Automobile use is obviously related to a variety of advantages such as on-demand mobility, comfort, status, speed and convenience. These advantages jointly illustrate why automobile ownership continues to grow worldwide, especially in urban areas. When given the choice and the opportunity, most individuals will prefer using an automobile. Several factors influence the growth of the total vehicle fleet, such as sustained economic growth (increase in income and quality of life), complex individual urban movement patterns (many households have more than one automobile), more leisure time and suburbanization. The acute growth in the total number of vehicles also gives rise to congestion at peak traffic hours on major thoroughfares, in business districts and often throughout the metropolitan area.

Cities are important generators and attractors of movements, which have created a set of geographical paradoxes that are self-reinforcing. For instance, specialization leads to additional transport demands while agglomeration leads to congestion. Over time, a state of automobile dependency has emerged which results in a diminution in the role of other modes, thereby limiting still further alternatives to urban mobility. In addition to the factors that contribute to the growth of driving, two major factors contribute to automobile dependency:

- **Underpricing and consumer choices**. Most road infrastructures are subsidized as they are considered a public service. Consequently, drivers do not bear the full cost of automobile use. Like the "Tragedy of the Commons", when a resource is free of access (road), it tends to be overused and abused (congestion). This is also reflected in consumer choice, where automobile ownership is a symbol of status, freedom and prestige, especially in developing countries. Single home ownership also reinforces automobile dependency.
- **Planning and investment practices**. Planning and the ensuing allocation of public funds aim towards improving road and parking facilities in an ongoing attempt to avoid congestion. Other transportation alternatives tend to be disregarded. In many cases, zoning regulations impose minimum standards of road and parking services and *de facto* impose a regulated automobile dependency.

There are several levels of automobile dependency with their corresponding land use patterns and alternatives to mobility. Among the most relevant indicators of automobile dependency are the level of vehicle ownership, *per capita* motor vehicle mileage and the proportion of total commuting trips made using an automobile. A situation of high automobile dependency is reached when more than three-quarters of commuting trips are done using the automobile. For the United States, this proportion has remained around 88 percent over recent decades. Automobile dependency is also served by a cultural and commercial system promoting the automobile as a symbol of status and personal freedom, namely through intense advertising and enticements to purchase new automobiles. Even if automobile dependency is often negatively perceived, and favors market distortions such as the provision of roads, its outcome reflects the choice of individuals who see the automobile more as an advantage than an inconvenience.

The second half of the twentieth century saw the adaptation of many cities in North America and Europe to automobile circulation. Motorized transportation was seen as a powerful symbol of modernity and development. Highways were constructed, streets were enlarged, and parking lots were created, often disrupting the existing urban fabric. However, from the 1980s, motorization started to be seen more negatively and several cities implemented policies to limit automobile circulation by a set of strategies including:

- **Dissuasion**. Although automobile circulation is permitted, it is impeded by regulations and physical planning. For instance, parking space can be severely limited and speed bumps placed to force speed reduction.
- **Prohibition of downtown circulation**. During most of the day the downtown area is closed to automobile circulation but deliveries are permitted during the night. Such strategies are often undertaken to protect the character and the physical infrastructures of a historic city. They do however, like most policies, have unintended consequences. If mobility is restrained in certain locations or during certain periods, people will simply go elsewhere (longer movements) or defer their mobility for another time (more movements).
- **Tolls**. Imposition of tolls for parking and entry to some parts of the city. Most evidence underlines however that drivers are willing to bear additional toll costs, especially when commuting is concerned since it is linked with their main income. Still, congestion pricing is a measure which is increasingly being considered.

Tentative solutions have been put forth such as transport planning measures (synchronized traffic lights, regulated parking), limited vehicle traffic in selected areas, the promotion of bicycle paths and public transit. In Mexico City, vehicle use is prohibited according to license plate numbers and the date (even–uneven). Affluent families have solved this issue by purchasing a second vehicle, thus worsening the existing situation. Singapore is the only country in the world which has successfully controlled the amount and growth rate of its vehicle fleet by imposing a heavy tax burden and the purchase of permits on automobile owners. Such a command-based approach is unlikely to be possible in other contexts.

There are many alternatives to automobile dependency such as intermodality (combining the advantages of individual and collective transport) or carpooling (strengthened by policy and regulation by the US Government). These alternatives, however, can only be partially executed as the automobile remains the prime choice for providing urban mobility. There are however powerful countervailing forces that can influence modal choice, namely congestion.

Congestion

Congestion occurs when transport demand exceeds transport supply in a specific section of the transport system. Under such circumstances, each vehicle impairs the mobility of others.

Recent decades have seen the extension of roads in rural but particularly in urban areas. Those infrastructures were designed for speed and high capacity, but the growth of urban circulation occurred at a rate higher than often expected. Investments came from diverse levels of government with a view to provide accessibility to cities and regions. There were strong incentives for the expansion of road transportation by providing high levels of transport supply. This has created a vicious circle of congestion which supports the construction of additional road capacity and automobile dependency. Urban congestion mainly concerns two domains of circulation, often sharing the same infrastructures:

- **Passengers**. In many regions of the world incomes have significantly increased to the point that one or more automobile per household is common. Access to an automobile conveys flexibility in terms of the choice of origin, destination and travel time. The automobile is favored at the expense of other modes for most trips, including commuting. For instance, automobiles account for the bulk of commuting trips in the United States.
- **Freight**. Several industries have shifted their transport needs to trucking, thereby increasing the usage of road infrastructure. Since cities are the main destinations for freight flows (either for consumption or for transfer to other locations) trucking adds to further congestion in urban areas. The "Last Mile" problem remains particularly prevalent for freight distribution in urban areas. Congestion is commonly linked with a drop in the frequency of deliveries, tying additional capacity to insure a similar level of service.

Infrastructure provision was not able to keep up with the growth in the number of vehicles, even less with the total number of vehicles-km. During infrastructure improvement and construction, capacity impairment (fewer available lanes, closed sections, etc.) favors congestion. Important travel delays occur when the capacity limit is reached or exceeded, which is the case for almost all metropolitan areas. In the largest cities such as London, road traffic is actually slower than it was 100 years ago. Marginal delays are thus increasing and driving speed becomes problematic with the level of density. Large cities have become congested most of the day, and congestion is getting more acute. Another important consideration concerns parking, which consumes large amounts of space. In automobile dependent cities, this can be very constraining as each economic activity has to provide an amount of parking space proportional to their level of activity. Parking has become a land use that greatly inflates the demand for urban land.

Daily trips can be either "mandatory" (workplace–home) or "voluntary" (shopping, leisure, visits). The former is often performed within fixed schedules while the latter comply with variable and discretionary schedules. Correspondingly, congestion comes in two major forms:

- **Recurrent congestion**. The consequence of factors that cause regular demand surges on the transportation system, such as commuting, shopping or weekend trips. However, even recurrent congestion can have unforeseen impacts in terms of its duration and severity. Mandatory trips are mainly responsible for the peaks in circulation flows, implying that about half the congestion in urban areas is recurring at specific times of the day and on specific segments of the transport system.
- **Non-recurrent congestion**. The other half of congestion is caused by random events such as accidents and unusual weather conditions (rain, snowstorms, etc.), which are unexpected and unplanned. Non-recurrent congestion is linked to the presence and effectiveness of incident response strategies. As far as accidents are concerned, their randomness is influenced by the level of traffic as the higher the traffic on specific road segments the higher the probability of accidents.

Behavioral and response time effects are also important; in a system running close to capacity, simply breaking suddenly may trigger what can be known as a backward traveling wave. It implies that as vehicles are forced to stop the bottleneck moves up from where the incident initially took place, often leaving drivers puzzled about its cause. The spatial convergence of traffic causes a surcharge on transport infrastructures up to the point where congestion can lead to the total immobilization of traffic. Not only does the massive use of the automobile have an impact on traffic circulation and congestion, but it also leads to the decline in public transit efficiency when both are sharing the same roads.

In some areas, the automobile is the only mode for which infrastructures are provided. This implies less capacity for using alternative modes such as transit, walking and cycling. At some levels of density, no public infrastructure investment can be justified in terms of economic returns. Longer commuting trips in terms of average travel time, the result of fragmented land uses and congestion levels are a significant trend. So is convergence of traffic at major highways that serve vast low density areas with high levels of automobile ownership and low levels of automobile occupancy. The result is energy (fuel) wasted during congestion (additional time) and supplementary commuting distances. In automobile dependent cities, a few measures can help alleviate congestion to some extent:

- **Ramp metering**. Controlling the access to a congested highway by letting automobiles in one at a time instead of in groups. The outcome is a lower disruption on highway traffic flows.
- **Traffic signal synchronization**. Tuning the traffic signals to the time and direction of traffic flows.
- **Incident management**. Making sure that vehicles involved in accidents or mechanical failures are removed as quickly as possible from the road.

- **HOV lanes**. High Occupancy Vehicle lanes insure that vehicles with two or more passengers (buses, vans, carpool, etc.) have exclusive access to a less congested lane.
- **Congestion pricing**. A variety of measures aimed at imposing charges on specific segments or regions of the transport system, mainly as a toll. The charges can also change during the day to reflect congestion levels so that drivers are encouraged to consider other time periods or other modes.
- **Public transit**. Offering alternatives to driving that can significantly improve efficiency, notably if it circulates on its own infrastructure (subway, light rail, buses on reserved lanes, etc.).

All these measures only partially address the issue of congestion; they alleviate, but do not solve, the problem. Fundamentally, congestion remains a failure at reconciling mobility demands and acute supply constraints.

The urban transit challenge

As cities continue to become more dispersed, the cost of building and operating public transportation systems increases. For instance, only about 80 large urban agglomerations have a subway system, the great majority of them being in developed countries. Furthermore, dispersed residential patterns characteristic of automobile dependent cities makes public transportation systems less convenient to support urban mobility. In many cities additional investments in public transit did not result in significant additional ridership. Unplanned and uncoordinated land development has led to rapid expansion of the urban periphery. Residents, by selecting housing in outlying areas, restrict their potential access to public transportation. Over-investment (when investments do not appear to imply significant benefits) and under-investment (when there is a substantial unmet demand) in public transit are both complex challenges.

Urban transit is often perceived as the most efficient transportation mode for urban areas, notably large cities. However, surveys reveal a stagnation or a decline of public transit systems, especially in North America. The economic relevance of public transit is being questioned. Most urban transit developments had little, if any, impacts to alleviate congestion in spite of mounting costs and heavy subsidies. This paradox is partially explained by the spatial structure of contemporary cities which are oriented to servicing the needs of the individual, not necessarily the needs of the collectivity. Thus, the automobile remains the preferred mode of urban transportation. In addition, public transit is publicly owned, implying that it is a politically motivated service that provides limited economic returns. Even in transit-oriented cities such as in Europe, transit systems depend massively on government subsidies. Little or no competition is permitted as wages and fares regulated, undermining any price adjustments to changes in ridership. Thus, public transit often serves the purpose of a social function ("public service") as it provides accessibility and social equity, but with limited relation to economic activities. Among the most difficult challenges facing urban transit are:

- **Decentralization**. Public transit systems are not designed to service low density and scattered urban areas that are increasingly dominating the landscape. The greater the decentralization of urban activities, the more difficult and expensive it becomes to serve urban areas with public transit. Additionally, decentralization promotes long-distance trips on transit systems.
- **Fixity**. The infrastructures of several public transit systems, notably rail and subway systems, are fixed, while cities are dynamical entities, even if the pace of change can take decades. This implies that travel patterns tend to change and that

a transit system built for servicing a specific pattern may eventually face "spatial obsolescence".

- **Connectivity**. Public transit systems are often independent from other modes and terminals. It is consequently difficult to transfer passengers from one system to the other.
- **Competition**. In view of cheap and ubiquitous road transport systems, public transit faced strong competition and loss of ridership in relative terms and in some cases in absolute terms. The higher the level of automobile dependency, the more inappropriate the public transit level of service. The public service being offered is simply outpaced by the convenience of the automobile. However, changes in energy prices are likely to impose a new equilibrium in this relationship.
- **Financing and fare structures**. Most public transit systems have abandoned a fare structure to a simpler flat fare system. This had the unintended consequence of discouraging short trips for which most transit systems are well suited, and encouraging longer trips. Information systems offer the possibility for transit systems to move back to a more equitable distance-based fare structure.

Method 1 – Traffic counts and traffic surveys

Traffic counting methods

Transport planning at all levels requires understanding of actual conditions. This involves determination of vehicle or pedestrian numbers, vehicle types, vehicle speeds, vehicle weights, as well as more substantial information such as trip length and trip purpose and trip frequency. The first group of data dealing with the characteristics of vehicle or people movement is obtained by undertaking traffic counts. Those related to measuring trips involving knowledge of origin and destination require more detailed surveys.

There is a wide range of counting methods available. It is useful to distinguish between intrusive and non-intrusive methods. The former include counting systems that involve placing sensors in or on the roadbed; the latter involve remote observational techniques. In general the intrusive methods are used most widely because of their relative ease of use and because they have been employed for decades. The only widely used non-intrusive method is manual counting, which enjoys wide application because of its ease. Intrusive methods, however, have evolved little over the last decade, but in the USA, with federal transport policy emphasis on IT solutions to traffic management, progress is being made in the development of non-intrusive methods. The major intrusive methods include:

- **Bending plate**. A weight pad is attached to a metal plate embedded in the road to measure axle weight and speed. It is an expensive device and requires alteration to the roadbed.
- **Pneumatic road tube**. A rubber tube is placed across the lanes and uses pressure changes to record the number of axle movements in a counter placed on the side of the road. The drawback is that it has limited lane coverage, may become displaced and can be dislodged by snow ploughs.
- **Piezo-electric sensor**. A device is placed in a groove cut into the roadbed of the lane(s) being counted. This electronic counter can be used to measure weight and speed. Cutting into the roadbed can affect the integrity of the roadbed and decrease the life of the pavement.

- **Inductive loop**. A wire is embedded in the road in a square formation which creates a magnetic field that relays the information to a counting device at the side of the road. This has a generally short life expectancy because it can be damaged by heavy vehicles, and is also prone to installation errors.

The major non-intrusive methods include:

- **Manual observation**. A very traditional method that involves placing observers at specific locations to record vehicle or pedestrian movements. At its simplest, observers use tally sheets to record numbers, on the other hand mechanical and electronic counting boards are available that the observer can punch in each time an event is observed. Traffic numbers, type and directions of travel can be recorded. Manual counts give rise to safety concerns, either from the traffic itself or the neighborhoods where the counts are being undertaken.
- **Passive and active infra-red**. A sensor detects the presence, speed and type of vehicles by measuring infra-red energy radiating from the detection area. Typically the devices are mounted overhead on a bridge or pylon. The major limitation is the performance during inclement weather, and limited lane coverage.
- **Passive magnetic**. Magnetic sensors that count vehicle numbers, speed, and type are placed under or on top of the roadbed. In operating conditions the sensors have difficulty differentiating between closely spaced vehicles.
- **Microwave Doppler/radar**. Mounted overhead, the devices record moving vehicles and speed. With the exception of radar, devices have difficulty in detecting closely spaced vehicles and do not detect stationary vehicles. They are not affected by weather.
- **Ultrasonic and passive acoustic**. These devices use sound waves or sound energy to detect vehicles. Those using ultrasound are placed overhead to record vehicle presence but can be affected by temperature and turbulence; the acoustic devices are placed alongside the road and can detect numbers and vehicle type.
- **Video image detection**. The use of overhead video cameras to record vehicle numbers, type and speed. Various software is available to analyze the video images. Weather may limit accuracy.

A recent study that examined the use of the various traffic count methods by State Departments of Transport in the USA found that less than half use any non-intrusive techniques. Part of the reason is the level of technical expertise required to operate the devices. Inductive loops are in use in all states, with very high levels of use (> 90 percent) for pneumatic rubber tubes and piezo-electric road sensors. Manual counts were used by 82 percent of states. In terms of satisfaction with the methods, manual counts and inductive loops were rated highest. Despite the poor acceptance of the non-intrusive devices, their cost-effectiveness was shown to be higher than the inductive loops. This suggests that the newer devices may gain wider use once their cost-effectiveness becomes more widely appreciated.

Surveys

Traffic counts may provide some precise information about numbers of vehicles, their type, weight or speed, but they cannot provide other data that are essential in transport planning, such as trip purpose, routing, duration, etc. Collecting these data requires more extensive survey instruments. These instruments include:

- **Mailed questionnaires**. These can include a wide range of questions. They are relatively cheap to administer to large numbers of people, although preparation can be expensive. The main problem is the generally low response rate.
- **Travel diaries**. Respondents are asked to keep a diary of the trips undertaken, times, purposes, modes, etc. Extremely useful instrument constrained largely by the number of people willing to complete such a detailed inventory.
- **Telephone surveys**. With automated dialing this can achieve extensive coverage, but response rates are usually low.
- **Face-to-face home interviews**. These can overcome many of the errors based on misunderstanding of questions in mail surveys, but are extremely time-consuming and costly.

Extensive traffic surveys began to be developed in the 1950s. One of the earliest was the Chicago Area Transportation Study (CATS), undertaken in 1956; it provided detailed O/D data on trip length, purposes, modes of travel and travel patterns. This was followed in 1960 with the US Census's first attempt to collect journey to work (JTW) travel data in urban areas. Other metropolitan areas in the USA and Canada, including Detroit and Toronto, copied and extended the scope of such surveys in the 1960s. The growth of surveys was encouraged by the results that provided the first comprehensive snapshots of urban travel activities in a society rapidly adopting the automobile and undertaking new types of travel behavior. This was a boon to transport planning. Furthermore, much of academic understanding of travel activity in cities has been drawn from these surveys. Since then national censuses in many countries have included travel surveys in their decennial inventories, and many planning agencies update and extend the results from the national surveys with local investigations (see below).

All survey techniques represent a compromise between the objectives of the survey, the resources available, the coverage that is feasible and the amount of data to be collected. The survey instrument(s) that are employed depend largely on the resources available. Even national agencies find the costs of conducting national surveys onerous. Very common is the mail-back questionnaire. CATS, for example, uses a questionnaire along with a travel diary, which involves sending out a letter of introduction to selected households, distribution of the questionnaire and instructions, mailing out reminder letters and a telephone follow-up to selected individuals to verify their information.

The degree of detail required in most travel surveys means that even the largest agencies have to rely on sampling. It is usual to target households rather than individuals, since the household is a good predictor of travel behavior. Fixing the size of the sample is an extremely important issue. Sample size determines the degree of reliability of the results, but these have to be conditioned by the resources available and the survey instruments to be employed. In its household surveys, CATS determined that 400 completed household responses would be sufficient to provide a statistically significant sample for each of the geographic units, and because it expected a 20 percent rate of response, it could plan for the distribution of 2,000 questionnaires in each zone. A clustered random sample of approximately 2,000 addresses in each zone was taken. For national surveys in the USA, samples of 26,000 households are sought. Because national surveys may not provide a sufficiently reliable or detailed set of data for the needs of individual states or planning agencies, these agencies frequently "back-on" additional counts in their areas when national surveys are undertaken.

The main problems encountered in traffic surveys are:

- **Comparability between surveys**. It is usually very important to compare survey results over time. This is frequently very difficult because of different sample sizes, different questions, different response rates and different geographical collection units. These are usually major problems for studies trying to compare the results from different agencies.
- **Non-response bias**. There are significant variations in the response rates achieved by surveys. The larger the non-response rate, the less reliable will be the results. A 60 percent response rate is sometimes considered as a threshold. Many surveys fail to achieve high rates of response, for example the 2001 National Household Travel Survey (NHTS) only achieved 41 percent.
- **Coverage bias**. The survey instruments frequently contain hidden biases. For example automatic telephone surveys exclude cell phone users and those without a land line connection.
- **Unreporting of trips**. Research is now showing that surveys and travel diaries may be undercounting trips made. Some test surveys are using GPS devices to record trips and indicate that in the Kansas City survey 10 percent of trips were unreported and in the case of Laredo the figure was as high as 60 percent.

Method 2 – Transportation/land use modeling

Types of models

> Essentially, all models are wrong but some are useful.
>
> (George Box)

To gain a better understanding of the behavior of urban areas, several operational transportation/land use models (TLUM) have been developed. The reasons behind using TLUM are numerous, such as the ability to forecast future urban patterns based on a set of economic assumptions or to evaluate the potential impacts of legislation pertaining to environmental standards. Other uses of TLUM relate to testing theories, policies and practices about urban systems. With a simulation model, urban theories can be evaluated and the impacts of policy measures, such as growth management and congestion pricing, can be measured. It is not surprising that since TLUM are planning tools *per se*, their development and application has mainly been done by various government agencies related to transportation, regional planning and the environment.

Broadly taken, a model is an information construct used to represent and process relationships between a set of concepts, ideas and beliefs. Models have a language, commonly mathematics (expressed as functions in various computer programming languages), an intended use and a correspondence to reality. There are four levels of complexity related to the modeling transportation/land use relationships:

- **Static modeling**. Expresses the state of a system at a given point in time through the classification and arithmetic manipulation of representative variables. Measuring accessibility can be considered as static modeling.
- **System modeling**. Expresses the behavior of a system with a given set of relationships between variables. The gravity model is an example of system modeling as it tries to evaluate the generation and attraction of movements.
- **Modeling interactions between systems**. Tries to integrate several models to form a meta-system (a large and complex system). A transportation/land use model offers such a perspective.

- **Modeling in a decision-taking environment**. This not only implies the application of a transportation/land use model, but also the analysis and reporting of its results in order to find strategies and recommendations. Geographic Information Systems are useful tools for that purpose as they can include the modeling and its graphic display, as well as being the platform over which decision making can take place.

On average, models tend to be relevant for constrained and well-structured problems with a specified number of variables, well-defined goals and firmly established technical solutions. This in itself limits significantly the applicability of TLUM as urban systems are complex entities. Still, these models have pros and cons:

- **Advantages**. They incite a conceptualization of urban economic and spatial processes. Advances in urban and regional sciences are often linked with advances in modeling, notably conceptual representations such as land economics. Moving from an urban concept to an urban model is simply a step forward, albeit an important one. The data requirements of TLUM are often an incentive to perform surveys from which useful information about urban mobility and spatial structure can be gathered. This information has the additional advantage of triggering studies that are not necessarily related to TLUM, but that contribute to advances in the understanding of the dynamics of urban systems.
- **Drawbacks**. Models may gear towards a mechanistic approach of urban dynamics where processes are compartmentalized. This leads to difficulties about "thinking outside the box" and by its nature modeling often fails to grasp significant economic, technological and social changes. They may also give the impression that a system can effectively be controlled since all its major elements have been summarized; solving a problem is thus a matter of tweaking parameters.

A transportation/land use system can be divided in to three subcategories of models (Figure 7.11):

- **Land use models** are generally concerned about the spatial structure of macro- and micro-economic components, which are often correlated with transportation requirements. For instance, by using a set of economic activity variables, such as population and level of consumption, it becomes possible to calculate the generation and attraction of passengers and freight flows.
- **Spatial interactions models** are mostly concerned about the spatial distribution of movements, a function of land use (demand) and transportation infrastructure (supply). They produce flow estimates between spatial entities, symbolized by origin–destination pairs, which can be disaggregated by nature, mode and time of day.
- **Transportation network models** try to evaluate how movements are allocated over a transportation network, often of several modes, notably private and public transportation. They provide traffic estimates for any given segment of a transportation network.

To provide a comprehensive modeling framework, all these models must share information to form an integrated transportation/land use model. For instance, a land use model can calculate traffic generation and attraction, which can be inputted in a spatial interaction model. The origin–destination matrix provided by a spatial interaction model can be inputted in a traffic assignment model, resulting in simulated flows on the transportation network.

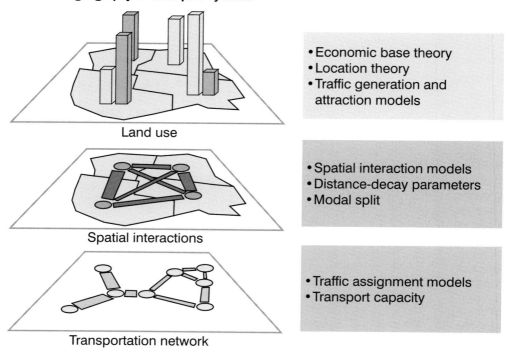

Land use

- Economic base theory
- Location theory
- Traffic generation and attraction models

Spatial interactions

- Spatial interaction models
- Distance-decay parameters
- Modal split

Transportation network

- Traffic assignment models
- Transport capacity

Figure 7.11 Components of the transportation/land use system

Four-stage transportation/land use modeling

The core foundation of TLUM involves two components. The land use component, which is based on the location of housing, industrial and commercial activities, tends to be more stable than the transportation component which is highly dynamic. Most TLUM has been applied regionally, mainly at the urban level, as a larger scale would be prohibitively complex to model. The modeling of the transportation components is particularly relevant and is divided in to four sequential stages for the estimation of travel demand: where movements originate, how they are allocated, what modes are used and finally what segments of the transport network are being used (Figure 7.12):

- The first stage is called **trip generation** and deals with trip rate estimates, usually at the zonal level. The most common methods for trip generation are cross-classification (also referred to as category analysis) and multiple regression analysis. Cross-classification seeks to identify specific socio-economic groups within the population that have common trip generation characteristics. The trip generation of a zone will thus be the outcome of its composition. Regression analysis estimates the number of trips generated by a zone (dependent variable) as a function of a series of independent variables.
- The second stage is referred to as **trip distribution** and deals with spatial movement patterns; the links between trip origins and destinations. The most common technique for estimating trip distribution is the gravity model. There are various forms

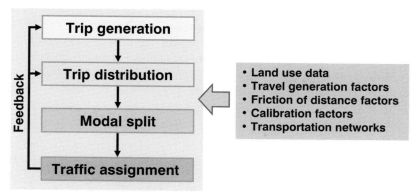

Figure 7.12 Four-stage transportation/land use model

of the gravity model and various calibration techniques as well. Cross-classification and multiple regression can also be used to estimate the number of trips a zone would attract.

- The third stage is **modal split**: the proportion of trips made by automobile drivers and passengers, transit passengers, cyclists and walkers. Logit modeling is commonly used as it evaluates the preference of each user in terms of probability of using a specific mode for a specific origin/destination pair.
- Finally, once the spatial patterns of movements by various modes are estimated, trips are **assigned** to the various transport links. This is done mostly by using operations research methods aiming at minimizing travel costs or time over a transport network.

Data requirements

Applying TLUM requires an extensive range of data, most of it related to spatial units, land use, spatial interactions and the transportation network. The most important information for TLUM is however origin–destination data. A variety of survey methods are used to collect these data including roadside questionnaires, telephone interviews and detailed activity modeling. Data availability and limitation are important factors behind the applicability of such models and there is a constant trade-off between the costs of fulfilling the data requirements and the benefits supplementary data may offer. Additionally, data need to constantly be updated as demographic, economic and technological changes take place. This is one of the major reasons why the transportation/land use modeling process, although theoretically and conceptually sound, has not been applied comprehensively. Among the major types of variables, it is possible to identify:

- **Land use data**. Include socio-economic variables pertaining to the area under investigation, such as population, employment, income level, commercial activity, etc. Such data are used to estimate or calibrate the amount of travel generated and attracted by each zone.
- **Travel generation factors**. Considering the available land use data, these factors estimate the number of trips, people and/or freight that each level of economic activity generates. They consider a multitude of issues such as income, modal preferences

and consumption levels. Most of this information can be gathered using surveys or inferring from observations made elsewhere.

- **Friction of distance factors**. They represent the difficulty of traveling between different locations of the area under investigation, commonly measured in terms of time, distance or cost. There is a significant variance according to mode and purpose of travel. Friction of distance factors enable the assessment of trip distribution and modal split.
- **Calibration factors**. It is uncommon for the results produced by an uncalibrated model to correspond to the reality. Calibration factors thus try to match the results produced by the model with data based on observations, surveys or common sense. Calibration can often be an obscure process, because it tries to incorporate factors that are not explained by the model itself.
- **Transportation networks**. A representation of the structure and geometry of transportation within the area under investigation, mainly composed of nodes and links. Transportation networks are commonly divided by modes. For road transportation, a node could represent an intersection, a stop or a parking lot, while a segment could be linked with attributes such as permitted speed, distance and capacity. For public transit, a node could represent a bus stop or a metro station, while a segment could have attributes such as capacity and frequency of service. Transportation networks, along with origin–destination matrices, are fundamental elements of the traffic assignment procedure.

The core of most transportation/land use models is some kind of regional economic forecast that predicts and assigns the location of the basic employment sector. As such, they are dependent on the reliability and accuracy of macro- and micro-economic forecasting. Traditionally, such forecasting tends not to be very accurate as it fails to assess the impacts of economic, social and technological changes. For instance, globalization and the emergence of global commodity chains have significantly altered the dynamics of regional economies.

Additionally, few TLUM deal with freight transportation. This can be explained by the fact that passenger transportation in urban areas tends to be highly regulated by governmental agencies (e.g. public transit) while freight transportation is dominantly controlled by private entities. Paradoxically, while freight-related activities such as terminals and distribution centers tend to occupy a large amount of space, they do not generate a large amount of passenger traffic.

CASE STUDY **City logistics**

City logistics is a relatively new field of investigation brought by the challenges of moving growing quantities of freight within metropolitan areas. While cities, particularly since the industrial revolution, have always been important producers and consumers of freight, many of these activities took place in proximity to major transport terminals, such as ports and railyards, with limited quantities of freight entering the city *per se*. The functional specialization of cities and the global division of production, as well as increasing standards of living, are all correlated with larger quantities of freight coming from, bound to or transiting through urban areas. According to the Institute of City Logistics, city logistics is "the process for totally optimizing the logistics and transport activities by private companies in urban areas while considering the traffic environment, the traffic congestion and energy consumption within the framework of a market economy". In simple terms, it concerns the means to achieve freight distribution in urban areas, by improving the efficiency of urban freight transportation, reducing traffic congestion and mitigating environmental impacts. Addressing city logistics requires an understanding of urban geography as well as supply chain management. Urban freight distribution thus has a unique set of challenges (Table 7.2).

By its characteristics, urban freight distribution reflects many dimensions of contemporary logistics and exacerbates many of its constraints because of the nature of congestion in urban areas. Particularly, urban freight distribution is subject to smaller volumes with time-sensitive freight necessary to replenish a constant demand. Repetitiveness is a salient issue as a regular flow of deliveries must be maintained in spite of peak hour congestion, and therefore many freight distribution activities take place during the night if possible. Since urban areas are large consumers of final goods, the issue of reverse logistics deserves attention in the form of the collection of wastes and recycling. The diffusion of e-commerce has also created new forms of demands and new forms of urban distribution with a growth in the home deliveries of parcels. From a regulatory perspective urban areas are highly constrained with a variety of rules related to zoning, emissions and even access conditions to road and terminals. High population densities imply a low tolerance for infringements and disturbances.

City logistics, as a distributional strategy, can take many forms. For instance, a high density and congested central city can be serviced by an independent freight distribution system calling from a terminal located at the margin of the area. The vehicles used to service the customers (either for deliveries or pickups along a flexible route) are likely to be smaller and thus better adapted for distribution in an urban environment. There is also the possibility of using the existing public transit system to move freight but this implies several challenges in terms of the adaptation of modes, the usage of existing passenger terminals and scheduling issues. The urban terminal itself could be a neutral facility interfacing with a set of distribution centers, each being connected to their respective supply chains. Thus, a wide array of supply chains connected to the city can achieve a better distributional efficiency within the central city.

Table 7.2 Key issues in urban freight distribution

Issue	Challenge
Increasing volume of freight moving in urban areas	Capacity of urban freight transport systems
Changes in the nature of freight distribution	Smaller volumes and time-sensitive freight
Repetitiveness	Many urban activities (retail, groceries and catering) require daily deliveries
Environmental issues	Growing demand for reverse logistic flows (waste and recycling)
Emergence of e-commerce	Growth in home deliveries
Congestion	Lower driving speeds and frequent disruptions (reliability). Peak hour interferences
Regulation	Emissions, access and zoning

Bibliography

Barry, M. (1991) *Through the Cities: The Revolution in Light Rail Transit*, Dublin: Frankfort Press.

Batty, M. and Y. Xie (1994) "From Cells to Cities", *Environment and Planning B*, 21, 531–48.

BTS (2001) Special Issue on Methodological Issues in Accessibility, *Journal of Transportation and Statistics*, 4(2/3), Bureau of Transportation Statistics (http://www.bts.gov), Sept./Dec.

Carter, H. (1995) *The Study of Urban Geography*, 4th edn, London: Arnold.

Cervero, R. (1998) *The Transit Metropolis: A Global Inquiry*, Washintgon, DC: Island Press.

Cox, W. (1998) "Light Rail in Minneapolis: A Bridge to Nowhere", *The Public Purpose, Urban Transport Fact Book*. http://www.publicpurpose.com/ut-mspsp.htm.

Dimitriou, H. (1993) *Urban Transport Planning*, New York: Routledge.

Environmental Protection Agency (1997) *Evaluation of Modeling Tools for Assessing Land Use Policies and Strategies*, EPA420-R-97-007, Ann Arbor, MI: EPA.

Ewing, R. (1993) "Transportation Service Standards – As If People Matter", *Transportation Research Record*, 1400 (www.trb.org), pp. 10–17.

FHWA (2001) *Transportation Performance Measures Toolbox, Operations*, Federal Highway Administration (www.ops.fhwa.dot.gov/travel/deployment_task_force/perf_measures.htm).

Gwilliam, K. (ed.) (2001) *Cities on the Move: A World Bank Urban Transport Strategy Review*, Strategy Paper, Washington, DC: World Bank. http://wbln0018.worldbank.org/transport/utsr.nsf.

Hanson, S. (ed.) (1995) *The Geography of Urban Transportation*, 2nd edn, New York: Guilford Press.

Harvey, J. (1996) *Urban Land Economics*, Houndsmills: Macmillan.

Hugill, P.J. (1995) *World Trade Since 1431*, Baltimore, MD: Johns Hopkins University Press.

Isard, W. (1956) *Location and Space-Economy*, Cambridge, MA: MIT Press.

Kauffman, R.J. (2001) *Paving the Planet: Cars and Crops Competing for Land*, Alert, Worldwatch Institute.

Moore, T. and P. Thorsnes (1994) *The Transportation/Land Use Connection*, Report no. 448/449, Washington, DC: American Planning Association.

Muller, P.O. (1995) "Transportation and Urban Form: Stages in the Spatial Evolution of the American Metropolis", in S. Hanson (ed.) *The Geography of Urban Transportation*, 2nd edn, New York: Guilford, pp. 26–52.

Newman, P. and J. Kenworthy (1996) "The Land Use–Transport Connection", *Land Use Policy*, 1, 1–12.

Newman, P. and J. Kenworthy (1999) *Sustainability and Cities: Overcoming Automobile Dependence*, Washington, DC: Island Press.

Rietveld, P. (2000) "Nonmotorized Modes in Transport Systems: A Multimodal Chain Perspective for The Netherlands", *Transportation Research D*, 5(1), 31–6.

Schafer, A. (2000) "Regularities in Travel Demand: An International Perspective", *Journal of Transport Statistics*, 3(3), http://www.bts.gov/jts/V3N3/schafer.pdf.

Taaffe, E.J., H.L. Gauthier and M.E. O'Kelly (1996) *Geography of Transportation*, Upper Saddle River, NJ: Prentice Hall.

Texas Transportation Institute (2008) *The 2007 Annual Mobility Report*, College Station, Texas. http://mobility.tamu.edu/.

Thomson, J.M. (1977) *Great Cities and Their Traffic*, London: Victor Gollancz.

Torrens, P.M. (2000) *How Land Use–Transportation Models Work*, Working Paper 20, Centre for Advanced Spatial Analysis, University College London, http://www.casa.ucl.ac.uk/working_papers/paper20.pdf.

TRB (1994) *Highway Capacity Manual*, Special Report 209, Transportation Research Board (www.trb.org).

Victoria Transport Policy Institute (2002) "Automobile Dependency", *Transport Demand Management Encyclopedia*, http://www.vtpi.org/tdm/tdm100.htm.

⬤8 Transportation, energy and environment

Until the 1990s, environmental concerns played a small role in transport infrastructure planning and operations. This situation has changed. The future of the transport industry is likely to be compromised without an understanding of environmental sustainability. The opportunities to participate in the sustainable development of transportation are likely to be manifold in the future. Paradoxically, geographers' capacity for understanding the rapidly changing environment has not been growing at the same pace as the provision of a knowledge base which the transport industry requires. Thus, there is a pressing need to re-equip a new generation of transport geographers with the necessary skills to apply sustainability issues to transportation. Three concepts are at the heart of the relationship between transport and the environment: transport and energy, the reciprocal influence of transport and the physical environment, and sustainable transport.

Concept 1 – Transport and energy

Energy

Human activities are closely dependent on the use of several forms and sources of energy to perform work. Energy is the potential that allows movement and/or modification of matter. Energy content is the available energy per unit of weight or volume from an energy source. Thus, the more energy consumed the greater the amount of work realized. Wood, coal, petroleum oils and natural gas are fossil fuels, whereas human and animal power, wind and water power and solar radiation are actual sources of energy. There are enormous reserves of energy able to meet the future needs of mankind. Unfortunately, one of the main contemporary issues is that many of these reserves cannot be exploited at reasonable costs or are unevenly distributed around the world. From the earliest time, people's choice of energy source depended on a number of utility factors. Since the industrial revolution, people have used fuels to provide steam power and electrical power. This has considerably improved industrial productivity by having as much work as possible being performed by machines. The development of the steam engine and the generation and distribution of electric energy over considerable distances have also altered the spatial pattern of manufacturing industries by liberating production from direct connection to a fixed power system. Industrial development generates enormous demands for fossil fuels.

At the turn of the twentieth century, the invention and commercial development of the internal combustion engine, notably in transport equipment, made possible the efficient movement of people, freight and information and stimulated the development of the global trade network. With globalization, transportation accounts for a growing share of the total amount of energy spent for implementing, operating and maintaining the international range and scope of human activities. At the beginning of the twenty-first century, despite growing supply and pricing uncertainties, fossil fuels, notably petroleum, remain the world's chief sources of energy with a production level estimated at 85 million barrels per day. Out of the world's power consumption of about 14 trillion watts a year, approximately 85 percent is derived from fossil fuels.

Transportation and energy consumption

Energy consumption has become a major focus of the global economy. There exists a strong correlation between energy consumption and level of economic development. Historically, high per capita energy consumption is associated with high income, relatively low energy prices and the need to move people, commodities and information. Among developed countries, transportation now accounts for 20–25 percent of all the energy being consumed. With less than 5 percent of the world's population, the United States consumes approximately 65 percent of all the transportation energy among G8 countries. The increasing motorization and the concomitant rise in land and air traffic in countries such as China, India, Russia and Brazil are stimulating growth in all aspects of the transportation industry.

The impact of transport on energy consumption is diverse, including many factors necessary for the provision of transport facilities:

- **Vehicle manufacture, maintenance and disposal**. The energy spent for manufacturing and recycling vehicles is a direct function of vehicle complexity, material used, fleet size and vehicle life cycle.
- **Vehicle operation**. Mainly involves the energy used to provide power for vehicles.
- **Transportation infrastructure construction and maintenance**. The building of roads, railways, bridges, tunnels, terminals, ports and airports and the provision of lighting and signaling equipment require a substantial amount of energy. They have a direct relationship with vehicle operations since extensive networks are associated with large amounts of traffic.
- **Administration of transport business**. The expenses involved in planning, developing and managing transport infrastructures and operations involves time, capital and skill that must be included in the total energy consumed by the transport sector.
- **Energy production and trade**. The processes of exploring, extracting, refining and distributing fuels or generating and transmitting energy also require power sources. The transformation of 100 units of primary energy in the form of crude oil produces only 85 units of energy in the form of gasoline. Any changes in transport energy demands influence the pattern and flows of the world energy market.

This close relationship between transport and energy is subject to different interpretations. As a generalization, it is possible to compare the costs of hauling passengers or commodities by ship, rail, road and air by expressing the costs to a common unit such as energy use per unit of transport production. Such comparison must be handled with care however as the actual passenger or ton-kilometers cost of an individual transport operation is influenced by a variety of factors such as distance, route characteristics, load factors, cargo value or value of service, rate structures, terminal charges, etc. Further, in these comparisons note has to be taken of fuel efficiency, congestion levels and environmental externalities.

- **Land transportation** accounts for the great majority of energy consumption. Road transportation alone consumes almost 80 percent of the total energy used by the transport sector. Fuel costs for the North American trucking industry account for a third of its expenses. In land transport, road is almost the mode mainly responsible for additional energy demands over the last 25 years because its market share of freight and passenger transport has increased. Despite a falling market share, rail transport, on the basis of 1 kg of oil equivalent, remains four times more efficient for passenger and twice as efficient for freight movement as road transport. Rail transport accounts for 6 percent of global transport energy demand.
- **Maritime transportation** accounts for 90 percent of cross-border world trade as measured by volume. The nature of water transport and its economies of scale make it the most energy efficient mode. This mode uses 7 percent of all the energy consumed by transport activities.
- **Air transportation** plays an integral part in the globalization of the transportation network. The aviation industry accounts for 8 percent of the energy consumed by transportation. Air transport has high energy consumption levels, linked to high speeds. Fuel is the second most important budget for the air transport industry accounting for 13–20 percent of total expenses.

Further distinctions in the energy consumption of transport can be made between passenger and freight movement:

- **Passenger transportation** accounts for 60–70 percent of energy consumption from transportation activities. The private car is the dominant mode. There is a close relationship between rising income, automobile ownership and distance traveled by vehicle. The United States has one of the highest levels of car ownership in the world with 488 cars per 1,000 persons in 1999. About 60 percent of all American households owned two or more cars, with 19 percent owning three or more. A more disturbing trend has been the increasing rise in ownership of minivans, sport utility vehicles and light-duty trucks for personal use and the corresponding decline in fuel economy.
- **Freight transportation** is dominated by rail and shipping, the two most energy efficient modes. Coastal and inland waterways provide an energy efficient method of transporting passengers and cargoes. A tow boat moving a typical 15-barge tow holds the equivalent of 225 rail car loads or 870 truck loads. The grounds for favoring coastal and inland navigation are also based on lower energy consumption rates of shipping and the general overall smaller externalities of water transportation. The United States Marine Transportation System National Advisory Council has measured the distance that one ton of cargo can be moved with 3.785 liters of fuel. A tow boat operating on the inland waterways can move one ton of barge cargo 857 kilometers. The same amount of fuel will move one ton of rail cargo 337 kilometers or one ton of highway cargo 98 kilometers.

A powerful trend that has emerged in the 1950s has been the growing share of transportation in the total oil consumption of developed countries. Transportation accounts for approximately 25 percent of world energy demand and for more than 55 percent of all the oil used each year. Transportation is almost completely reliant (95 percent) upon petroleum products with the exception of railways using electrical power. While the use of petroleum products from other economic sectors, such as industrial and electricity generation, has remained relatively stable, the growth in oil demand is mainly attributed to the growth in transportation demand. In such a context, the price of oil increased substantially, leading to a set of potential impacts on transport systems (Figure 8.1).

For the transportation sector, six major types of interdependent impacts of high oil prices can be expected:

- **Usage level**. This trend is clear and straightforward as users of a specific mode generally respond to higher prices by limiting or rationalizing (e.g. speed) their usage level. Transport operators, such as airline companies, can reduce the frequency of their services. It is a matter of price elasticity where an increase of price P will result in a usage level change of Q. This function is rarely linear. At first, price increases may have limited effects as they are simply absorbed with the expectation that they are a temporary condition. Once a specific price threshold is reached, then significant changes will result as marginal and extraneous usage will be cut until a new equilibrium is reached. Usage for this mode is said to have reached a paradigm shift.
- **Modal shift**. In conjunction with a drop in usage level for a mode, an alternative mode may capture the traffic of that change, in whole or in part, through a modal shift. Again, this process is commonly not linear and a modal balance (A/B) can shift rapidly once a price threshold is reached. Thus, an increase of price P may result in a substantial shift, $Q(A/B)$, in the modal balance. A modal shift commonly takes place towards a mode which is less energy intensive (less elasticity) than the other. It can thus be expected that with higher oil prices some trucking will shift towards rail and that public transit will gain in market share.
- **Service area changes**. Under a specific price level, each mode has an optimum service area; a distance at which it provides mobility in a cost-effective fashion. Since

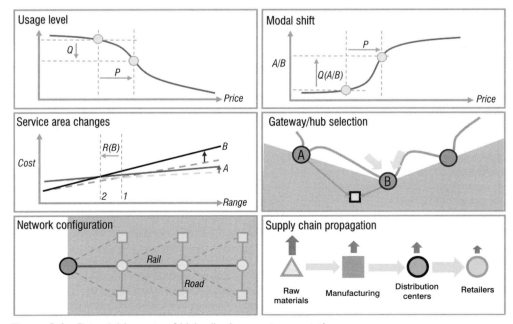

Figure 8.1 Potential impacts of high oil prices on transportation

each mode has a different elasticity, an increase in prices will have different impacts on the cost/distance function. For two modes, *A* and *B*, the same increase in energy price would create a different inflection of the cost/distance function where the range of mode *B* would reduce by *R(B)*. Thus, mode *A* gains in market share. An example is trucking versus rail in North America where about 700 miles (1,120 km) was considered to be a standard cutting point, but a rise in fuel prices has placed this range around 500 miles (800 km).

- **Gateway/hub selection**. A gateway is a point of interface between two systems of circulation. Since these systems have different elasticities, a rise in energy prices can eventually change their relationships, particularly the locations where intermodal transportation takes place. A shipping service using gateway port A and taking advantage of faster (but more energy intensive) hinterland connections may instead switch to gateway port B which is closer to customers. Although this change may result in longer total shipping times, the cost trade-off would make it acceptable. Higher energy prices are thus likely to reinforce gateways that have the most efficient hinterland connections, notably in terms of modal choice.
- **Network configuration**. Enduring high energy prices are likely to trigger shifts in the configuration of transportation networks in terms of gateways, hubs, routing and corridors. For instance an inland corridor may experience a change of the linkages with inland load centers that minimize road use and maximize rail. An airline may decide to abandon less profitable routes and offer more direct (point-to-point) services. An air transport network may experience a reconfiguration and an abandonment of marginal services, namely at small airports.
- **Supply chain propagation**. A supply chain is composed of a series of inputs and outputs having a complex geographical and functional structure. Rising energy prices imply a wide variety of changes in the cost structure within a supply chain, namely a propagation of those costs. Procurement, manufacturing and distribution costs are all impacted to various degrees. It is possible for some of these costs to be absorbed through reduced profit margins and higher efficiencies, but they do eventually propagate and end up in higher consumer prices. The functional and geographical structure of a supply chain is a key element of the nature and extent of its costs propagation. At a certain price level, some supply chains cease to be profitable and a reconfiguration becomes necessary.

It is assumed here that the origins and destinations remain relatively constant. However, it is clear that locational choices are significantly impacted as well. For instance, many comparative advantages in global trade are mainly based on low transport costs. In a higher energy prices environment, locational practices may change in several manufacturing sectors with sites closer to final markets being advantaged, even if characterized by higher labor (or input) costs.

Transportation and alternative fuels

All other things being equal, the energy with the lowest cost will always be sought. The dominance of petroleum fuels is a result of the relative simplicity with which it can be stored and efficiently used in the internal combustion engine vehicle. The transportation sector is heavily dependent on the use of petroleum fuels. Other fossil fuels (natural gas, propane and methanol) can be used as transportation fuels but require a more complicated storage system. The main issue concerning the large-scale uses of these alternative vehicle fuels is the large capital investments required in distribution facilities as compared with conventional fuels. Another issue is that in terms of energy density, these alternative fuels have lower efficiency than gasoline and thus require a greater volume of on-board storage to cover the equivalent distance as a gasoline propelled vehicle.

Alternative fuels in the form of non-crude oil resources are drawing considerable attention as a result of shrinking oil reserves, increasing petroleum costs and the need to reduce pollutant emissions:

- **Biogas** such as ethanol, methanol and biodiesel can be produced from the fermentation of food crops (sugar cane, corn, cereals, etc.) or wood-waste. Their production however requires large harvesting areas that may compete with other types of land use. Besides, it is estimated that one hectare of wheat produces less than 1,000 liters of transportation fuel per year which represents the amount of fuel consumed by one passenger car traveling 10,000 kilometers per year. This limit is related to the capacity of plants to absorb solar energy and transform it through photosynthesis. This low productivity of the biomass does not meet the energy needs of the transportation sector. In 2007, the US government proposed to reduce oil consumption by 20 percent by using ethanol. As the USA currently produces 26 billion liters of ethanol each year, this objective would require the production of nearly 115 billion liters of ethanol by 2017, which amounts to the total annual US maize production. Besides, the production of ethanol is an energy-intensive process. The production of one thermal unit of ethanol requires the combustion of 0.76 unit of coal, petroleum or natural gas. Biodiesel can be obtained from a variety of crops. The choice of biomass fuel will largely depend on the sustainability and energy efficiency of the production process.
- **Hydrogen** is often mentioned as the energy source of the future. The steps in using hydrogen as a transportation fuel consist in: (1) producing hydrogen by electrolysis of water; (2) compressing or converting hydrogen into liquid form; (3) storing it on-board a vehicle; and (4) using fuel cells to generate electricity on demand from the hydrogen to propel a motor vehicle. Hydrogen fuel cells are two times more efficient than gasoline and generate near-zero pollutants. But hydrogen suffers from several problems. A lot of energy is wasted in the production, transfer and storage of hydrogen. Hydrogen manufacturing requires electricity production. A hydrogen-powered vehicle requires 2–4 times more energy for operation than an electric car which does not make it cost-effective. Besides, hydrogen has a very low energy density and requires a very low temperature and very high pressure storage tank which adds weight and volume to a vehicle. This suggests that liquid hydrogen fuel would be a better alternative for ship and aircraft propulsion.
- **Electricity** is being considered as an alternative to petroleum fuels as an energy source. A pure battery electric vehicle is considered a more efficient alternative to a hydrogen fuel-propelled vehicle as there is no need to convert energy into electricity since the electricity stored in the battery can power the electric motor. Besides an electric car is easier and cheaper to produce than a comparable fuel-cell vehicle. The main barriers to the development of electric cars are the lack of storage systems capable of providing driving ranges and speed comparable to those of conventional vehicles. The low energy capacity of batteries makes the electric car less competitive than internal combustion engines using gasoline. An electric car has a maximum range of 100 kilometers and a speed of less than 100 kph, requiring 4–8 hours to recharge.
- **Hybrid** vehicles with a propulsion system that uses an internal combustion engine as well as an electric motor and batteries provide interesting opportunities to

combine the efficiency of electricity with a long driving range. A hybrid vehicle still uses liquid fuel as the main source of energy but the engine provides the power to drive the vehicle or is used to charge the battery via a generator. Alternatively the propulsion can be provided by the electricity generated by the battery. When the battery is discharged, the engine starts automatically without intervention from the driver. The generator can also be fed by using the braking energy to recharge the battery. Such a propulsion design greatly contributes to overall fuel efficiency. Given the inevitable depletion of oil resources, the successful development and commercialization of hybrid vehicles appears the most sustainable option to conventional gasoline engine-powered vehicles.

Transportation and peak oil

The extent to which conventional non-renewable fossil fuels will continue to be the primary resources for nearly all transportation fuels is subject to debate. Some studies estimate global resources for oil at about a trillion barrels. This represents 30 years of reserves at the present rate of consumption. But the gap between demand and supply, once considerable, is narrowing, an effect compounded by the peaking off of global oil production. The steady surge in demand from China and India requires an additional output of 2–3 million barrels a day. This raises concern about the capacity of major oil producers to meet this rising world demand. The producers are not running out of oil, but the existing reservoirs may not be capable of producing on a daily basis the increasing volumes of oil that the world requires. Reservoirs do not exist as underground lakes from which oil can easily be extracted. There are geological limits to the output of existing fields. This suggests that an additional 4–5 million barrels a day need to be found to compensate for the declining production of existing fields. Additional reserves in Alaska, off-shore West Africa and the Caspian Sea basin are not enough to offset this growing demand. The bitumen reserves in Alberta, Canada, for instance are estimated at 170 billion barrels, second in the world in terms of oil reserves, behind Saudi Arabia. But extracting heavy oil from bitumen sands necessitates much energy and water. The production of one barrel of bitumen requires burning 10–20 percent of the energy content of the resulting crude oil in the form of natural gas.

Other studies argue that the history of the oil industry is marked by cycles of shortages and surplus. The rising price of oil will render cost-effective oil recovery in difficult areas. Deep water drilling or extraction from tar sands should increase the supply of oil that can be recovered and extracted from the surface. But there is a limit to the capacity of technological innovation to find and extract more oil around the world. Technological development does not keep pace with surging demand. The construction of drilling rigs, power plants, refineries and pipelines designed to increase oil exploitation is a complex and slow process. The main concern is the amount of oil that can be pumped to the surface on a daily basis. Some studies predict that carbon sequestration in the form of CO_2 capture and storage, if technically and economically viable, could enhance the recovery of oil from conventional wells and prolong the life of partially depleted oil fields well into the next century.

The penetration of non-fossil fuels in the transportation sector has serious limitations. As a result, the price of oil will certainly continue to increase as more expensive fuel-recovery technologies will have to be utilized with soaring demand for gasoline. But high oil prices are inflationary leading to recession in economic activity and the search for alternative sources of energy. Already, the peaking of conventional oil production is leading to the implementation of coal-derived oil projects. Coal liquefaction technology allows the transformation of coal into refined oil after a series of processes in an environment of high temperature and high pressure. While the cost-effectiveness of this technique has yet to be demonstrated, coal liquefaction is an important measure in the implementation of transportation fuel strategies in coal-rich countries, such as China and South Africa.

The costs of alternative energy sources to fossil fuels are higher in the transportation sector than in other types of economic activities. This suggests higher competitive advantages for the industrial, household, commercial, electricity and heat sectors to shift away from oil and to rely on solar, wind or hydro-power. Transportation fuels based on renewable energy sources might not be competitive with petroleum fuels unless future price increases are affected by different fuel taxes based on environmental impacts. Excessive fuel price could stimulate the development of alternatives. But economists have demonstrated that automotive fuel oil is price inelastic. Higher prices result in very marginal changes in demand for fuel. While $100 per barrel was for a long time considered a threshold that would limit demand for automotive fuel and lead to a decline in passenger-km and freight-km, evidence suggests that higher oil prices had limited impact on the average annual growth rate of world motorization. The analysis of the evolution of the use of fossil fuels suggests that in a free market economy the introduction of alternative fuels is leading to an increase in the global consumption of both fossil and alternative fuels and not to the substitution of crude oil by bio-based alternative fuels. This suggests that in the initial phase of an energy transition cycle, the introduction of a new source of energy complements existing supply until the new source of energy becomes price competitive enough to be an alternative. The presence of both renewable and non-renewable types of fuels stimulates the energy market with the concomitant result of increasing greenhouse gas emissions. The production of alternative fuels is added to the existing fossil fuels and does not replace it. World market consumption of all primary energy forms has grown by 40 percent during the period 1980–2000.

In a context where petroleum prices are relatively low, substitution to alternative fuels in the transportation sector will require very strong government interventions forcing energy suppliers to purchase available green energies on the market at a fixed price. Without strong regulatory controls conventional oil substitution by renewable vehicle fuel requirements (ethanol and biodiesel) will be relative and marginal. Only under the conditions of price equilibrium between conventional and alternative fuels could the market become an effective transitional force. Answering the energy demand of the transportation sector will rest on a delicate balance between technological improvements, behavioral changes and environmental policies. Without presuming on the outcome, a major trend is already apparent. The energy crisis imposes capital rationing with greater emphasis on quality of transport infrastructures.

Concept 2 – Transport and environment

The impacts of the environment on transport

The relationships between transport and the environment are complex. It is clear that over time the impacts of transport on the environment have been increasing and this brings into question how the relationships may be better managed. The environment has always played a constraining effect on the mobility of people and goods. The main elements are physical distance, topography, hydrology, climate and natural hazards:

- **Physical distance** has had a paramount effect limiting spatial interactions. The cost of overcoming the friction of distance is one of the most important features of transport geography (see Chapter 1). As has been discussed elsewhere, human progress has been marked by technological developments in transport that have shrunk distance barriers, while never completely overcoming them, except for telecommunications.

- **Topography** continues to play an important role in shaping transportation. Several transport modes, such as rail and canals, are still very constrained by slopes, and while other land modes, such as road and pipelines, have the capacity to adjust to gradients, their routings are influenced by topography. Transport systems seek the paths of least resistance, and for this reason valleys and plains have usually the densest networks. Topography also influences water transport, since the configuration of river basins may or may not coincide with trade flows; and port sites may not be available where needed. Historically, mountains and deserts have served as barriers to interaction, serving to isolate regions.

- **Hydrology** has a particular effect on water transport. The depth of channels may not be adequate to support the size of vessels required. Fluctuations in water levels because of seasonality may interrupt navigation. Tidal conditions influence access to ports and impede loading operations in ports where the tidal ranges are large. Land transport modes may also be influenced by hydrological conditions. Rivers and estuaries may serve as barriers to interactions between the different shores, requiring transfers to ferries or necessitating long detours to bridgeheads. Permafrost makes construction of routes and transport infrastructures extremely difficult in arctic and sub-arctic regions.

- **Climate** affects transport systems in both direct and subtle ways. In the days of sailing ships, trade patterns were dictated by prevailing winds. Among many examples, there was the triangular trade between Europe, Africa and the Americas during the seventeenth to nineteenth centuries. Contemporary air travel is very much influenced by the air circulation of the upper atmosphere. Winter conditions, particularly the freezing of water bodies, such as the St. Lawrence Seaway and the Baltic Sea, interrupt regular shipments. Climate also has many more subtle impacts that are of shorter duration. Snow, fog, wind and rain events may cause delays and disruptions. Air transport is particularly prone to such impacts, although other modes may be affected as well by such events.

- **Natural hazards** have the potential to impact greatly on transport systems, as indeed they affect all human activities. Geological events such as earthquakes, tsunamis and volcanic eruptions can be catastrophic in their human impacts and along with more localized occurrences such as landslides affect the operations of all modes. The 1995 Kobe earthquake in Japan brought about such a disruption to the port that shipping lines transferred their business to other ports, and Kobe has still not recovered this lost business. Hurricanes and tornadoes have the power to at least disrupt activities and at worse to destroy infrastructures. Exceptional rain storms can produce flooding, and because transport routes tend to follow low-level paths, severe dislocations to transport systems can occur. In 2005, Hurricane Katrina led to the closure of the port of New Orleans, the main transit for American grain exports.

Environmental conditions can complicate, postpone or prevent the activities of the transport industry. Technological developments have permitted to overcome the obstacles of the physical environment. The physical relief has been changed, and these changes generated costs. Environments have been transformed, destroyed or even artificially created to such an extent that it is extremely difficult to identify a pristine reference. More importantly, transport operations, freight and passenger movements, maintenance activities and the construction of equipments have led to major environmental impacts.

The impacts of transport on the environment

The issue of transportation and the environment is paradoxical in nature. Transportation activities support increasing mobility demands for passengers and freight, notably in urban areas. But transport activities have resulted in growing levels of motorization and congestion. As a result, the transportation sector is becoming increasingly linked to environmental problems. The most important impacts of transport on the environment relate to climate change, air quality, noise, water quality, soil quality, biodiversity and land take.

- **Climate change**. The activities of the transport industry produce several million tons of pollutants each year into the atmosphere. These include the emission of lead (Pb), carbon monoxide (CO), carbon dioxide (CO_2), methane (CH_4), nitrogen oxides (NO_x), nitrous oxide (N_2O), chlorofluorocarbons (CFCs), perfluorocarbons (PFCs), silicon tetraflouride (SF_6), benzene and volatile components (BTX), heavy metals (zinc, chrome, copper and cadmium) and particulate matters (ash, dust). The road transport sector is responsible for 74 percent of global CO_2 emissions, while aviation, shipping and railways account for 12 percent, 10 percent and 4 percent respectively. There is an ongoing debate to what extent these emissions (labeled as "greenhouse gases") may prevent the wavelengths of electromagnetic radiation from leaving the Earth's surface and thus contribute to global warming. This could lead to an increase in the average temperature at the Earth's surface, reducing snow cover of polar regions, which in turn could contribute to sea level rise and an increase in ocean heat content. Some of these gases also participate in depleting the stratospheric ozone (O_3) layer which naturally screens the Earth's surface from ultraviolet radiation.
- **Air quality**. Highway vehicles, marine engines, locomotives and aircraft are the sources of pollution in the form of gas and particulate matter emissions that affect air quality, causing damage to human health. Toxic air pollutants are associated with cancer, cardiovascular, respiratory and neurological diseases. Carbon monoxide (CO) when inhaled affects bloodstream, reduces the availability of oxygen and can be extremely harmful to public health. An emission of nitrogen dioxide (NO_2) from transportation sources reduces lung function, affects the respiratory immune defense system and increases the risk of respiratory problems. The emissions of sulphur dioxide (SO_2) and nitrogen oxides (NO_x) in the atmosphere form various acidic compounds that when mixed in cloud water creates acid rain. Acid precipitation has detrimental effects on the built environment, reduces agricultural crop yields and causes forest decline. The reduction of natural visibility by smog has a number of adverse impacts on the quality of life and the attractiveness of tourist sites. Particulate emissions in the form of dust emanating from vehicle exhaust as well as from non-exhaust sources such as vehicle and road abrasion have an impact on air quality. The physical and chemical properties of particulates are associated with health risks such as respiratory problems, skin irritations, eyes inflammations, blood clotting and various types of allergies.
- **Noise**. Noise represents the general effect of irregular and chaotic sounds. It is traumatizing for the hearing organ and may affect the quality of life by its

unpleasant and disturbing character. Long-term exposure to noise levels above 75 dB seriously hampers hearing and affects human physical and psychological well-being. Transport noise emanating from the movement of transport vehicles and the operations of ports, airports and railyards affects human health, through an increase in the risk of cardiovascular diseases. Increasing noise levels have a negative impact on the urban environment reflected in falling land values and loss of productive land uses.

- **Water quality**. Transport activities have an impact on hydrological conditions. Fuel, chemical and other hazardous particulates discarded from aircraft, cars, trucks and trains or from port and airport terminal operations, such as de-icing, can contaminate rivers, lakes, wetlands and oceans. Globally, world seaborne trade grew from 2.6 billion tons in 1970 to over 7 billion tons of loaded goods in 2006. Because demand for shipping services is increasing, marine transport emissions represent the most important segment of water quality inventory of the transportation sector. The main effects of marine transport operations on water quality predominantly arise from dredging, waste, ballast waters and oil spills. Dredging is the process of deepening harbor channels by removing sediments from the bed of a body of water. Dredging is essential to create and maintain sufficient water depth for shipping operations and port accessibility. Dredging activities have a two-fold negative impact on the marine environment. They modify the hydrology by creating turbidity that can affect the marine biological diversity. The contaminated sediments and water raised by dredging require spoil disposal sites and decontamination techniques. Waste generated by the operations of vessels at sea or at ports cause serious environmental problems, since they can contain a very high level of bacteria that can be hazardous for public health as well as marine ecosystems when discharged in waters. Besides, various types of garbage containing metals and plastic are not easily biodegradable. They can persist on the sea surface for a long time and can be a serious impediment for maritime navigation in inland waterways and at sea, and affecting berthing operations as well. Ballast waters are required to control a ship's stability and draught and to modify the center of gravity in relation to cargo carried and the variance in weight distribution. Ballast waters acquired in a region may contain invasive aquatic species that, when discharged in another region, may thrive in a new marine environment and disrupt the natural marine ecosystem. There are about 100 non-indigenous species recorded in the Baltic Sea. Invasive species have resulted in major changes in nearshore ecosystems, especially in coastal lagoons and inlets. Major oil spills from oil cargo vessel accidents are one of the most serious problems of pollution from maritime transport activities. The *Erika*, *Prestige*, and *Sea Empress* oil spills that occurred in the European Atlantic generated a significant amount of pollution that destroyed aquatic species including algae, mollusks, crustaceans, marine mammals, fish and invertebrates.
- **Soil quality**. The environmental impact of transportation on soil consists of soil erosion and soil contamination. Coastal transport facilities have significant impacts on soil erosion. Shipping activities are modifying the scale and scope of wave actions leading to serious damage in confined channels such as river banks. The removal of earth for highway construction or lessening surface grades for port and airport developments have led to important loss of fertile and productive soils. Soil contamination can occur through the use of toxic materials by the transport industry. Fuel and oil spills from motor vehicles are washed off road sides and enter the soil. Chemicals used for the preservation of railroad ties may enter into the soil. Hazardous materials and heavy metals have been found in areas contiguous to railroads, ports and airports.

- **Biodiversity**. Transportation also influences natural vegetation. The need for construction materials and the development of land-based transportation has led to deforestation. Many transport routes have required draining land, thus reducing wetland areas and driving out water plant species. The need to maintain road and rail rights of way or to stabilize slopes along transport facilities has resulted in restricting the growth of certain plants or has produced changes in plants with the introduction of new species different from those which originally grew in the area. Many animal species are becoming extinct as a result of changes in their natural habitats and reduction of ranges.

- **Land take**. Transportation facilities have an impact on the urban landscape. The development of port and airport infrastructure is a significant feature of the urban and peri-urban built environment. Social and economic cohesion can be severed when new transport facilities such as an elevated train and highway structures cut across an existing urban community. Arteries or transport terminals can define urban borders and produce segregation. Major transport facilities can affect the quality of urban life by creating physical barriers, increasing noise levels, generating odors, reducing the urban aesthetic and affecting the built heritage.

A comprehensive assessment of the environmental impacts of the transportation system is not restricted to these issues. Additional effects such as accidents and the movement of hazardous materials need to be included. It is also possible to break down the total environmental impact of the transport industry into contribution from downstream and upstream requirements for the provision of transport infrastructures. Another issue is that the scale of the impacts that may vary from the local to the global. Transportation impacts can fall within three categories:

- **Direct impacts**. The immediate consequence of transport activities such as pollutant emissions and respiratory diseases. The cause-and-effect relationship is generally clear and well understood.

- **Indirect impacts**. These are the secondary (or tertiary) effects of transport activities. They are often of higher consequence than direct impacts, but the involved relationships are often misunderstood and difficult to establish, such as congested traffic and stress.

- **Cumulative impacts**. These are the additive, multiplicative or synergetic consequences of transport activities. They take into account the varied effects of direct and indirect impacts on an ecosystem, often unpredicted such as physical barriers and the migration of exotic species.

As shown on Figure 8.2, the environmental dimensions of transportation include:

- **Causes**. Two major factors contribute to the level of transport activities. Land use refers to the spatial structure and location of the transport demand. The most important driving factor for transport pollution emission is economic growth which is in turn related to higher incomes, increased motorization and travel activity expressed in passenger-kilometers or ton-kilometers.

- **Activities**. A wide array of factors are involved in the use of transportation infrastructures and related services. All these activities have environmental outputs.

- **Outputs**. Several factors are to be considered. The first outcome of transportation activities are pollutant emissions. According to the geographical characteristics of the area where emissions are occurring (e.g. wind patterns) ambient levels are created. Once these levels are correlated with population proximity, a level of

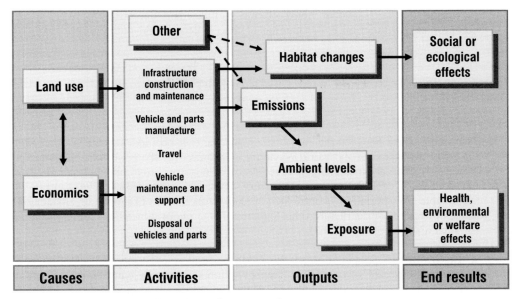

Figure 8.2 Environmental dimensions of transportation

exposure to harmful pollutants can be calculated. This exposure is likely to have negative consequences.

● **End results**. They include all the health, environmental and welfare effects of exposure to emissions from transportation activities.

Transportation gives rise to considerable environmental impacts in its construction and operation and through the commercial, industrial and residential activities it encourages. These changes to the environment generate important costs that are often not borne by transport users. Placing monetary values to environmental externalities is increasingly being considered to justify the implementation of policies. The economic valuation of the cost of pollution is difficult. Global external costs of transportation have been estimated at more than $1 trillion. The most important external costs are climate change, air pollution, congestion and traffic accidents. Much uncertainty underlies the calculation of transport externalities. The process involves quantifying emissions, estimating their spatial concentration, assessing their impact and equating monetary values. The most common approaches are to measure the loss of wealth (production, human capital), to assess the cost of restoration and to estimate the avoidance costs (cost of prevention measures). The difficulties in quantifying external environmental costs should not justify inaction. The objective is not to converge on a single value for a given cost, but rather to help rationalize choices with regards to environmental policies. The economic assessment is then applied to transport demand management through the use of fiscal incentives or taxes related to environmental externalities with a view that the resulting environmental protection outweighs the utility loss of price increases in transport services. The relationships between transport and the environment are complex and multidimensional. The spatial accumulation of transportation has become a dominant factor behind the emission of most pollutants and their impacts on the environment. With growth in transport and unbalanced modal split, business as usual in the transport sector is no longer a viable option. Controlling the negative externalities of transportation facilities and operations is likely to be compromised without an understanding of the challenges and policy implications of sustainability.

Concept 3 – Transport and sustainability

Sustainable transportation

The relationships between the environment and the transport industry are strong. With the rapid expansion of the world economy, concerns over the environmental impacts of transportation are increasing. The reduction of negative environmental externalities has become a central theme for transport development strategies. The concept of sustainable transportation is intricately linked with the development of sustainable transport modes, infrastructures and logistics. Three major dimensions are considered for such a purpose (Figure 8.3):

- **Environment**. A reduction of the environmental impacts of transportation is a likely strategy for sustainability. Transportation significantly contributes to harmful emissions, noise and climate changes. However, vehicles are becoming more environmentally efficient, although there are more of them around, so it often evens out. An improvement of the land use impacts of transportation, especially the pressures infrastructures create, is also a strategic goal to achieve. The transportation system is also a generator of wastes (vehicles, parts, packaging, etc.) that must be reduced.

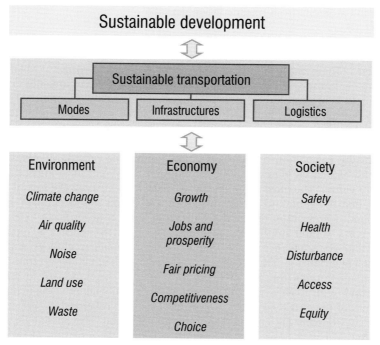

Figure 8.3 Sustainable transportation (Source: adapted from UK Department of the Environment, Transport and the Regions, 1999)

- **Economy**. Transportation is a factor of economic growth and development. A sustainable strategy would aim to efficiently use transportation for the purpose of growth and for the creation of jobs. Transportation should also have a fair pricing strategy, meaning that users bear the full costs (direct and indirect) of their usage of the transport system. A system where competition is fair and open is likely to promote the choice and efficiency of distribution systems. In a system where transport is a public or private monopoly, price distortions and misallocations of capital are created which in the long run are likely to render the system unsustainable.
- **Society**. Sustainable transportation should benefit society. It should be safe, should not impair human health and should minimize disturbance on communities. Access and equity are also two important principles as transportation should promote access to goods and services for as many people as possible.

Most studies agree that automobile dependence is related to an unsustainable urban environment. However, such an observation is at odds with the choice and preference of the global population where the automobile is adopted when income levels reach a certain threshold. It can be seen that other transport alternatives are limited in comparison to the convenience of the automobile. Private and flexible forms of transportation, such as the automobile, are thus fundamental to urban mobility and should not be discarded for the sake of "sustainability". Freight transportation must also been considered in this process, considering the substantial growth of raw materials and goods being traded in a global economy. In fact, freight transportation relies on much more environmentally sound modes such as rail and maritime.

Despite the apparent and projected success of measures to promote transport sustainability, they have their limits. For instance, the formation of compact and accessible cities must be allowed to contend with the already existing built environment, while obeying several limits to development and urban renewal through temporal constraints and lack of available capital. Indeed, the built environment cannot change quickly enough to solve the bulk of problems related to unsustainable transport. Most of the investment that is already in place will remain in place for 50 years or more and new investments (in additional or improved infrastructure) will not represent much more than a few percentage points of change in terms of reducing traffic congestion and ease of access to strategic urban locations. It is necessary, therefore, to favor conditions that would influence modal choice, the supply of transport services and attractiveness of walking and cycling within the constraints of the built environment. While policies, rules and regulations have a tendency to favor a misallocation of resources (such as compliance), users tend to instinctively react to price signals and discard modes that are becoming costly (unsustainable).

The main dilemma in environmental protection is the conflict between a top-down and a bottom-up approach. Top-down decision making recognizes the leadership of the international community and governments in focusing efforts on international regulations and their implementation. Bottom-up decision making acknowledges the role of transport firms in reducing the environmental impact of their activities based on their analysis of operating conditions, environmental assessment, established priorities and organizational capabilities. The two approaches are not mutually exclusive as they are influenced by the concept of sustainability.

Top-down approach to sustainable transport

Since the mid-1980s, many of the changes that have occurred in transport systems have been undertaken in parallel with the search for a balance between the economic, social and environmental dimensions of development. The concept of sustainable development was popularized in 1987 with the publication of the Bruntland report which defined sustainable development as the ability to meet the needs of the current generation without compromising the needs of future generations. In June 1992, the Rio Earth Summit declared that sustainable development involves the equitable sharing of the benefits of economic progress by focusing on the conservation and preservation of natural resources, and by tackling the reciprocal influences of environmental, social and economic issues. Efforts to promote sustainable development were further enhanced with the Johannesburg Summit in 2002, with commitments on poverty reduction, and on protecting the Earth's biodiversity and ecosystems. Sustainable development is concerned with seeking an optimal balance between environmental, economic and social objectives.

The emergence of a consensus on the necessity of implementing strategies of sustainable development applied to the evolution, performance and organization of transport systems led to the recognition of the concept of environmentally sustainable transport. In 1996, the Organization for Economic Cooperation and Development (OECD) designated environmentally sustainable transport as one that does not endanger public health and ecosystems and meets access needs, while using renewable resources below their rate of generation, and using non-renewable resources below the development rates of renewable substitutes.

Sustainable development applied to transport systems requires the promotion of linkages between environmental protection, economic efficiency and social progress. Under the environmental dimension, the objective consists in understanding the reciprocal influences of the physical environment and the practices of the industry, and that environmental issues are addressed by all aspects of the transport industry. Under the economic dimension, the objective consists of orienting progress in the sense of economic efficiency. Transport must be cost-effective and capable of adapting to changing demands. Under the social dimension, the objective consists in upgrading standards of living and quality of life.

The environmental, economic and social dimensions of sustainable transport are interdependent and lead to various trade-offs and opportunities for transport decision makers. Policies to improve accessibility will increase motorization and environmental externalities. Economic policies increasing transport tariffs to reflect real costs will affect individual incomes and be detrimental to the poor. Social policies favoring the informal transport sector with a view to answering the mobility needs of the poor may give rise to a significant increase in polluting emissions. Solutions to achieve a relative balance between these trade-offs can be found in health and safety improvements, efficient transport pricing, infrastructure maintenance and land use design.

In 2001, the United Nations proposed that sustainable development when applied to transport refers to its role in securing a balance between equity, efficiency and the capacity to answer the needs of future generations. More specifically, this implies: (1) securing energy supply; (2) reflecting the costs of non-renewable resources in transport vehicle operations; (3) creating responsive and effective markets; and (4) adopting production processes that respect the environment by eliminating negative externalities detrimental to future generations.

In reviewing sustainable development strategies, the OECD developed ten guidelines for the management of sustainable transport for future-oriented policy making and practices:

- Develop **long-term vision** of a desirable transport future that is sustainable for the environment and health and provides the benefits of mobility and access.

- Assess **long-term transport trends** considering all aspects of transport, their health and environmental impacts, and the economic and social implications of continuing with business as usual.
- Define **health and environmental quality objectives**, based on health and environmental criteria, standards and sustainability requirements.
- Set quantified **sector-specific targets** derived from the environmental and health quality objectives and set target dates and milestones.
- Identify **strategies to achieve environmentally sustainable transport** and combinations of measures to ensure technological enhancement and changes in transport activity.
- Assess the **social and economic implications** of the vision and ensure that they are consistent with social and economic sustainability.
- Construct **packages of measures and instruments** for reaching the milestones and targets of environmentally sustainable transport.
- Develop an **implementation plan** that involves the well-phased application of packages of instruments capable of achieving environmentally sustainable transport, taking into account local, regional and national circumstances.
- Set provision for **monitoring implementation** and for public reporting on the environmentally sustainable transport strategy.
- Build **broad support and cooperation** for implementing environmentally sustainable transport.

Governments have a very important role to play in achieving the environmental objectives of the sustainable development of the transport industry but cannot implement its policies without some form of cooperation and alliances with the private sector. For the past decade, governments have introduced a variety of rules in different sectors that constitute steps towards attaining a sustainable environment. Examples of legislation included energy consumption, transport development, polluting emissions, protection of ecosystems, etc. These laws and rules increase the amount and strength of measures to protect and improve the environment. Environmental legislation is placing increasing restrictions on transport activity and operators have to respond by developing management systems enabling them to meet regulatory requirements. All the partners of the transport industry (i.e. carriers, terminal operators, shippers, stevedores, etc.) must answer these new regulatory requirements. Terminal operators and carriers must be responsible for the damage they cause to the environment. This can be imposed by sanctions, financial obligations or withdrawal of permits. The planning and implementation processes of investments in transport infrastructures around the globe increasingly include an environmental impact assessment (EIA) satisfying minimum standards of analysis.

An **environmental impact assessment** is a process for carrying out an appraisal of the full potential effects of a development project on the physical environment.

This suggests that the most efficient means to implement strategies of environmental sustainability defined by governments consist in the elaboration of a policy framework giving responsibilities of a sustainable development blueprint to the transport operators.

Bottom-up approach to sustainable transport

There is a wide range of responses to environmental sustainability. The various trajectories for a sustainable environment involve three steps: (1) transport operations must conform to local, national and international regulations; (2) environmental costs of transport operations must be built into the price of providing transport facilities and services; (3) environmental performance must be introduced into the organization's management. Environmental sustainability represents a growing area of responsibility for transport companies, one that is forcing them to acquire expertise in environmental management. The most important challenge for the industry is to implement environmentally sustainable transport within competitive market structures. As a result, before implementing a systematic approach to managing environmental performance, senior management of transport firms consider key cost issues: inventory and resources:

- **Inventory**. The first obstacle common to all enterprises that wish to adopt and implement an environmental policy is to collect transport data on environmental output. Planning environmentally sustainable transport rests on numerous components linked to environmental conditions, legislation and transport operations. As a result, many transport firms identify one or a few selected issues pertaining to transport operations, legislation and the physical environment. They will search for a proper answer that would fulfill their objectives in terms of sustainable environment without the formal use of an environmental impact assessment. Transport companies need flexibility and do not necessary have to adopt formal procedures. Often, transport companies need to decide the rate of their environmental performance in relation to their management capacity.
- **Resources**. The adoption of an environmentally sustainable transport policy involves costs in terms of time, personnel and resources that are not always available for small and medium-size transport companies. The continuous process of data collection and evaluation depends on the presence within the corporate structure of experts responsible for surveying, initiating and coordinating all the measures for protecting the environment. As a result, some transport enterprises invest in the development of information technology pertaining to performance indicators of environmental impact assessment. Other companies use available commercial software. Any environmental impact assessment must consider using the best available technology at an appropriate cost. Given these inherent costs, the adoption and the implementation of a set of techniques of environmental management can contribute but do not by themselves guarantee optimal environmental results.

The practices of environmentally sustainable transport increasingly affect the competitiveness of the transport industry around the world. However, there are many reasons why environmental management should be integrated with the traditional economic considerations of transport enterprises.

- **Investments**. The transport industry needs to invest in clean technologies, improving energy efficiency, increasing renewable energy use, waste reduction and recycling. These various methods and techniques may reduce the financial costs of the transport industry in meeting emission reduction requirements.
- **Credit rates**. The financial sector is increasingly concerned about environmental sustainability. As a result, merchant banks are implementing credit programs charging different interest rates to terminal operators and carriers in relation to their environmental performance.
- **Insurance premiums**. Environmental costs influence insurance premiums. Transport firms that undertake activities that pose a potential risk to the environment pay an annual premium for any damage that might occur to the environment and must prepare a contingency plan. The conditions and insurance premiums are a function

of the past environmental performance of a transport operator. Insurance companies reduce insurance premiums of firms that have a green certification.

- **Market capitalization**. Increasingly, environmental components are accounted in the evolution of the value of shares for firms registered in the stock market. A growing number of investors are seeking socially responsible investment opportunities.
- **Revenues**. Several factors permit the transport industry to increase revenues. State instruments such as fiscal advantages, government purchasing policies and government subsidies that favor through fiscal measures the purchase of green technologies permit firms that make use of these programs promoting environmental sustainability, to achieve important savings. Equally revenue sensitive are environmentally differentiated fairway charges that permit carriers to increase their revenues.
- **New markets**. It is true that environmental management imposes restrictions that could result in more costly solutions. However, a transport enterprise that acquires expertise in environmental management also secures a marketable knowledge. This expertise can be harnessed to become a statutory and even a commercial advantage.
- **Strategic alliances**. There are an increasing number of transport operators attesting in their annual reports their compliance to environmental legislation. Legal and compensation costs related to transport project development have an influence in global transactions of mergers and acquisitions. This suggests that environmental risk reduction represents an asset for active stakeholders.

Method 1 – Transport environmental management

Environmental management systems

All transport infrastructures vary in terms of property, investment provisions, types of activities and volume of traffic. As a result, it is not possible to provide a unique model of environmental management as problems are mode specific and there are no agreed common international standards. Nevertheless, there are several environmental management systems (EMS) that provide procedures and specifications in a structured and verifiable manner to meet environmental objectives.

An **environmental management system** is a set of procedures and techniques enabling an organization to reduce environmental impacts and increase its operating efficiency.

Obviously, transport firms can only manage environmental issues on which they can exert a controlling influence. The best environmental practices include procedures that:

- Match transport facilities, operations or projects with environmental components.
- Link environmental components with regulatory requirements.
- Assess risks, impacts and responsibilities.
- Identify environmental issues to be addressed.
- Consider commercial strategies and operations of private and public sector organization.
- Introduce best practices.
- Undertake continuous monitoring and auditing.

These issues must be clearly understood and addressed before designing a particular framework of environmental management for a transport organization. There exist numerous environmental management systems. Obviously, the choice of a system is specific to each transport enterprise in relation to the problems, risks, impacts and responsibilities identified and the geographical environment in which the enterprise must operate. The most often mentioned environmental management systems are EMAS and ISO 14001:

● **Eco-Management and Audit Scheme (EMAS).** In 1993, the European Union created the norm EMAS, conceived to provide European firms with a framework and operational tools that would permit to better protect the environment. EMAS has developed a handbook entitled *Identification of Environmental Aspects and Evaluation of their Importance*. This approach rests on the necessity to identify environmental impacts and the various types of environment that are affected by the operations and activities of any types of organizations including transport enterprises. The impacts are evaluated according to a step-by-step procedure that examines each activity of an enterprise and its impacts on the environment. Each impact is then assessed in relation to criteria developed by the organization. These criteria must evaluate the potential damage to the environment, the fragility of the environment, the size and frequency of the activity, the importance of that activity for the organization, the employees and the local community, and the legal obligations emanating from environmental legislation.

● **ISO 14001.** The International Organization for Standardization has developed a set of norms that represent the main industrial reference in terms of environmental management systems and sustainability. ISO 14001 offers three categories of indicators to measure the environmental performance that could be applicable to the transport industry. (1) The indicators of environmental conditions (IEC) present the information on the environmental conditions permitting a better understanding of the impacts or the potential impacts of transport operations. (2) The indicators of management performance (IMP) present information on the management efforts that are being made to influence the environmental performance of transport operations. (3) The indicators of operational performance (IOP) present information on the environmental performance of transport operations. Generally, these indicators permit the identification of the most significant environmental impacts that are associated with transport operations. They include the evaluation, review and increase of the environmental performance of transport corporations, the identification of new practices and opportunities for better management of transport operations, and the provision of constant, credible and measurable information and data on the relationship between the environmental performance of the firm and its environmental objectives, targets and policies.

EMAS has been developed to stimulate and synchronize European environmental policies. EMAS mainly addresses manufacturing and transportation issues and is site specific. EMAS has a focus on internal corporate activities (as ISO) but also on external stakeholders. As a result EMAS holders are required to publish environmental statements for the public, while ISO 14001 has no such provision. In contrast, ISO is global in scope and is company specific. The corporate benefits do not differ between the two systems and studies suggest that the two standards have no practical effects on environmental performance. The most important issue is that both environmental management systems have strength and areas to improve, but it is the corporate environmental outlook that is the real engine to a high level of environmental performance and therefore a strong EMS.

Environmental impact assessment is a key instrument in the elaboration of environmentally sustainable transport. The process of environmental impact assessment implies numerous activities. There are many criteria to use to determine the interrelationships between environmental components and a given transportation activity or project. Any environmental impact assessment must be designed to accommodate these conditions.

As shown in Figure 8.4, a model of environmental management systems (EMS) should be flexible and be adapted to suit the needs of a particular industry. An EMS developed for ports and maritime transport should focus on issues such as water quality, air quality, waste management, habitat conservation, noise, dredging, contaminated soils, anti-fouling paints and energy consumption. For all these issues, compliance with legislation affecting shipping and port operations should be considered. An environmental management system implies interdependence and information flows. This is not to say that it is not possible to implement sustainable development through a piecemeal approach, but it is preferable to recognize the complementarities and dynamic feedback of the various dimensions composing the model. A key feature of environmental management consists in maintaining a balance between the environmental, legislative and commercial dimensions. Evaluating the trade-offs is one of the main challenges facing decision makers.

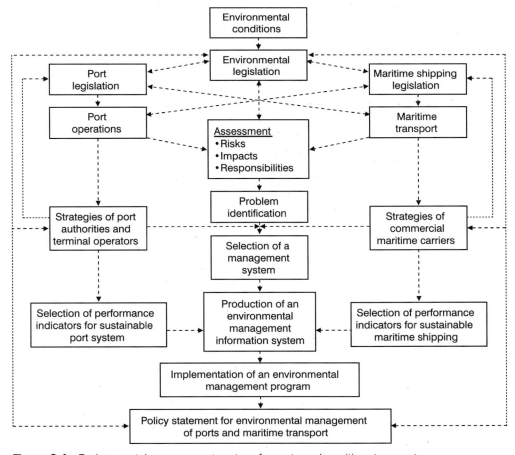

Figure 8.4 Environmental management system for ports and maritime transport

Thus, to undertake an environmental impact assessment three main tasks are necessary: (1) match transport operations with environmental components; (2) link them with regulations; (3) assess risks, impact and responsibilities. This is particularly true with a transport project. These tasks allow environmental issues to be identified, commercial operations of the industry to be considered, best practices to be introduced, and measures of control to be planned. All EIA are constructed, albeit with more details, under this framework.

Operational constraints

Obviously the adoption of an EMS favors the conformity and the adaptability of transport operations to environmental legislation. The issue of responsibility is at the heart of the processes of environmental sustainability. The modern history of environmental legislation reveals that different laws have been promulgated on a wide range of physical components of the environment. In the field of transport, several measures have been adopted to reflect the objectives of sustainable environment. While many of these measures are perfectible, international environmental laws and legislation tend to put pressure within and beyond national boundaries. The growth in the number and strength of environmental policies, rules and practices has increased the number of standards and has permitted the development of a wide range of techniques of environmental sustainability such as voluntary remediation programs, flexible standards and procedures, financial and technical support. The main benefit resides in the reduction of legal costs that affect the profits and the productivity of the transport industry. There exist four functions that can anchor the administrative responsibility of the transport industry:

- **Quantify the terms of references**. Everywhere, transport companies have to adapt their environmental objectives in relation to a great variety of geographical conditions, commercial factors, technological changes and environmental policies, legislation and regulations. Henceforth, transport corporations cannot limit themselves to simply enunciate principles or policies in the field of environment. Environmental management systems applied to transportation require a massive amount of information on environmental conditions and the dynamics of the transport system. The best practices are those adopting an analytical framework disaggregating environmental objectives. This implies homogenous, exploitable and credible units of measurement that are time referenced with a view to observe the evolution and comparison by sector and at different geographical scales. Data must permit the environmental impacts of transport activities to be quantified continuously. Reasonable objectives need to be fixed that in turn represent benchmarks for defining strategies of environmental sustainability in different sectors and at different levels. For instance, this could be the reduction of polluting emissions by a certain percentage over a given period of time in relation to a given benchmark.
- **Devise a calendar of operations**. The objectives of sustainable environment can be very complex for several reasons: (1) the lack of data for evaluating the impact and the cost of environmental measures; (2) the importance of strategies and actions at the international level; (3) the lack of procedures or methods to solve the problems; (4) conflicts of jurisdictions; and (5) the growth in the production of polluting emissions. New problems may need to be controlled or solved while our understanding of environmental problems improves and new environmental technologies are introduced. The most efficient strategies are those that target short-, medium- and long-term objectives with precise values. The best strategies are dynamic and integrated within a continuous evaluation process.

- **Establish benchmarking**. There is a need to establish the minimum standards of quality that are sought from transport operations. Since it is almost impossible to establish a pristine reference, the standards must permit to specify the state of environmental quality that is sought. These standards must express the specific environmental status with regards to water, air, soil and all the other components of the physical environment within a precise geographical area. The standards will clarify the level of pollution or other impacts that can be supported by people and the environment without any risks. The best practices are those that: (1) establish standards on the basis of scientific criteria; (2) engage public administration in the development of procedures for writing, applying and controlling legislation; and (3) integrate these standards within the practices of territorial planning.
- **Implement measures of control**. It is important to impose standards of quality and parameters for the different components of the physical environment. The objective must be the eradication of any toxic substances that may present a risk for people's health and the environment. The practices and the policies of environmental sustainability within the transport industry demonstrate the need for flexibility and adaptability of transport systems to the challenges of protecting the environment through the adoption of appropriate technologies and material. Notwithstanding the criteria of analysis in the environmental management plan, it is important to undertake frequent assessments with a view to control the respect of transport operations to the existing environmental legislation. Environmental certification represents the best instrument of control of the transport industry.

Governance strategies

Implementing an environmental management system requires a broad range of instruments. Six instruments are conducive to the implementation of strategies for environmental sustainability applied to the transport industry:

- **Strategic instruments**. Any strategy of environmental sustainability must rest on a vision of development that defines general orientations and interacts with existing policies. Corporate leadership plays a key role in the success of practices of sustainable environment. A company-wide vision of sustainability facilitates the integration of sustainable environment goals within management practices. Furthermore, it may help to obtain government support and encourage the participation of stakeholders.
- **Legal instruments**. Legislation remains one of the most important instruments to achieve sustainable environment. The best practices are associated with different legislation emanating from public administrations at the local, regional, national and international level.
- **Geographic instruments**. Geographic and cartographic tools are fundamental for environmental sustainability planning. For transport enterprises, these tools permit the construction of data bases on the physical characteristics of land uses, inventory and mapping of freight and passenger flows, trip length and frequencies.
- **Economic instruments**. Cost-benefit analyses are important in the elaboration of pricing and fiscal policies and fixing quotas to protect the environment from transport activities. Economic instruments can further be modified to assess more accurately the costs of environmental damage. The most efficient "green taxes" in terms of environmental sustainability rest on the establishment of dues that reflect the marginal costs of environmental damage.
- **Communication instruments**. Personnel training, research and development activities, dissemination of impact assessment and risk evaluation reports are extremely

important in influencing the behavior of transport users and corporate decision making. The best performances in sustainable environment are achieved in transport firms that have adopted measures of knowledge growth and environmental responsibilities among all the personnel working in the transport organization.

- **Cooperation instruments**. These instruments aim at increasing the institutional capacity of the transport industry by integrating all the elements of environmental sustainability in corporate strategies. Cooperation and voluntary alliances between governments and transport industry stimulate and facilitate the identification of objectives and the elaboration of strategies for a sustainable environment.

CASE STUDY Environmental practices in Swedish maritime transport

Port operation permit

In 1999, the Swedish government adopted an environmental code that requires all national ports to acquire an exploitation permit before 2006. To obtain the permit necessitates ports to complete a comprehensive environmental impact assessment of all maritime activities related to transport (navigation, rail, road and pipeline), energy consumption and waste disposal. In addition they are required to have identified the measures by which they intend to improve environmental performance. All the study costs required to obtain the permit have to be paid for by the port operators themselves. The Swedish environmental management framework targeted at the port industry contains three principles. First, port operators have to identify the environmental impacts to be controlled. Port operators have to possess the relevant knowledge of the area in which terminal operations are taking place with a view to prevent damage and detriment. They have to adopt the precautionary principle to limit the risk of emissions and all necessary measures to combat adverse health and environmental effects. They are also subject to the appropriate location principle where the choice of site for land and marine operations are compatible with the environment. Second, port operators must introduce good environmental practices on the ground. National port operators are subdued to the polluter pays principle to protect the environment. The code stipulates that port operators must conduct their operations in accordance with resource management and ecocycle principles to insure efficient use of raw materials and energy and minimize consumption and waste. It must be possible to use, reuse, recycle and dispose of all materials extracted from the environment in a sustainable manner with the minimum use of resources and without damaging the environment. Port operators are subject to the product choice principle which consists in restraining the use of chemical products that may involve hazards to human health and the environment. Such an approach has compelled port operators to adopt the best possible environmental technology. Third, port operators must incorporate the concept of continual improvement. Port operators must therefore adopt a result-based management in the light of benefits and costs in applying environmental quality standards that must be achieved by sector, region or date. They must demonstrate that their operations are continuously undertaken in an environmentally acceptable manner.

Marine fairway charges

Since 1998, Sweden has established differentiated fairway and port charges for any passenger and cargo ships calling at Swedish ports, regardless of the ships' flag state.

This system was part of a tripartite agreement between the Swedish Maritime Administration, the Swedish Shipowners' Association and the Swedish Ports' and Stevedores' Association to employ vigorous measures to decrease ship-generated air pollution. The system is based on two charging components. One portion corresponds to the gross tonnage of the ship and the other portion is based on the volume of cargo being transported by the ship. The charging levels for the gross tonnage portion are differentiated with respect to emissions of nitrogen oxides (NO_x) and sulphur oxides (SO_x) by the ship. The charging scheme does not represent an increase in shipping costs, nor an increase in port incomes. Ships which have taken environmentally protective measures will be charged reduced dues, while ships with higher emission levels will pay higher dues. The differentiated charging scheme is completed with a system of subsidies for the installation of catalytic converters. The Swedish Maritime Administration reimburses the fairway dues being paid for a five-year period to shipowners that have invested in the installation of catalytic converters. This reimbursement can be as high as 40 percent of the installation costs. This subsidy applies to any flagged ship notwithstanding its port of registry or nationality, registered at the Swedish Maritime Administration. The differentiated fairway charges scheme has permitted significant reductions in nitrogen oxides (NO_x) and sulphur oxides (SO_x) emissions by ships calling at Swedish ports. It is important to underline that the system does not lead to reduction in global incomes for the Swedish Maritime Administration.

Bibliography

André, P., C.E. Delisle and J.P. Reveret (2004) *L'évaluation des Impacts sur l'Environnement*, Montréal: Presses Internationales Polytechnique.

Attali, J. (1975) *La Parole et l'Outil*, Paris: PUF.

Barke, M. (1986) *Transport and Trade*, Edinburgh: Olivier & Boyd.

Biondi, V., M. Frey and F. Iraldo (2000) "Environmental Management Systems and SMEs. Motivations, Opportunities and Barriers related to EMAS and ISO 14001 Implementation", *Greener Management International*, 29: 55–69.

Bonnafous, A. and C. Raux (2003) "Transport Energy and Emissions: Rail", in D.A. Hensher and K.J. Button (eds) *Handbook of Transport and the Environment*, vol. 4, Amsterdam: Elsevier, pp. 293–307.

Comtois, C. and P.J. Rimmer (2004) "China's Complete Push for Global Trade: Port System Development and the Role of COSCO", in D. Pinder and B. Slack (eds) *Shipping and Ports in the 21st Century*, London: Routledge, pp. 40–62.

Delucchi, M.A. (2003) "Environmental Externalities of Motor Vehicle Use", in D.A. Hensher and K.J. Button (eds) *Handbook of Transport and the Environment*, vol. 4, Amsterdam: Elsevier, pp. 429–49.

Evans, R.L. (2007) *Fueling our Future*, Cambridge: Cambridge University Press.

Freimann, J. and M. Walther (2001) "The Impacts of Corporate Environmental Management Systems. A Comparison of EMAS and ISO 14001", *Greener Management International*, 36: 91–103.

Gilbert, R. and A. Perl (2008) *Transport Revolutions. Moving People and Freight without Oil*, London: Earthscan.

Gwilliam, K.M. and Z. Shalizi (1996) *Sustainable Transport: Sector Review and Lessons of Experience*, Washington, DC: World Bank.

Holmen, B.A. and D.A. Niemeier (2003) "Air quality", in D.A. Hensher and K.J. Button (eds) *Handbook of Transport and the Environment*, vol. 4, Amsterdam: Elsevier, pp. 61–79.

Jakob, A., J.L. Craig and G. Fisher (2006) "Transport Cost Analysis: A Case Study of the Total Costs of Private and Public Transport in Auckland", *Environmental Science and Policy*, 9: 55–66.

Johansson, B. (2003) "Transportation Fuels – A System Perspective", in D.A. Hensher and K.J. Button (eds) *Handbook of Transport and the Environment*, vol. 4, Amsterdam: Elsevier, pp. 141–57.

Khare, M. and P. Sharma (2003) "Fuel options", in D.A. Hensher and K.J. Button (eds) *Handbook of Transport and the Environment*, vol. 4, Amsterdam: Elsevier, pp. 159–83.

Lawrence, D.P. (2003) *Environmental Impact Assessment. Practical Solutions to Recurrent Problems*, Hoboken, NJ: Wiley-Interscience.

Leinbach, T.R. and J.J. Bowen (2004) "Airspaces: Air Transport, Technology, and Society", in S.D. Brunn, S.L. Cutter and J.W. Harrington (eds) *Geography and Technology*, Dordrecht: Kluwer Academic, pp. 285–313.

Lenzen, M., C. Dey and C. Hamilton (2003) "Climate Change", in D.A. Hensher and K.J. Button (eds) *Handbook of Transport and the Environment*, vol. 4, Amsterdam: Elsevier, pp. 37–60.

Leppäkoski, E., S. Gollasch, P. Gruszka, H. Ojaveer, S. Olenin and V. Panov (2002) "The Baltic – A Sea of Invaders", *Canadian Journal of Fisheries and Aquatic Sciences*, 59(7): 1175–88.

Marine Transportation System National Advisory Council (MTSNAC) (2001) *Challenges and Opportunities for the US Marine Transportation System*, Washington, DC: US Maritime Administration (MARAD).

Mass, P. (2005) "The Breaking Point", *New York Times Magazine*, August 21.

Organization for Economic Co-operation and Development (1988) *Transport and the Environment*, Paris: OECD.

Organization for Economic Co-operation and Development (2002) *OECD Guidelines towards Environmentally Sustainable Transport*, Paris: OECD.

Potter, S. and I. Bailey (2008) "Transport and the Environment", in R. Knowles, J. Shaw and I. Docherty (eds) *Transport Geographies. Mobilities, Flows and Spaces*, Oxford: Blackwell Publishing, pp. 29–48.

Raymond, K. and A. Coates (2001a) *Guidance on EIA Screening*, Luxembourg: Office for Official Publications of the European Communities.

Raymond, K. and A. Coates (2001b) *Guidance on EIA Scoping*, Luxembourg: Office for Official Publications of the European Communities.

Rubenstein, J.M. (2004) "Motor Vehicles on the American Landscape", in S.D. Brunn, S.L. Cutter and J.W. Harrington (eds) *Geography and Technology*, Dordrecht: Kluwer Academic, pp. 267–83.

Schipper, L.J. and L. Fulton (2003) "Carbon Dioxide Emissions from Transportation: Trends, Driving Factors and Forces for Change", in D.A. Hensher and K.J. Button (eds) *Handbook of Transport and the Environment*, vol. 4, Amsterdam: Elsevier, pp. 203–25.

Sperling, D. (2003) "Cleaner Vehicles", in D.A. Hensher and K.J. Button (eds) *Handbook of Transport and the Environment*, vol. 4, Amsterdam: Elsevier, pp. 185–99.

Talley, W.K. (2003) "Environmental Impacts of Shipping", in D.A. Hensher and K.J. Button (eds) *Handbook of Transport and the Environment*, vol. 4, Amsterdam: Elsevier, pp. 279–91.

UK Department of the Environment, Transport and the Regions (1999) *Indicators of Sustainable Development*, http://www.sustainable-development.gov.uk/sustainable/quality99/.

United Nations (2001) *Sustainable Transport Pricing and Charges. Principles and Issues*, New York: United Nations.

United Nations Commission on Trade and Development (2007) *Review of Maritime Transport 2007*, New York: United Nations.

United Nations Economic and Social Commission for Asia and the Pacific (2002) *Report of the Regional Seminar on Liberalisation of Maritime Transport Services under WTO 2002*, New York: United Nations.

Valcic, I. (1980) *Le Bruit et ses Effets Nocifs*, Paris: Masson.

Vance, J.E. (1990) *Capturing the Horizon*, Baltimore, MD: Johns Hopkins University Press.

Velas, F. (1991) *Le Transport Aérien*, Paris: Economica.

Walker, L.J. and J. Johnston (1999) *Guidelines for the Assessment of Indirect and Cumulative Impacts as Well as Impact Interactions*, Luxembourg: Office for Official Publications of the European Communities.

⑨ Transportation planning and policy

Since transportation is such an important component of contemporary society, capable of producing significant benefits, yet giving rise to many negative externalities, appropriate policies need to be devised to maximize the benefits and minimize the inconveniences. At the same time the allocation, design and construction of such transport infrastructure and services must be subject to careful planning, both by public and private agencies. In this chapter a distinction is drawn between policy and planning. The major features of the policy and planning processes are examined. Because they both have to reflect the fundamental changes in society and contemporary issues and problems, policies and planning change. The changing orientation of public policy is described and the chapter goes on to explore the evolving nature of urban transport planning and intervention methods.

Concept 1 – The nature of transport policy

Defining policy and planning

The terms "policy" and "planning" are used very loosely and are frequently interchangeable in many transport studies. Mixing them together is misleading. Policy and planning represent separate parts of an overall process of intervention. There are circumstances where policy may be developed without any direct planning implications, and planning is frequently undertaken outside any direct policy context. However, precise definitions are not easily come by. For example here are two definitions of policy:

> A set of principles that guide decision-making or the processes of problems' resolution.
>
> (Studnicki-Gizbert, 1974)

> The process of regulating and controlling the provision of transport.
>
> (Tolley and Turton, 1995)

Transport planning is facing a similar issue related to its definition:

> Transport planning is taken to be all those activities involving the analysis and evaluation of past, present and prospective problems associated with the demand for the movement of people, goods and information at a local, national or international level and the identification of solutions in the context of current and future identification of economic, social, environmental, land use and technical developments and in the light of the aspirations and concerns of the society which it serves.
>
> (Transport Planning Society, UK)

A programme of action to provide for present and future demands for movement of people and goods. Such a programme is preceded by a transport study and necessarily includes consideration of the various modes of transport.

(European Environment Information and Observation Network)

In this chapter the following definitions are used:

Transport policy: the development of a set of constructs and propositions that are established to achieve particular objectives relating to social, economic and environmental development, and the functioning and performance of the transport system.

Thus, transport policy can be concomitantly a public and private endeavor, but governments are often the most involved in the policy process since they either own or manage many components of the transport system. Governments also often perceive that it is their role to manage transport systems due to the important public service they provide.

Public policy is the means by which governments attempt to reconcile the social, political, economic and environmental goals and aspirations of society with reality. These goals and aspirations change as the society evolves, and thus a feature of policy is its changing form and character. Policy has to be dynamic and evolutionary.

Transport planning deals with the preparation and implementation of actions designed to address specific problems.

A major distinction between planning and policy is that the latter has a much stronger relation with legislation. Policies are frequently, though not exclusively, incorporated into laws and other legal instruments that serve as a framework for developing planning interventions. Planning does not necessarily involve legislative action, and is more focused on the means of achieving a particular goal.

Why transport policy?

Transport policies arise because of the extreme importance of transport in virtually every aspect of national life. Transport is taken by governments of all types, from those that are interventionalist by political credo to the most liberal, as a vital factor in economic development. Transport is seen as a key mechanism in promoting, developing and shaping the national economy. Many regional development programs, such as the Appalachia Project in the USA in the 1960s and the contemporary Trans-European Networks (TENs) policy in the EU are transport based. Governments also seek to promote transportation infrastructure and services where private capital investment or services may not be forthcoming. Paradoxically, academics question the directness of the links between transport and economic development.

Transport frequently is an issue in national security. Policies are developed to establish sovereignty or to insure control over national space and borders. The Interstate Highway Act of 1956, that provided the United States with its network of expressways, was formulated by President Eisenhower on the grounds of national security (see Figure 9.1). Security was at the heart of the 2002 imposition of requirements on document clearance prior to the departure of freight from foreign countries to the USA.

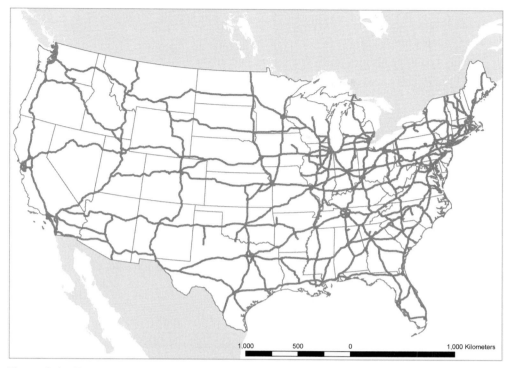

Figure 9.1 The Interstate highway system

Transport raises many questions about public safety and the environment. Issues of public safety have for a long time led to the development of policies requiring driving licenses, limiting the hours of work of drivers, imposing equipment standards, establishing speed limits, mandating highway codes, seat belts and other accident controls. More recently, environmental standards and control measures are being instituted in response to the growing awareness of the environmental impacts of transport. Examples include banning leaded gasoline and mandating catalytic converters in automobiles.

Transport policy has been developed to prevent or control the inherent monopolistic tendency of many transport modes. Unrestrained competition leads to market dominance by a company thereby achieving monopoly power. Such dominance brings into question many issues affecting the public interest such as access (in a port would smaller shipping lines be excluded?), availability (would smaller markets continue to receive air service by a monopoly carrier?) and price (would the monopolist be in a position to charge high prices?).

Other reasons for policy intervention include the desire to limit foreign ownership of such a vital industry. For example, the USA limits the amount of foreign ownership of its domestic airlines to a maximum of 49 percent, with a maximum of 25 percent control. Other countries have similar restrictions.

Policy instruments

Governments have a large number of instruments at their disposal to carry out transport policy. An extremely important instrument is public ownership. The direct control by the state of transportation is very widespread. Most common is the provision by public agencies of transport infrastructure such as roads, ports, airports, canals. Public ownership also extends to include the operation of transport modes. In many countries airlines, railways, ferries and urban transit are owned and operated by public agencies.

Subsidies represent an important instrument used to pursue policy goals. Many transport modes and services are capital intensive, and thus policies seeking to promote services or infrastructure that the private sector are unwilling or unable to provide may be made commercially viable with the aid of subsidies. Private railroad companies in the nineteenth century received large land grants and cash payments from governments anxious to promote rail services. In the USA, the Jones Act, which seeks to protect and sustain a US-flagged merchant fleet, subsidizes ship construction in US shipyards. Indirect subsidies were offered to the air carriers of many countries in the early years of commercial aviation through the awarding of mail contracts. Dredging of ship channels and the provision of other marine services such as pilotage and navigation aids are subsidies to facilitate shipping.

Both public ownership and subsidies represent instruments that require the financial involvement of governments. Regulatory control represents a means of influencing the shape of transportation that is very widely employed. By setting up public agencies to oversee particular sections of the transport industry, governments can influence the entire character and performance of the industry. The agencies may exert control on entry and exit, controlling which firms can offer transportation services, at what prices, to which markets. Thus, while the actual services may be offered by private firms, the regulator in fact plays a determining role. Regulatory agencies in the USA such as the Civil Aeronautics Board played a critical role in shaping the US airline industry for decades.

Other policy instruments are less direct, although in many cases can be as equally important as the three discussed above. Many governments are major promoters of research and development in transportation. Government research laboratories are direct products of state investments in R&D, and much university and industry R&D is sustained by government contracts and programs. The fruits of this research are extremely important to the industry. It is a vital source for innovation and the development of new technologies such as intelligent vehicles and intelligent highway systems.

Labor regulations pertaining to conditions of employment, training and certification may not be directed purposefully at influencing transport, but as a policy they may exert a significant effect over the industry. Safety and operating standards, such as speed limits, may have a similar effect. The restrictions on limiting the number of hours a truck driver may work may be instituted for safety reasons and for enhancing the working conditions of drivers, but they shape the economics of truck transport. In the same fashion speed limits help fix the distance of daily trips that one driver may undertake, thereby shaping the rate structure of the trucking industry.

Trends in policy development

Public policies reflect the interests of decision makers and their approaches to solving transport problems. These interests and approaches are both place specific (they apply to a particular area of jurisdiction) and time specific (they are established to reflect the conditions of transport and the intended solutions at a point in time). Policies change and evolve, therefore, as the conditions change and as the different sets of problems are recognized. Policies are dynamic.

The dynamic nature of policy is reflected in the way the policy instruments have been employed over the years. In the nineteenth century, when many of the modern transport systems were being developed, the prevailing political economy was one of *laissez-faire*, in which it was believed that the private sector should be the provider of transport services and infrastructure. Examples of private transport provision include:

- **Turnpikes**. The first British modern roads in the eighteenth century were the outcome of private trusts aiming to derive income from tolls on roads they built and maintained. It was likely the first massive private involvement in transport infrastructure provision (see Chapter 1, Concept 3).
- **Canals**. Many of the earliest canals were built with private capital. One of the first canals that helped spark the industrial revolution in Britain was the Bridgewater Canal, built by the Duke of Bridgewater between 1761 and 1765 to haul coal from his mines to the growing industrial city of Manchester.
- **Urban transit**. In most North American cities public transit was operated by private firms. The earliest examples were horsecars that followed rail lines laid out on city streets. With electrification at the end of the nineteenth century, the horse-cars were converted to streetcars and the network was greatly expanded. In the twentieth century buses were introduced by private companies operating on very extensive route systems.
- **Ships**. Most maritime companies were private enterprises. Many were family businesses, some of which became large companies, such as the Cunard Line in the UK. The main government involvement concerns military navies and ferries.
- **Railways**. Railways were developed by private companies during the nineteenth century, including such famous companies as Canadian Pacific and Union Pacific. In the USA this has continued to the present day.

This situation was not completely without public policy involvement, however. The massive subsidies that were granted to US and Canadian railroads are an example of state intervention. In the early twentieth century the overprovision of rail lines, competition between carriers and market failures led to a crisis in many parts of the transport industry, particularly after 1918. This led to a growing degree of government involvement in the transport industry, both to offset market failures, jurisdictional conflicts and to ensure that services could be maintained for the sake of the "public good":

- The failure of the Canadian Northern Railway and the threat of a Canadian Pacific monopoly led the Canadian federal government to establish Canadian National in 1921.
- In many cities private bus companies were taken over by municipally controlled transit commissions in the 1930s and 1940s.
- The airline industries in many countries were placed under the control of a national public carrier, for example Air France, Trans Canada Airlines, British Overseas Airways Corporation.
- Railways were nationalized in Europe after World War II, and in the USA in 1976 after the collapse of the Penn Central Railroad and several other lines, when a publicly funded passenger system (Amtrak) was set up, and a publicly owned freight railroad was established (Conrail).

In addition to the public ownership of transport modes, there emerged in the twentieth century a growing amount of regulatory control. The airline and the trucking industries saw entry limited by permits, and routes and rates were fixed by regulatory boards that had been set up to control the industries. At the same time greater safety regulations were being imposed and working conditions were increasingly being shaped by labor legislation. By the 1960s therefore, transportation had come under the sway of public policy initiatives that exerted an enormous influence on the industries and their spatial structures.

By the 1960s, however, there was a growing body of evidence that indicated that public ownership and regulation were not always in the public interest. Transportation costs that were fixed by the regulatory authorities were maintained at higher levels than were necessary. Research demonstrated that many regulatory boards had been "captured" by those they were supposedly regulating, so that they were frequently acting to protect the industries rather than the public. At the same time there was a crisis of public finances in many countries, where the costs of operating the state-owned transportation industry were seen to be unsustainable. Some economists espoused the theory of contestability, which repudiated traditional economic theory concerning monopoly power. Contestability theory argued that the threat of entry of a new actor was sufficient to thwart a monopolist's ability to impose monopoly pricing. The key, therefore, is to relax entry thresholds, by allowing new firms to start up, something the regulatory boards were impeding.

This evidence was brought into the public policy arena by politicians who espoused market-oriented views, notably President Reagan in the USA and Prime Minister Thatcher in the UK. Although President Carter had initiated the first steps towards deregulation in the USA in the mid-1970s, it was in the 1980s during the Reagan presidency that trucking, the airline industry and the railways were largely deregulated. In the UK, there was a massive move to privatize most sectors of the transport industry, including state and most municipally owned bus companies, the national airline, trucking, the railway, airports and most seaports.

Deregulation and privatization policies have spread, unequally, to many other parts of the world. New Zealand has perhaps the most open transport policy, but many others, such as Canada and Australia, have made significant steps in this direction. In the EU, the pace of deregulation and privatization is proceeding unevenly. Subsidies to state-owned transport companies have been terminated, and many airlines have been privatized. Government-owned railroads still exist in France, Germany, Italy and Spain, but the tracks have been separated from the traction and rail service operations, and have been opened up to new service providers. In Latin America, most of the state-owned transport sector has been deregulated. While the former centrally planned states have had to make the furthest adjustments to a more open market economy, several, such as China, have opened up large sections of the transport industry to joint ventures with foreign private enterprises. In China, many new highways and most of the major ports are being developed with private capital. Thus, at the beginning of the twenty-first century, transportation is under less direct government economic control worldwide than at any period over the last 100 years.

Changing nature of policy interventions

Recent trends in transport policy towards liberalization and privatization have not necessarily weakened government interventions. Controls over monopoly power are still in place, and even in the most liberal of economies there is still strong evidence of public policy intervention even in such capitalist countries as the USA, for example:

- **Ownership of ports and airports** in the United States. Terminals continue to be largely under State or municipal ownership. Thus the Port of Los Angeles is a department of the City of Los Angeles; the port of Hampton Roads is owned by the Virginia Port Authority; New York's port and three major airports are owned by the Port Authority of New York and New Jersey.

- **Highway provision, upgrade and maintenance** remains one of the most significant and enduring commitments of public funds.
- The **Surface Transportation Board**, the new regulatory agency controlling the railways, refused to sanction the proposed merger between Canadian National and Burlington Northern and Sante Fe Railroads in 2002. This was the first time in 20 years that the regulator had turned down an application for merger. It cited concerns of concentration of ownership.

Government policy orientations have changed, however (Table 9.1). The private sector plays a significant role in the change in public policy, which is mimicking the changes that have taken place in the strategies of private transport corporations. The public policy environment is thus shifting towards the consideration of transport as a set of interacting modes instead of independent modes. There is a growing recognition that the scale at which economic processes are taking place is both regional and global, and that seeking consensus is beneficial to all those involved. Coalitions where private and public stakeholders related to a transport system are established to insure consensus about proper policies are increasingly common. Yet, the public financing of transportation infrastructure is getting problematic, implying that different forms of public/private partnerships are being sought. Each case involves a different balance of public and private actors so different financing approaches have to be considered. Since privatization has been the dominant paradigm users of the system are increasingly seen as customers to whom a mobility subject to market forces is being offered. This implies a certain level of service at a fair price subject to accountability. Users reflect a public subsidy perspective while customers are elements of a revenue-generation strategy. A plan-based policy approach, which has commonly failed to capture correct market and technological trends, is being replaced by a more liberal market approach where deregulation leaves transportation more subject to price signals.

Governments are beginning to exert greater control over environmental, safety and security concerns, issues that are replacing former preoccupations with economic matters. The environment is becoming a significant issue for government intervention (see Chapter 8). Coastal zone legislation has made it increasingly difficult for ports to develop new sites in the USA. Air quality is a major factor influencing the allocation of US federal funds for urban transport infrastructure. In Europe, environmental issues are having an even greater influence on transport policy. The EU Commission

Table 9.1 Shift in public transport policy perspective

Conventional	*Emerging*
Independent modes	Intermodal systems
Local economies	Regional/global economies
Independent jurisdictions	Coalitions/consensus
Publicly funded	Public/private partnerships
Users (public subsidy)	Customers (revenue generation)
Build (infrastructure provision)	Manage (optimization of existing resources)
Plan (regulations; political signals)	Market (deregulations; price signals)

is promoting rail and short sea shipping as alternatives to road freight transport. Projects are assessed on the basis of CO_2 reduction. All transportation projects are subject to extensive environmental assessments, which may lead to a rejection of proposals, despite strong economic justification, such as the case of the Dibden Bay proposal for expanding the port of Southampton in the UK. As a major source of atmospheric pollution and environmental degradation, the transportation industry can anticipate many further government environmental policy interventions.

Safety has always been a policy issue. Legislation imposing speed limits, mandating seat belts and other measures often enshrined in official highway codes have sought to make travel safer. However, it is in the area of security that the most recent set of policy initiatives have been drawn. Screening of people and freight has become a major concern since 9/11. Both the US government and such international organizations as the International Maritime Organization (IMO) and the International Civil Aviation Organization (ICAO) have instituted new measures that impact on operations, and represent additional costs to the transport industry.

Concept 2 – The policy process

Problem definition

Policies are developed in response to the existence of a perceived problem or an opportunity; they never exist in a vacuum. The context is extremely important because it will shape the kinds of actions considered. For example:

● Who has **identified the problem**? Is it widely recognized by society as a whole or is it limited in scope to a local pressure group for example? In the case of the former there may be a greater willingness to intervene than in the latter, depending on the political power exerted by the pressure group.
● Do the public authorities have the **interest or will to respond**? There are usually many more problems than the policy makers are willing to address. Many issues remain unaddressed.
● Do the public authorities wish to wield the **instruments necessary to carry out a policy response**? The problem may be recognized, but public authorities may have little ability to effect change. Such is the problem of many environmental problems that require global solutions.
● What is the **timescale**? How pressing is the problem, and how long would a response take? Policy makers are notoriously prone to attempt only short-term interventions, since their mandates are usually of relatively short duration. Long-term issues may not attract policy makers because the results of any policy intervention may be decades away.

These questions lie at the heart of the need to correctly identify the problem or opportunity. No policy response is likely to be effective without a clear definition of the issue. The following elements need to be considered in defining a problem:

● Who has **identified the problem**, and why should it be seen to be a problem? Many problems exist, but few are taken up because they are not brought before a wide audience.
● Is there **agreement on the problem**? If there is no agreement that a problem exists, it is unlikely that a strong policy response will be forthcoming. Effective policies are more likely to be formulated if there is widespread recognition of a

problem and its causes. A problem for the Kyoto Accord on global warming was that decision makers in the USA were not convinced that the problem is due to human-induced carbon dioxide emissions.

- Is it an issue that can be **addressed by public policy**? The price of oil is regarded by many as a problem, but individual countries have no power to affect the price of this commodity.
- Is it **too soon to develop a policy**? This argument was used by the lobby in California that opposed stricter emission controls on vehicles in the early 1990s, based on the argument that the technology of alternative energy for vehicles was not sufficiently advanced.
- Is the problem **seen differently by groups with different values**? Environmentalists see many transport issues differently than many other interest groups. Divergence of opinions may affect how the problem is addressed.
- Is the problem **fully understood**? Do we know the causal relationships that may be necessary to provide a solution? Transport and development, and the role of transport in global warming are issues around which there is a debate.
- Can the **relationships** between the factors that make up the problem **be quantified**? Problem definition is better when it is possible to measure the scale and scope of the issues involved.

In defining the problem or opportunity and to help address the questions above, background studies are required. The state of affairs needs to be provided which will identify the actors, the issues and the possible means that are available. It is also important to forecast trends in order to identify whether the issue is likely to change.

Policy objectives and options

The eventual success of a policy depends on establishing clear goals. If there are multiple objectives they must be consistent. They must be flexible enough to change over time as the circumstances evolve. In simple terms the objectives must:

- Identify the present **conditions and situation**.
- Indicate what the **goals** are.
- Identify the **barriers** to achieving the goals.
- Identify what is **needed** from other agencies.
- Determine how **success** will be judged.
- Identify what **steps** are required to achieve success.

Having defined the problem and objectives, policy options must be formulated and evaluated. In many cases more than one solution has to be considered for policy adoption. The objectives may be realized in many different ways. Best practices from other jurisdictions may be considered, and all other possible solutions need to be considered. By evaluating the options it may be possible to identify the one that best meets the goals that have been established and at the same time is the best fit for local circumstances. These types of evaluations are referred to as *ex ante*, because the outcomes are being assessed even before the policy is put into practice. Although one can never completely anticipate the outcome of different prospective policy options, *ex ante* evaluations are capable of bringing to light what problems may develop when the preferred option is implemented. Thus, when the future policy is to be evaluated (*ex post*), problems of data, reporting and identification of success criteria may have been already anticipated and resolved through an earlier *ex ante* assessment.

Many types of evaluation methods are employed in both *ex ante* and *ex post* assessments. These include cost-benefit analysis, multi-criteria analysis, economic impact and Delphi forecasting. Because evaluation takes place at several of the steps in the policy process, it is now regarded as a critically important issue. New ideas involving managing the policy process include performance-based management, where evaluation is built into the entire process. This means that in the policy process a great deal of attention has to be paid to how the goals, results and beneficiaries are to be measured. The selection of indicators has to be agreed upon by policy managers from the inception.

Policy implementation

The implementation of the selected option represents a critical aspect of the policy process. The most carefully crafted policy that is widely accepted by those it affects can flounder because of improper implementation. It is impossible to define an optimal implementation procedure because of the wide range of socio-economic circumstances that policies are applied to, and also because of the diversity of policies themselves. However, a ten-step model of policy implementation can be considered:

1 Policies must not face **insurmountable external constraints**. This means that the policy must not exceed the jurisdictional or constitutional limits of the agency. This is a common issue in federal states, where different transport modes may be under different jurisdictions. One of the factors that impeded the success of Montreal's second airport at Mirabel was that the Provincial government, which had opposed the site selected by the Federal government, refused to build an expressway to provide better access to the city. Other examples include cases where the transport issue cannot be resolved because of international borders. However, transnational agreements, especially within the European Union, have considerably reduced external constraints in transport policy implementation.

2 In implementing the policy there must be an **adequate time frame and resources**. The policy may be appropriate, but may fail because its implementation took longer or was more expensive than budgeted. A recent example is that of airport and port divestiture in Canada, where the two policies had similar goals but different implementation procedures. Airports had access to much greater financial assistance to carry out the transfer process; that of the ports was much smaller. As a result the port policy took much longer to be carried out.

3 The implementing agency must have **adequate staff and resources** to carry out the policy. A growing problem with environmental legislation is that agencies do not have the means to insure guidelines and standards are enforced. This has been a particular problem for many of the East European countries admitted into the EU in 2004, which have to adopt stricter standards than before.

4 The premises of **policy and theory must be compatible**. At one time public ownership was seen as a valid policy alternative. Today it may be a valid option in theory in some circumstances, but in most countries it is not politically acceptable.

5 **Cause-and-effect relationships** in the policy must be direct and uncluttered. A successful policy must be seen to be based on clear and unambiguous relationships. Complex policies are more likely to be misunderstood. It took many years for the new urban transport policy of the USA to be implemented. The Intermodal Surface Transportation Efficiency Act of 1991 (ISTEA) was an extremely complex piece of legislation that left many local agencies that were required to carry out the Act quite perplexed. It required simplification under the 1998 Transportation Efficiency Act for the Twenty-first Century (TEA21).

6 **Dependency relationships** should be **kept to a minimum**. If the agency in charge of implementing the policy has to rely on others to carry it out, the more fragmented will become the authority. The implementing agency will become more dependent on others with not necessarily the same interests.

7 The **basic objectives** of the policy need to be **agreed upon and understood**. All actors in the policy process must possess a clear understanding of the policy and what is required to carry it out. It goes without saying that all those involved must understand the policy and have knowledge about their roles in carrying it out. Information and training are essential elements in the policy process.

8 **Tasks** must be specified in an **appropriate sequence**. Implementation is a process with connected steps from conception to the end. If the steps are not carried out in the correct sequence the policy may fail. Difficulties may arise, for example, if evaluation is completed without the indicators of success being agreed upon beforehand, or if another agency is involved before necessary pre-conditions for its participation have been completed.

9 **Communication and coordination** need to be on the same wavelength. Those implementing the policy have to possess the same information base, have to interpret it in the same way and to communicate well with each other.

10 There must be **compliance**. Those agencies involved in implementing the policy must work towards total compliance. Many times policies are formulated but their compliance is lacking (see 3 and 7 above).

Policy evaluation and maintenance

The implementation stage is not the final step in the policy process. The effectiveness of the policy needs to be assessed after a certain period of time, and steps must be taken to insure that there are resources and means to maintain a successful policy. In the past, this tended to be overlooked, and after a while policies would be sidetracked by other newer initiatives. The long-term effect was the presence of many different policy initiatives frequently with conflicting goals. Prior to the ISTEA, US federal highway policy was marked by an accumulation of interventions, the so-called "entitlements" that were added one after the other, with little thought as to compatibility or integration with other funding. The result was that policies in place frequently conflicted with each other in terms of goals or implementation measures.

Ongoing program evaluation is thus central to the maintenance of policy. This has tended to be a difficult issue for managers who today find their programs being assessed by methods and data requirements that were never built into the policy initially. Performance-based management has become an essential tool in the policy process as a result. Under this system evaluation is built into all stages of the policy process, and indicators are agreed upon by the managers who carry out the programs as well as the units that undertake evaluation.

Concept 3 – Transport planning

The traditional transport planning process

Transport planning is usually focused on specific problems or on broad transport concerns at a local level. It has been traditionally a preoccupation of lower tier governments, such as the state or municipality. Because of this fact, transport planning is most developed in the urban sphere, and it is there where most experience has been gathered. The planning process, however, has a number of similarities with the policy process. Identifying a problem, seeking options and implementing the chosen strategy are essential steps in planning too. Because it tends to deal with localized problems, the solutions adopted in transport planning tend to be much more exact and specific than policy directives.

Many aspects and issues involved in urban transport planning have already been covered in Chapter 7. For a long time it was a field dominated by traffic engineers who gave it a distinctly mechanistic character, in which the planning process was seen as a series of rigorous steps undertaken to measure likely impacts and to propose engineering solutions. There were four major steps: trip generation, trip distribution, modal split and route selection. They involved the use of mathematical models, including regression analysis, entropy-maximizing models and critical path analysis.

There are many reasons why the results of these models should be treated with caution:

- They are only as good as the **data they manipulate** and many times the data is inaccurate or incomplete.
- They are based on **assumptions** that the mathematical relationships between variables remain constant.
- They can be **manipulated** to produce the outcome that the analyst knows the client prefers.
- Because the predictions were **rarely subjected to subsequent evaluation**, their validity is largely questioned, and the modeler is happy to predict the future.

The predictions of future traffic flows produced by the four-stage sequence are then used to identify planning options. Since the most common prediction of the modeling is that present capacities will be unable to cope with traffic growth, the tendency has been to produce planning solutions that call for an expansion of capacity. This has been referred to as predict and accommodate. It is the solution that has typified so much urban transport planning from the 1940s to the 1980s. It has given rise to the enormous expansion of highway construction that reinforces the dominance of the automobile. Rarely are there postmortems of the prediction models, and as has been learned by empirical observation, the issue of induced demand has distorted the actual traffic.

Contemporary transport planning

In cities traffic problems have increased significantly over the last 50 years, despite a great deal of urban transport planning. There is a growing realization that perhaps planning has failed and that the wrong questions have been asked. Rather than estimate traffic increases and then provide capacity to meet the expected growth, it is now accepted that what is required is better management of the transport system through new approaches to planning. Just as urban planning requires the inputs of many specialists, so transport planning is beginning to utilize multidisciplinary teams in order to broaden the scope of the planning process. Planning is still a multi-step process, but it has changed considerably over the last 20 years.

- **Goals and objectives**. While the goal of traditional transport policy, improving accessibility, is still useful, it has to be considered in the context of other desirable goals. For instance improving safety and health, reducing emissions from vehicles, improving equity, enhancing economic opportunities, improving community livability and promoting mobility are all valid. But which goal(s) are pursued results in a very different planning process. Defining goals becomes a much more complicated stage in contemporary planning. Increasingly goals have turned to consider managing demand, rather than trying to build capacity.

- **Options**. Given the possible range of goals that transport planners have to consider, it becomes necessary to provide a set of possible options. Several objectives may be desirable, and thus it is important to consider what they imply. Several scenarios may have to be considered, and they must become important components of the planning process.

- **Identification of actors, institutions, stakeholders**. Given that transport planning has the potential to influence so many elements of society – economic wellbeing, environmental conditions, social integration – it is important that those affected by the transport problem and its potential resolution should be identified so that they can be engaged. This would be a much wider list of affected parties than simply those involved in transportation activity itself, and requires recognizing a role for citizen participation.

- **Predicting outcomes, identifying benefits and assessing costs**. The stage of predicting the outcomes for each of the options is a critical step in the process. Models continue to play an important role, but whereas the traditional models were based on the number of trips, increasingly modeling is becoming more activity based. Transport is seen in the context of scheduling household decisions in time and space. Demographic and social data are used extensively, and the mathematical models have become more sophisticated. Nevertheless there are roles for other types of analyses, including non-objective forecasts. The predicted outcomes must then be assessed as to their benefits and costs. These may be expressed in monetary terms, but many transport planning situations call for measurement in other terms, such as visual impacts, environmental dislocations and employment impacts.

- **Choosing course of action**. Evaluation of the scenarios has to consider the costs and benefits from the frequently conflicting perspectives of the stakeholders and actors. Extensive public consultation may be required. The information has to be disseminated and explained so that an informed public can participate in the debate. Ultimately it will be the politicians who decide, but they are swayed by the strength of the arguments presented by the transport professionals, and in publicly contentious cases by pressure brought to bear by citizen groups.

Transport demand management

In rejecting the former paradigm of building capacity, transport planners have turned increasingly to managing both demand and the transport system. Building roads has produced a car-oriented society in which the other modal alternatives have little opportunity to coexist.

Car ownership is beyond the ability of the transport planner to control directly and the question remains whether it should be. But car use and ownership is affected by land use and density, both elements that planners can affect. High population densities, in particular, favor walking, bicycling and public transit use. It is for this reason that a great deal of attention in planning is being paid to densification and integration. This includes concentrating development along well-served transport corridors (transit-oriented development) and increasing densities in areas undergoing rehabilitation.

Managing the demand for transport is made up of a large number of small interventions that cumulatively can impact car use, but in particular improve the livability of cities. A sample of well-practiced and successful interventions includes:

- **Park and ride**. Parking spaces are provided, usually close to an expressway, where drivers can board buses that provide service to the city center. This has become a staple feature in the outer zones of many US and British cities. Its success is variable, however, and there is some evidence from the UK that park and ride may actually increase car use, as people who may have used regular bus services now use their cars to drive to the car parks.
- **Traffic calming**. Measures that seek to reduce the speed of vehicles in urban areas, such as speed bumps and street narrowing. For residential streets the goal is to make their use by car drivers unattractive because of the obstacles; for thoroughfares the objective is to reduce average speeds. The measures indicate the need for much greater attention to street design and layout.
- **Priority lanes for buses and high occupancy vehicles**. Lanes on major thoroughfares and expressways that are reserved for buses, taxis and passenger vehicles with several occupants. This has become an important feature of transport planning in North America, where major highway expansion projects offer priority lanes. The goal is to encourage use of buses and high occupancy vehicles that can be seen to travel at higher speeds along the reserved lanes by other drivers who may be stuck in traffic jams.
- **Alternate work schedules**. Encouraging work hours other than the dominant 9 to 5 schedule. One of the great problems in transport planning is that demand is concentrated in two main peak periods. In the past, efforts were made to meet this demand by increasing road capacity, which was never sufficient, and resulted in an underuse of the capacity the other 20 hours each day. Promoting flexible schedules and encouraging telecommuting are policies that seek to spread out the demand for transport over more hours and even to reduce the demand altogether.
- **Promoting bicycle use**. In some countries, particularly the Netherlands, the bicycle is an important mode of travel. It is a green and healthy mode, but in automobile-dependent cities the bicycle does not share the roads easily with trucks and cars. Encouraging greater use of the bicycle requires significant planning adjustments, such as the provision of bicycle lanes and bike stands.
- **Car sharing**. Encouraging drivers to share car use with neighbors or co-workers.
- **Enhancing pedestrian areas**. In most cities vehicles dominate the streets. In many areas of high population density, the quality of life (enhanced safety, less pollution, etc.) and the visual attractiveness of streetscapes can be enhanced by excluding vehicles from streets altogether, or limiting access to public transport vehicles. In Europe this has become a distinctive feature of the historic cores of many cities.
- **Improving public transit**. For 50 years or more public transit use has declined in most cities. Yet it is the only major alternative to the car in these cities, and thus enhancing the use of transit has become a major planning objective. Improvements include making transit more attractive, by improving bus schedules and improving the appearance and comfort of transit vehicles and stations. At the same time efforts are underway to widen the range of transit alternatives. These include extending commuter rail services, and constructing new systems such as light and heavy rail modes.
- **Parking management**. Restricting on-street parking, charging higher rates for parking.

Pricing

While planning interventions may have a positive cumulative effect in shaping transport demand, some economists suggest that a more direct approach involving the imposition of more stringent cost measures on car users is necessary. It is widely accepted that car users pay only a small proportion of the actual costs of their vehicle use. Economists argue that the external costs should be borne by the users. As intuitively rational as this argument may be, there are several problems with its application.

- First, there are **difficulties in measuring externalities**, with considerable variations in estimates between different studies. Different types of use, speeds, engines, vehicle weight, driving conditions, etc. make it difficult to produce broadly accepted values. Decision makers have difficulty in agreeing to impose charges when there is a diversity of evidence about external costs.
- Second, there are **practical difficulties in collecting these costs**. One of the easiest (and most widely used) methods is a gasoline tax. It is a crude approach, however, because it imperfectly distinguishes between driving conditions and engine type – a fuel-efficient vehicle may have just as high a consumption in heavy urban traffic as a gas-guzzler in a rural setting.
- Third, is the **political difficulty of imposing such additional costs on the public**. In North America, in particular, access to "free" roads is regarded as a birthright, and it is intensely unpopular to propose any new forms of revenue generation that hints at additional taxation.

The effectiveness of economic controls is evident by the experience of Hong Kong where, despite high incomes, car ownership and use remains at a very low level. This is due in the main to the high cost of parking. An even more drastic example is Singapore, where extreme measures limiting car purchases, high vehicle licenses, electronic tolls on highways and cordon pricing in the downtown area have restrained car use.

The use of pricing mechanisms may be less in other countries, but the trend towards greater application of some forms of tolling is accelerating. Cordon pricing has been applied in a number of jurisdictions, especially in Norway in Oslo, Bergen and Trondheim. Under cordon pricing, access to certain areas, usually the CBD, is tolled. The most famous application was the decision to charge private vehicles for entry into Central London in early 2003, a program that has proved to be successful, despite a great deal of opposition.

Another form of charging is the imposition of tolls on new highways and bridges. In North America, the public had become used to the notion that highways are "free", a legacy of the Interstate Highways Act, funded largely by Congress. In both Canada and the US legislation now permits private companies to build and operate private roads and bridges, and to collect tolls to cover costs. In Canada, Highway 407 outside Toronto and the Confederation Bridge linking Prince Edward Island to the mainland are examples of tolled facilities developed and operated by private corporations. The same trend applies to developing countries such as China where many new roads and bridges are toll based.

Another form of pricing is congestion or fair pricing. Here certain lanes of a highway are tolled, but at variable rates. When traffic is moving freely, the charges for the tolled lanes are nil. But as traffic builds up and speeds are reduced, the costs of using the reserved lanes increase. Collection of the tolls is electronic, and drivers are informed of the current charges by large signs. Drivers are given a choice, therefore, to stay in the slower lanes for free, or move to the tolled lanes at a cost that is proportionate to the speed on the congested lanes. This system is now in place in several US states, after successful tests in California and Texas.

Intelligent vehicles and intelligent highways

Technology is seen by many transport planners as a solution to a wide range of transport problems. This is an approach that has achieved wide acceptance in the USA, where there has always been a strong emphasis on seeking engineering solutions to urban transport problems. It involves using information technologies (ITS) to provide better information and control over traffic flow and individual vehicle use. Many of the solutions involve the application of remote sensing techniques along with ITS.

One of the most promising approaches involves interactive highways. They are a means of communication between the road and driver that warn of approaching road conditions. Warnings include electronic message boards that suggest alternate routes to approaching motorists, and designated radio frequencies that give updated traffic reports. It is based on a closed-circuit TV system (CCTV) that records lane-by-lane occupancy, volume and speed. At the same time ramp meters record in real time the amount of traffic entering the highway. This information is analyzed and processed at a control center that can dispatch emergency equipment to accidents as they happen, and can inform other drivers of road conditions, accidents, construction and delays.

A further technology is emergency signal priority. This is a means of providing emergency vehicles and public transport buses priority at traffic lights in congested areas. The system allows a vehicle equipped with a system emitter to send a coded infrared message to the system detector, installed at the traffic intersection. When activated, the detector receives the coded message and then either holds the existing green light until the vehicle passes through or changes the existing red light to a green light.

ITS is being applied in many further innovative ways to improve the efficiency of emergency vehicles. For example, in Montreal mathematical models are being used to predict where road accidents are likely to occur given the time of day, traffic volumes and weather conditions. Ambulances can be assigned to these zones. Once deployed and assigned to a specific event, optimal routing is determined and relayed to drivers. When the first responders have identified the extent and type of injuries, the information is relayed to a control center which determines availability of doctors and nurses at which hospital emergency room, and suggests a routing for the ambulance using a least-time model estimation.

ITS is providing many solutions to the problems of road pricing. Toll collection is increasingly using electronic means to collect tolls without requiring vehicles to stop at toll booths. In its simplest form, vehicles equipped with a transponder that emits details of the vehicle are allowed to pass through toll lanes without stopping to pay. Receptors at the booth record the passage and debit the account. This is at the heart of cordon pricing and of most other new toll systems.

This technology, however, is being wedded to global positioning systems (GPS), which is likely to produce radical changes in the way vehicular traffic is priced. This combination of technologies will permit a more effective means of applying road pricing than the road tax. Vehicles will be required to have an on-board unit that includes a GPS receiver, a set of digital maps showing jurisdictional boundaries, an odometer feed, a set of distance rate charges and a wireless communication system to report billing data. During each trip the GPS determines the jurisdictional zones, the odometer calculates the distance traveled in each zone, the computer tabulates the running total of fees and then periodically signals the data to the billing agency. These systems are presently being evaluated in several states in the USA. A comparable system is already in place in Germany, where since late 2004 all truck movements are charged an environmental tax based on distance traveled and vehicle characteristics.

Freight planning

The vast preponderance of transport planning, certainly at the urban level, has been devoted to passengers. The automobile and public transit issues have preoccupied planners since individual mobility can be a highly political issue (drivers are also voters). Yet freight traffic represents a significant part of the many problems that planning seeks to address. The models and data inputs used in transportation planning are of little relevance when applied to freight movements. For example, demographic data, such as household size, the backbone of passenger analysis, are irrelevant for freight. The bi-polar daily peak of traffic movements applies only to passengers, freight movements being distributed in a different profile over a 24-hour period.

While trucks account for approximately 10 percent of vehicles on the road, their size, low maneuverability, noisiness and high pollution output make their presence particularly objectionable. Truck pickup and delivery in city centers is particularly problematic because of limited parking. At the same time trucks are vital to the economy and wellbeing of society. Commerce is dominated by trucking, and the logistics industry in particular is dependent on road transport for pickup and delivery. Garbage pickup, snow removal and fire protection are among many essential services that are truck-oriented.

Planning for freight movements is still in its infancy. As a largely private sector activity freight transport is difficult to control, and many of the decisions that affect trucking are made by the industry itself. The emergence of large logistics/distribution centers on the outer fringes of metropolitan areas is taking place without public control or oversight. In Europe, some attempt to manage such development by establishing publicly promoted freight villages has had only limited success.

Several cities are seeking to limit trucking as pressures keep mounting up. In many jurisdictions limits on heavy trucks in urban areas are in place, and there are restrictions on times of delivery and pickup, which in some European cities extend to the exclusion of all trucks in the urban core during daytime hours. The question remains about how to constrain urban freight circulation while not impairing the economy.

All these steps are tackling the problem at the edges. In many cities there are no census data on freight traffic, so that planning in the few cases where it takes place is inevitably hit and miss. There needs to be a much greater focus on freight planning overall, since it is almost universally recognized that freight transport is important.

Concept 4 – Transport safety and security

A new context in transport security

While issues of safety and security have been before transport planners and managers for many years, it is only recently that physical security has become an overriding issue. Concerns were already being raised before the Millennium, but the tragic events of 9/11 thrust the issue of physical security into the public domain as never before and set in motion responses that are reshaping transportation in unforeseen ways. In addition, threats to health, such as the spread of pandemics, present significant challenges to transport planning and operations.

As locations where passengers and freight are assembled and dispersed, terminals have particularly been a focus of concern about security and safety. Because railway stations and airports are some of the most densely populated sites anywhere, crowd control and safety have been issues that have preoccupied managers for a long time. Access is monitored and controlled, and movements are channeled along pathways that provide safe access to and from platforms and gates. In the freight industry security concerns have been directed in two areas: worker safety and theft. Traditionally, freight terminals have been dangerous work places. With heavy goods being moved around yards and loaded onto vehicles using large mobile machines, accidents are systemic. Significant improvements have been made over the years, through worker education and better organization of operations, but freight terminals are still comparatively hazardous. The issue of thefts has been one of the most severe problems confronting all types of freight terminals, especially where high value goods are being handled. Docks in particular, have been seen as places where organized crime has established control over local labor unions. Over the years access to freight terminals has been increasingly restricted, and the deployment of security personnel has helped control thefts somewhat.

Physical security of passengers

Airports have been the focus of security concerns for many decades. Hijacking aircraft came to the fore in the 1970s, when terrorist groups in the Middle East exploited the lack of security to commandeer planes for ransom and publicity. Refugees fleeing dictatorships also found taking over aircraft a possible route to freedom. In response, the airline industry and the international regulatory body, ICAO, established screening procedures for passengers and bags. This process seems to have worked in the short run at least, with reductions in hijackings, although terrorists changed their tactics by placing bombs in unaccompanied luggage and packages, as for example in the Air India crash off Ireland in 1985 and the Lockerbie, Scotland, crash of Pan Am 103 in 1988.

The growth in passenger traffic and the development of hub-and-spoke networks placed a great deal of strain on the security process. There were wide disparities in the effectiveness of passenger screening at different airports, and because passengers were being routed by hubs, the numbers of passengers in transit through hub airports grew significantly. Concerns were raised by some security experts, but the costs of improving screening and the need to process ever larger numbers of passengers and maintain flight schedules caused most carriers to oppose tighter security measures.

The situation was changed irrevocably by the events of September 11, 2001. The US government created the Department of Homeland Security which in turn established a Transportation Security Authority to oversee the imposition of strict new security measures on the industry. Security involves many steps, from restricting access to airport facilities, fortifying cockpits, to the more extensive security screening of passengers. Screening now involves more rigorous inspections of passengers and their baggage at airports. For foreign nationals inspection employs biometric identification, which at present involves checking fingerprints but in the future may include retinal scans and facial pattern recognition. A new system, the Computer Assisted Passenger Prescreening System (CAPPS II), requires more personal information from travelers when they book their flights, which is used to provide a risk assessment of each passenger. Passengers considered as high risk are further screened.

The imposition of these measures has come at a considerable cost. In the USA alone, it is estimated that the expense of additional airport security is $6 billion. A significant factor has been the integration of screeners into the federal workforce, with important increases in salaries and training costs. The purchase of improved screening machines, and the redesigning of airport security procedures have been important cost additions. These measures have also had a major influence on passenger throughputs. Clearing security has become the most important source of delays in the passenger boarding process. Passengers are now expected to arrive two hours before departure at the terminal in order to clear security.

Security issues have had a negative effect on the air transport industry. As reviewed above, not only have costs increased, but delays and inconveniences to passengers. Business travel, the most lucrative sub-market for the regular airlines, has suffered particularly sharp declines. Anecdotal evidence suggests that these passengers are switching to other modes for shorter trips so as to avoid the time delays and aggravation caused by the security process and to executive jets for longer trips.

Freight security

Security in the freight industry has always been a major problem. Illegal immigrants, drug smuggling, piracy and the deployment of substandard vessels have been some of the most important concerns. However, as in the air passenger business, the events of 9/11 highlighted a new set of security issues. The scale and scope of these problems in freight is of an even greater magnitude. The less-regulated and greater international dimensions of the shipping industry, in particular, have made it a vulnerable target in an era of global terrorism. The number of ports, the vast fleet of global shipping and the range of products carried in vessels, and the difficulty of detection has made the issue of security in shipping an extremely difficult one to address. The container, which has greatly facilitated globalization, makes it extremely difficult to identify illicit and/or dangerous cargoes. In the absence of scanners that can X-ray the entire box, manual inspection becomes a time-consuming and virtually impossible task. Hubbing compounds the problem, as large numbers of containers are required to be handled with minimum delays and inconvenience.

In the USA the response was to enact the Maritime Transportation and Security Act in 2002. The basic elements of this legislation were adopted by the International Maritime Organization (IMO) in December 2002 as the International Ship and Port Security code (ISPS). There are three important features of these interventions. First, is the requirement of an automated identity system (AIS) for all vessels between 300 and 50,000 dwt. AIS requires vessels to have a permanently marked and visible identity number, and there must be a record maintained of its flag, port of registry and address of the registered owner. Second, each port must undertake a security assessment. This involves an assessment of its assets and facilities and an assessment of the effects of damages that might be caused. The port must then evaluate the risks, and identify the weaknesses to its physical security, communication systems, utilities, etc. Third, is that all cargoes destined for the USA must receive customs clearance prior to the departure of the ship. In addition, it is proposed that biometric identification for seafarers be implemented and that national databases of sailors be maintained.

The ISPS code is being implemented in ports around the world. Without certification, a port would have difficulty in trading with the USA. Security is thus becoming a factor in a port's competitiveness. The need to comply with ISPS has become an urgent issue in ports large and small around the world. The costs of securing sites, of undertaking risk assessments and of monitoring ships all represent an additional cost of doing business, without any commercial return. US ports have been able to tap funding from the Department of Homeland Security, but foreign ports have to comply or risk the loss of business.

In 2008 legislation in the USA required all containers being shipped to the USA to undergo screening. Foreign ports will be expected to purchase very expensive X-ray scanners, and undertake screening of all US-bound containers, regardless of the degree of security threat. This is a further financial and operational headache foreign ports have to contend with. Security has become an additional element in determining competitive advantage.

Method 1 – Cost–benefit analysis

The Framework

Cost–benefit analysis (CBA or COBA) is a major tool employed to evaluate projects. It provides the researcher with a set of values that are useful to determine the feasibility of a project from an economic standpoint. Conceptually simple, its results are easy for decision makers to comprehend, and therefore enjoys a great deal of favor in project assessments. The end product of the procedure is a benefit/cost ratio that compares the total expected benefits to the total predicted costs. In practice CBA is quite complex, because it raises a number of assumptions about the scope of the assessment, the time frame, as well as technical issues involved in measuring the benefits and costs.

Before any meaningful analysis can be pursued, it is essential that an appropriate framework be specified. An extremely important issue is to define the spatial scope of the assessment. Transport projects tend to have negative impacts over short distances from the site, and broader benefits over wider areas. Thus extending a runway may impact severely on local residents through noise generation, and if the evaluation is based on such a narrowly defined area, the costs could easily outweigh any benefits. On the other hand defining an area that is too broad could lead to spurious benefits.

> The aim of the study area definition should be to include all parts of the transport network which are likely to include significant changes in flow, cost or time as a result of the project.
>
> (UN, 2003, p. 17)

Because transport projects have long-term effects, and because the analysis is carried out on a real-term basis, the benefits and costs must be assessed using specific and predetermined parameters. For example: when is the project start date, when will it be completed, over what period of time will the appraisal run, and what discount rate will be used to depreciate the value of the costs and benefits over the appraisal period? These and other parameters must be agreed upon. Costs and benefits are presented in nominal values, that is monetary values of the start year and discounted for inflation over the project period. Because most transport projects are assessed for a 30-year period employing different discount rates may greatly influence the outcomes.

Costs and benefits

Costs associated with the project are usually easier to define and measure than benefits. They include both investment and operating costs. Investment costs include the planning costs incurred in the design and planning, the land and property costs in acquiring the site(s) for the project, and construction costs, including materials, labor, etc. Operating costs typically involve the annual maintenance costs of the project, but may include additional operating costs incurred, as for example the costs of operating a new light rail system.

Benefits are much more difficult to measure, particularly for transport projects, since they are likely to be diffuse and extensive. Safety is a benefit that needs to be assessed, and while there are complex issues involved, many CBA studies use standard measures of property savings per accident avoided, financial implications for reductions in bodily injury or deaths for accidents involving people. For example, Transport Canada uses $1.5 million in 1991 dollars for each fatality saved. One of the most important sets of benefits are efficiency gains as a result of the project. These gains might be assessed by estimating the time savings or increased capacity made possible by the project.

Measurement of other costs and benefits

Many other elements relating to social impacts, aesthetics, health and the environment are more difficult to assess. The latter, in particular, is a major factor in contemporary project assessment, and usually separate environmental impact analyses are required. Where possible these factors must be considered in CBA, and a variety of measures are used as surrogates for environmental benefits and costs. For example the commercial losses of habitat destruction and property damage can be estimated; the difference in the values of properties adjacent to airports and those further away are used to assess the costs of noise.

Three separate measures are usually obtained from CBA to aid decision making:

- **Net present value (NPV)**. This is obtained by subtracting the discounted costs and negative effects from the discounted benefits. A negative NPV suggests that the project should be rejected because society would be worse off.
- **Benefit–cost ratio**. This is derived by dividing the discounted costs by the discounted benefits. A value greater than 1 would indicate a useful project.
- **Internal rate of return (IRR)**. The average rate of return on investment costs over the life of the project.

The first two measures are broadly similar, though with significant differences. A project may have a high B/C ratio but still generate a smaller NPV. The results should be subjected to a sensitivity analysis. This would include considering the robustness of the predictions of costs and benefits, and usually involves the identification of aspects that would introduce uncertainty into the predictions. If certain elements are shown to be subject to variations (inflation, higher fuel charges, etc.) various scenarios would be prepared, and the cost/benefit values re-evaluated.

CASE STUDY Security, transport and health planning: the challenge of pandemics[1]

A pandemic is an epidemic of infectious disease that spreads through human populations across a large area, even worldwide. Over the last 300 years, ten major influenza pandemics have occurred. The 1918 pandemic (Spanish Flu) is considered to be the yet most severe. Of the world's population, 30 percent became ill and between 50 and 100 million died. One important factor why the Spanish Flu spread so quickly and so extensively was through modern transportation, which at the beginning of the twentieth century offered a global coverage. The virus was spread around the world by infected crews and passengers of ships and trains and severe epidemics occurred among shipyard and railway personnel.

---

1 Dr. Tom Luke, Department of Virology, Naval Medical Research Center, and Dr. Michael Osterholm, Director of the Center for Infectious Disease Research and Policy (CIDRAP), University of Minnesota, contributed to this section.

Concerns about the emergence of a new pandemic are salient, particularly in light of recent outbreaks such as SARS (Severe Acute Respiratory Syndrome) in 2002–3, which quickly spread because of the convenience of air travel. The next influenza pandemic could be equally severe and widespread illness or absenteeism in freight transportation sectors can cause cascading disruptions of social and economic systems.

The relationships between transportation and pandemics involve two major sequential dimensions:

- **Transportation as a vector**. With ubiquitous and fast transportation comes a quick and extensive diffusion of a communicable disease. From an epidemiological perspective, transportation can thus be considered as a vector, particularly for passenger transportation systems. The configuration of air transportation networks shapes the diffusion of pandemics. This issue concerns the early phases of a pandemic where transportation systems are likely to spread any outbreak at the global level.
- **Continuity of freight distribution**. Once a pandemic takes place, or immediately thereafter, the major concerns shift to freight distribution. Modern economic activities cannot be sustained without continuous deliveries of food, fuel, electricity and other resources. However, few events can be more disruptive than a pandemic as critical supply chains essentially shut down. Disruptions in the continuity of distribution are potentially much more damaging than the pandemic itself.

The more efficient the transportation, the more efficient is the vector that can transmit an infectious disease. International and long-distance transport such as air and rail, modes and terminals alike, concentrate passengers and increase the risk of exposure. Since the incubation time for the average influenza virus is between one and four days, there is ample time for someone infected to travel to the other side of the world before noticing symptoms.

Once symptoms have developed, there is also a "denial phase" where an infected individual will continue traveling, particularly if going back to his place of origin. An infected individual beginning to show symptoms is likely to cancel an outbound travel, but will do the utmost, even breaking quarantine (or warnings), to go back home. Thus, in a window of a few days before an outbreak could become apparent to global health authorities, a virus could have easily been transposed in many different locations around the world. At this point, the vector and velocity of the modern transport system would insure that an epidemic becomes a pandemic. In some cases, the velocity of global transportation systems is higher than at the regional level, which paradoxically implies that a virus can spread faster at the global level – between major gateways – than at the regional level.

Once an outbreak becomes apparent, the global passenger transportation system, such as air travel and passenger rail, can quickly be shut down in whole or in part, either voluntarily (more likely if the outbreak is judged to be serious) or by the unwillingness of passengers to be exposed to risks. The latter is what happened during the SARS outbreak in 2003. For instance, while the public transportation systems of several large Chinese cities were still operated, the number of users precipitously dropped because of risk avoidance. The SARS outbreak also had a substantial impact on the global airline industry. After the disease hit, flights in Pacific Asia decreased by 45 percent from the year before. During the outbreak, the number of flights between Hong Kong and the United States fell 69 percent. It is quite clear that this impact would pale in comparison to that of a 12- to 36-month worldwide influenza pandemic.

However dramatic the impacts of modern transportation as a high velocity vector for a pandemic, a potential greater risk resides in the geographical and functional structure of supply chains because the continuity of freight distribution could be compromised. The interconnectedness of the global economy, while being a net advantage from a supply chain standpoint, could make the next influenza pandemic more devastating than the ones before it. Even the slightest disruption in the availability of parts, finished goods, workers, electricity, water and petroleum could bring many aspects of contemporary life to a halt. The global economy has been favored by the exploitation of comparative advantages and a more tight management of supply chains. Inventories are kept to a minimum. Virtually no production surge capacity exists. As a consequence, most markets depend on the timely delivery of many critical products (such as pharmaceuticals, medical supplies, food and equipment parts) and services (such as communications support). A pandemic could quickly exhaust available food, energy and medical resources, and replacements would not be forthcoming. Thus, supply chain issues are expected to seriously compound the impacts of an influenza or other pandemic.

Many pandemic preparation plans fail to account the full importance and ramifications of global supply chains. They are essentially designed with the assumption that national economies are independent and are mostly self-reliant. The geographic and functional realities of the global economy are quite different. Cascading disruptions in vulnerable freight transportation systems and strategic supply chains can compound the difficulties of maintaining social cohesion and critical infrastructures during a pandemic. Transportation workers must therefore receive a high priority for support – including vaccine, prophylactic antivirals, public health interventions and access to healthcare. Pandemic planners must cooperatively develop plans and obtain the agreements and resources necessary to conduct health assurance campaigns at major transportation chokepoints and corridors.

Bibliography

Bailey, E.E. and W.J. Baumol (1984) "Deregulation and the Theory of Contestable Markets", *Yale Journal on Regulation*, 1, 111–37.

Button, K. (1993) *Transport Economics*, 2nd edn, Aldershot: Edward Elgar.

Caddy, J. (1997) "Hollow Harmonisation? Closing the Gap in Central European Environmental Policy", *European Environment*, 7, 73–9.

Dion, S., B. Slack and C. Comtois (2002) "Port and Airport Divestiture in Canada: A Comparative Analysis", *Journal of Transport Geography*, 10, 187–94.

Ewing, R.H. (1999) *Traffic Calming: State of the Practice*, Washington, DC: Institute of Transportation Engineers.

Goetz, A.R. (2002) "Deregulation, Competition, and Antitrust Implications in the US Airline Industry", *Journal of Transport Geography*, 10, 1–19.

Goh, M. (2002) "Congestion Management and Electronic Road Pricing in Singapore", *Journal of Transport Geography*, 10, 29–38.

Hogwood, B. and L.A. Gunn (1984) *Policy Analysis for the Real World*, Oxford: Oxford University Press.

Janelle, D.G. and A. Gillespie (2004) "Space–Time Constructs for Linking Information and Communication Technologies with Issues in Sustainable Transportation", *Transport Reviews*, 24, 665–77.

Litman, T. (2005) *Rail Transit in America: A Comprehensive Evaluation of Benefits*, http://www.vtpi.org/railben.pdf.

Luke, T.C. and J.-P. Rodrigue (2008) "Protecting Public Health and Global Freight Transportation Systems during an Influenza Pandemic", *American Journal of Disaster Medicine*, 3(2), 99–107.

Muller, P.O. (2004) "Transportation and Urban Form: Stages in the Evolution of the American Metropolis", in S. Hanson and G. Giuliano (eds) *Geography of Urban Transportation*, 3rd edn, New York: Guilford.

Osterholm, M.T. (2005) "Preparing for the Next Pandemic", *Foreign Affairs*, July/August, pp. 24–37.

Paaswell, R.E. (1995) "ISTEA: Infrastructure Investment and Land Use", in D. Banister (ed.) *Transport and Urban Development*, London: E&FN Spon, pp. 36–58.

Parkhurst, G. and J. Richardson (2002) "Modal Integration of Bus and Car in Local UK Transport policy", *Journal of Transport Geography*, 10, 195–206.

Picciotto, R. (1997) "Evaluation in the World Bank", in E. Chelimsky and W. Shadrish (eds) *Evaluation for the 21st Century*, Thousand Oaks: Sage.

Sorenson, P. and B. Taylor (2005) "Paying for Roads: New Technology for an Old Dilemma", *Access*, 26, 2–9.

Studnicki-Gizbert, K.W. (1974) *Issues in Canadian Transport Policy*, Toronto: Macmillan.

Tolley, R. and B. Turton (1995) *Transport Systems, Policy, and Planning*, London: Longman.

United Nations (2003) *Cost Benefit Analysis of Transport Infrastructure Projects*, New York: United Nations.

Victoria Transport Policy Institute (2005) *Transportation Demand Encyclopaedia*, Victoria: VTPI, www.vtpi.org/tdm/index.php.

⟨10⟩ Conclusion
Issues and challenges in transport geography

In this final chapter some of the main issues confronting transportation today and which are likely to have an even greater impact in the future are reviewed. What are the major questions that will confront future transport geographers, the readers of this book? What role will transport geographers play in addressing future challenges? These are complex questions, and inevitably given the breadth of the field of transport geography there has had to be a selection of topics. In looking back over the other chapters of this book, different groups of issues stand out for their present-day importance and their potential to reshape future geographies. Congestion has been selected because it is a feature with profound consequences today, and it is almost certainly to be an issue that transport geographers will have to deal with in the future. Infrastructures are everywhere a prevalent issue, notably in terms of provision and funding. The environment is already influencing transportation, and because of increasing pressures from the environment, and also because of growing awareness of the problems, it will be a factor of growing importance. Another issue relates to the need to understand the management responses required to cope with future transport developments, and the way that an understanding of spatial relationships can contribute to better management of the system. Issues related to energy and security, which previously were of lesser importance, have been brought forward because of economic and geopolitical changes. Finally, the role of geographers in addressing these issues is discussed.

Congestion

The causes of congestion are well understood, even if the solutions are not. Congestion arises from two causes. Most important is when demand for mobility exceeds the capacity to support it. It can also occur when random events bring about a temporary disruption to service, such as an accident or a natural hazard such as flooding. In the case of the second set of causes, it is possible to mitigate their effects if the occurrence is frequent, such as accidents, or if the risks are great, as for example of flooding in a flood plain. In the first case a solution is to increase capacity. However, as has been shown, increasing capacity engenders a hidden demand, so that adding lanes to an expressway tends to attract even more cars. Furthermore, demand is increasing ceaselessly, so that the practicality of this solution may be questioned.

The issue of congestion is likely to remain one of the great ongoing issues in transport geography because there are unprecedented demands for transportation being generated by a global economy that is ever more dependent upon the transport industry. The growth of demand is likely to have major impacts on the nature and form of the future transport industry.

In the short term at least, road transport is likely to continue its domination of the transport industry. There are two basic reasons for this assertion. In the developed world automobiles and trucks already dominate the market, and the spatial patterns of people, industries and services have adjusted themselves somewhat to the demands of these modes. Such low density, space extensive patterns are pushing the traffic congestion ever further out, and make it very difficult for other higher capacity modes to compete. At the same time the demand for mobility is growing as a result of the rapid industrialization of countries such as China and India. There too a modal shift is occurring in favor of road transport. Increasing prosperity in these countries represents a great potential for growth in road transport.

Congestion is not limited to internal urban-generated traffic. International trade is likely to continue to be dominated by maritime transport (in terms of weight) and air transport (in terms of value). This has already led to a concentration of traffic at a relatively small number of hubs, which are capable of extracting scale economies. For example, the 20 largest container ports handled more than 52 percent of global traffic in 2005. The traffic concentration however is already producing capacity problems in many of these hubs. International trade has grown at a rate faster than the global economy in recent decades and there are expectations that hub congestion will remain an issue in the future.

For geographers there are a whole range of issues arising out of the growth of demand and the paralysis of congestion. Here, they are grouped into two categories: first, are a series of questions surrounding how to provide solutions; second are the effects on future spatial patterns.

In the past the solution to congestion was to provide more capacity by building more infrastructure. Such a response depended heavily on engineering solutions. As has been learned over the last few decades, the model of "predict and accommodate" has not worked well. It is now recognized that a multidisciplinary approach is required. It is recognized that there will still be a heavy reliance on engineering skills to design and construct infrastructure and systems, and to develop further technological innovations required for the "intelligent highway". However, transport policy and planning requires a broader perspective, one that considers different goals and alternatives, responds to different needs for mobility, and one that seeks ways to manage demand. Under what conditions and in what types of locations can travel demand be modified? Does the current emphasis on proposing densification as a solution to reducing car dependence work? How might freight transport be better integrated in the urban environment?

Congestion is a phenomenon that is spatially bound. It takes place in specific locations with impacts at a multitude of scales, from a particular highway intersection that may delay traffic over a few hundred meters, to blockage in a port that may disrupt the flow of goods over half a continent. Each event produces a spatial response, from the car driver who searches out an alternative route in future to the shipper who selects a different mode or point of entry for succeeding shipments.

Increased demand and the rising likelihoods of congestion will intensify new spatial responses and thus it appears very likely that new spatial flows and structures will come into being. What will be the effects? What kinds of impact will be evident at the local, regional or global scales? Will congestion be sufficient to counteract the strong forces favoring concentration? Already there is evidence in air transport for growth in passengers and freight in some smaller airports. Will congestion in the newly industrializing countries act as a break on development?

Infrastructure

Regardless of the specific solutions to congestion that are considered, increasing demand is placing unprecedented requests for investment in transport infrastructures.

A major question confronting all countries of the world is how to finance the construction and maintenance of transport infrastructures. Governments have traditionally been the primary source of funding in the transport sector, but the costs of keeping pace with the growth in demand are making it difficult for even the richest countries to countenance public funding on the scale required.

Capital requirements are particularly prevalent on both sides of the infrastructure life-cycle spectrum. Over this matter the highways in China and North America represent two salient cases. For China, the last decade has seen an impressive level of highway construction with the setting of a national highway network which totaled more than 53,000 km in 2007, the second longest in the world. Comparatively, the American Interstate highway system of about 75,000 km is nearing a phase of its life cycle where a substantial amount of capital investment will be required to upgrade the system and maintain its operability, including thousands of aging highway bridges. While most of Interstate is publicly funded, almost all Chinese highways were funded by private interests that are using tolls to recover their investments.

Public–private partnerships and completely private ventures are one set of solutions. For many developing countries this is the only solution, since public finances are inadequate to the task. Thus, in the future, a greater private involvement in the provision of transport infrastructure is to be expected. Several models are already well tested: BOT (Build–Operate–Transfer), where the private sector builds and operates a facility or system for a period of time, but then transfers it back to the government after an agreed period; BLT (Build–Lease–Transfer) where after building the facilities, it is leased for a fixed period for operation and finally transferred back; ROT (Rehabilitate–Operate–Transfer) where the private party refurbishes an existing facility to be operated for a term prior to being returned to the state.

Another approach that is gaining momentum is charging for use of transport infrastructure. Pricing is becoming an important feature of transport planning in urban areas. Whether it is cordon pricing, congestion pricing or tolling, drivers are being forced to pay for their use of roads and limited price elasticity has been observed so far. With the growing concerns over the environment, charging for the externalities of transport modes is becoming a reality in many jurisdictions. How effective are these alternatives? What effects do they have over travel behavior?

The difficulties are not to be underestimated, however. Most transport infrastructure projects are long term, but are typified by the heaviest capital investment requirements being incurred over a short initial phase. Most private enterprises cannot take a long-term perspective, because they need to cover their expenses over short period of time. With the growing unwillingness or inability of the public sector to fund and provide transport infrastructure, what form and structure in infrastructure provision, maintenance and operation will be achieved?

Environmental challenges

The issue of sustainability has become an increasingly important consideration for the transport industry. It is now broadly recognized that there needs to be a balance between economic efficiency, social factors and the environment. Of these three, the issue of economic efficiency has always been to the forefront, and governments have been important in regulating social conditions (safety, security and working conditions). Despite the strong historic relationship between transport and the environment, the latter has tended to be overlooked by the industry. This is changing, and environmental issues are likely to play an ever more important role in the transport industry, particularly over three core dimensions:

- **Transport and atmospheric pollution**. Air quality standards are being implemented with increasing rigor in more and more countries around the world. There are still striking differences between regions and between the modes. For example, most of the countries of the developing world still have to go a long way to fixing and enforcing standards. In North America, passenger vehicles are more rigorously controlled than trucks, and ships are much less controlled than other modes. However, the trend is towards greater control over emissions. What will be the modal effects and the impacts on modal competition? Despite international accords, climate change issues are unlikely to be overturned in the immediate future. Already a higher incidence of severe climatic events such as hurricanes and droughts are being ascribed by environmentalists to atmospheric warming, while others argue that such occurrences are part of natural cycles with anthropogenic factors being negligible. Which regions and what transport systems are most likely to be impacted?
- **Transport and water quality**. The contribution of transport to the pollution of rivers and oceans is considerable, and is only recently being addressed by international legislation. Considerable progress has been made in a number of areas such as ballast water, waste and oil spills. As the legislation increases in its comprehensiveness, the more the transport industry is impacted. This is particularly evident in matters relating to dredging, where environmental constraints are placing a growing financial burden on ports that are seeking to deepen channels in order to keep pace with the growth of vessel size. Will these constraints serve to reduce the competitiveness of some ports? Will increased dredging costs bring about a break in the growth of vessel size? Similar questions arise out of coastal zone legislation, especially the provisions for protecting wetlands.
- **Transport and land take**. Increased demand for transport is already placing enormous pressures for new infrastructures. Many of these transport facilities such as airports and ports require very large amounts of land for their own internal operations and for the external transport links that have to be provided. A fundamental question is can the environment and society afford to provide sites of the scale required by the transport industry? Will the transport industry have to move away from its preferred model of massive hubs and load centers?

Management of transport systems

The transportation industry is changing so significantly in form and function that it is easy to overlook the very important changes in the way it is organized and managed. Yet it is through different management practices that the spatial manifestations of the industry are expressed. It is perhaps easiest to see the changes in management through the lens of governance, where an industry that used to be largely managed and controlled by the state has become increasingly controlled by the private sector. The privatization of transport companies and infrastructures has been an important feature of the last decade, and is likely to continue further into the present century. However, there are still many questions about the role of the state in transportation. Under what conditions and in what circumstances should continued state control be maintained and even strengthened? What are the best models of public–private partnerships in the transport industry?

The growing role of the private sector over an industry that is becoming global and multi-functional has necessitated a shift in management and ownership relationships that are still evolving. They include:

- The emergence of **horizontally linked global corporations** that through a series of acquisitions and mergers have bought up similar operating companies in different markets. A good example is the port terminal operators, such as Hutchison Port Corporation, a Hong Kong-based firm with major investments in Europe, China and the Americas.
- The development of **vertically integrated corporations** that have grown by merger and acquisition to control several segments of the transport chain. Examples include A.P. Moller, a company that controls the world's largest container fleet, an airline, terminal operating activities, trucking, barge, railroad services and logistics services in many parts of the world.
- **Intermediaries** that provide transport services on a global scale, without direct ownership of infrastructure. Third-party logistics (3PL) companies such as Kuhne and Nagel, Schenker and Panalpina operate in many markets and are major actors in the transport chain.
- **Alliances**, informal groupings of transport providers that pool resources and offer joint services between major global markets. Examples include the One World and Star airline alliances.

At the same time transport is being increasingly integrated in global production systems. It is becoming an integral part of production and distribution chains. Can it still be considered a derived demand therefore? What are the reciprocal relations between transport and production/distribution systems? How are Walmart's distribution networks shaped by transport, and how is transport impacted by Walmart?

These management and business structures give rise to distinct patterns of spatial organization, with different operating practices. The impact of Southwest Airlines on the spatial structure of the US airline industry has been considerable, for example, and the operational interests of a vertically integrated enterprise is different than one horizontally linked. This highlights the need to understand the nature of the organization of the businesses involved in transport as a means of explaining existing patterns and predicting their future forms. The concentration of traffic (and resultant congestion) is as much explained by the organization of transport firms as it is by traditional explanations involving demand and capacity. In turn, the organization of the global firms themselves is shaped by conditions of local spatial markets. A distinct geography of transport firms exists, a geography that is still largely *terra incognita*.

Energy, safety and security

The macro-economic and policy environment in which the transport sector evolves has substantially changed in recent years, bringing concerns to the forefront which before were rather secondary. With increases in energy costs, significant adjustments in transport modes may be expected in the future. While technologies may make alternative fuel vehicles a commercial option to the internal combustion engine, the main question is the effect of higher prices on automobiles and trucks. As the costs are passed on to users, how will global production systems that depend upon cheap transport be impacted? How will the logistics industry that exploits the most energy inefficient modes be affected? Will a modal shift to more energy efficient modes, such as rail or shipping, take place? What forms of transport and mobility will take shape as the energy transition away from fossil fuels takes place?

Another prevalent matter concerns security practices that are now part of the business environment in which passenger and freight transport systems are evolving.

Most of these measures are imposed by regulatory agencies with consequences often difficult to assess, but always involving additional costs and delays for transport operators. A balance between security measures and the efficient flow of passengers and freight will need to be achieved through a variety of regulatory, operational and technological innovations.

The role of geographers

Geographers have played a relatively small role in the field of transport studies, a field that has been dominated by engineers and economists. This was due in part to the needs of the industry being focused on providing infrastructures and technologies, at what cost and benefits and at what level of pricing. The contemporary industry is much more complex, with issues as varied as safety, aesthetics, working conditions, equity, deprivation, the environment and governance being necessary considerations. A much broader set of skills are required therefore, and transport studies are essentially multidisciplinary today. Geographers thus have important opportunities to contribute to transport studies, transport planning and transport operations, in part because of the breadth of the approach and training. Still, transport geography, like the field of transportation in general, does not receive a level of attention in academia proportional to its economic and social importance.

It is also a fundamental fact that transport is a spatial activity. It has always been a space adjusting service, but over the last few decades it has become increasingly global in scope. Contemporary transport operates at a wider range of scales than ever before. There are complex interactions between the local and the global. For example, the issues surrounding the expansion of an airport are usually decided at the local level, and the impacts are likely to be felt locally, namely its externalities such as noise and congestion. However, the effects on passenger and freight flows may have a global impact. The spatiality of transport and the many scale levels at which it operates are elements that are the particular concerns of geographers. No other discipline has as its core interest the role of space in shaping human activities.

One reason for the success of engineers and economists in transport studies and applications is that their training has been rigorous in the application of mathematics and multivariate statistics. They have demonstrated the ability to provide precise answers to the questions that decision makers have to consider – what to build, at what cost, with what cost effects. There has evolved a culture in the transport industry that unless it can be quantified it is of little value. Many transport geographers have the quantitative skills that have made their work accepted by the broader scientific community. There is little doubt that training in modeling, graph theory and multivariate statistics is required. However, there are newer techniques that provide geographers with opportunities to contribute to transport studies. GIS-T, in particular, should be an essential element in the training of a transport geographer. The multi-scalar, multivariate nature of the transport industry makes GIS-T an invaluable tool, and one that will raise the profile of geographers in the transportation industry.

One of the great challenges in transport studies is data availability. In many cases, official census and survey data are inadequate or unavailable in the form required. Knowledge of survey techniques and their limitations are an important part of the transport geographer's toolkit. Many of the traditional tools and approaches of geographers are still relevant. They allow us to address problems that are frequently overlooked by other disciplines because of the lack of data. Questionnaires and interviews represent a vital source of information in many situations. Content analysis is extremely useful in providing quantified data from non-quantified sources. At the same time, fieldwork provides the opportunity to obtain detailed understanding of the particularities of the local conditions that cannot be obtained from reading texts and official documents.

The prospects for transport geography and transport geographers appear to be excellent. A look back at the subject matter and topics covered in this book indicates an industry that is growing in significance and changing. The kinds of issues that are achieving greater importance – sustainability, congestion, governance and management – are ones to which geographers have the opportunity to contribute. As the transport industry becomes more complex, old approaches, focusing on a narrow range of factors, have to be replaced by more nuanced analysis and solutions. In the transport industry itself, in public planning and in research institutions, the scope for geographers appears bright.

 **Glossary**

Many of the glossary terms are adapted from the Bureau of Transportation Statistics, the European Conference of Ministers of Transport, the Economic Commission for Europe, the Intermodal Association of North America and the Mineta Transportation Institute.

access
> The capacity to enter and exit a transport system. It is an absolute term implying that a location has access or does not.

accessibility
> The measure of the capacity of a location to be reached by, or to reach, different locations. The capacity and the structure of transport infrastructure are key elements in the determination of accessibility.

aerodrome
> A defined area on land or water (including any buildings, installations and equipment) intended to be used either wholly or in part for the arrival, departure and movement of aircraft. Aerodromes may include airports, heliports and other landing areas.

aframax
> A tanker of standard size between 75,000 and 115,000 dwt. The largest tanker size in the AFRA (Average Freight Rate Assessment) tanker system.

agglomeration economies
> See economies of agglomeration.

air cargo
> Total volume of freight, mail and express traffic transported by air. Includes: freight and express commodities of all kinds, such as small package counter services, express services and priority reserved freight.

air carrier
> Commercial system of air transportation, consisting of domestic and international scheduled and charter service.

air space
> The segment of the atmosphere that is under the jurisdiction of a nation or under an international agreement for its use. They include two major components, one being land-based (takeoffs and landings) and the other air-based, mainly composed of air corridors. These corridors can superimpose themselves to altitudes up to 22,500 meters. The geography of air transport is limited to the use of predetermined corridors.

air transportation
> Includes establishments that provide domestic and international passenger and freight services, and establishments that operate airports and provide terminal facilities.

airport

(1) An area of land or water that is used or intended to be used for the landing and takeoff of aircraft, and includes its buildings and facilities, if any. (2) A facility used primarily by conventional, fixed-wing aircraft. (3) A facility, either on land or water, where aircraft can take off and land. Usually consists of hard-surfaced landing strips, a control tower, hangars and accommodation for passengers and cargo. (4) A landing area regularly used by aircraft for receiving discharging passengers or cargo.

alternative fuels

Low-polluting fuels used to propel a vehicle instead of high-sulfur diesel or gasoline. Examples include methanol, ethanol, propane or compressed natural gas, liquid natural gas, low-sulfur or "clean" diesel and electricity.

Amtrak

Operated by the National Railroad Passenger Corporation of Washington, DC. This rail system was created by President Nixon in 1970, and was given responsibility for the operation of intercity, as distinct from suburban, passenger trains between points designated by the Secretary of Transportation.

arterial street

A major thoroughfare, used primarily for through traffic rather than for access to adjacent land, that is characterized by high vehicular capacity and continuity of movement.

Association of Southeast Asian Nations (ASEAN)

Free trade area established on August 8, 1967, in Bangkok, Thailand, with the signing of the Bangkok Declaration. The members of ASEAN are Brunei Darussalam, Indonesia, Laos, Malaysia, Myanmar, Philippines, Singapore, Thailand and Vietnam. The Secretariat of the Association is located in Jakarta, Indonesia.

average vehicle occupancy (AVO)

The number of people traveling by private passenger vehicles divided by the number of vehicles used.

average vehicle rideship (AVR)

The ratio of all people traveling by any mode, including cars, buses, trains and bicycles (or telecommuting), in a given area during a given time period to the number of cars on the road. A key measure of the efficiency and effectiveness of a transportation network – the higher the AVR, the lower the level of energy consumption and air pollution.

back haul

Traffic for the return movement of a car or container towards the point where the initial load originated or to handle a shipment in the direction of the light flow of traffic.

balance of payments

A record of receipts from and payments to the rest of the world by a country's government and its residents. The balance of payments includes the international financial transactions of a country for commodities, services and capital transactions.

balance of trade

The difference between a country's total imports and exports. If exports exceed imports, a positive balance of trade exists.

barge

A non-motorized water vessel, usually flat-bottomed and towed or pushed by other craft, used for transporting freight. Dominantly used on river systems.

barrel

A unit of volume equal to 42 US gallons (or 159 liters) at 60 degrees Fahrenheit, often used to measure volume in oil production, price, transportation and trade.

base fare

The price charged to one adult for one transit ride; excludes transfer charges, zone charges, express service charges, peak-period surcharges and reduced fares.

base period

The period between the morning and evening peak periods when transit service is generally scheduled on a constant interval. Also known as "off-peak period". The time of day during which vehicle requirements and schedules are not influenced by peak-period passenger volume demands (e.g. between morning and afternoon peak periods). At this time, transit riding is fairly constant and usually low to moderate in volume when compared with peak-period travel.

berth

A specific segment of wharfage where a ship ties up alongside at a pier, quay, wharf or other structure that provides a breasting surface for the vessel. Typically, this structure is a stationary extension of an improved shore and intended to facilitate the transfer of cargo or passengers.

bill of lading

A document that establishes the terms of a contract between a shipper and a transportation company. It serves as a document of title, a contract of carriage and a receipt for goods.

block

A group of railcars destined to the same location.

break-bulk cargo

Refers to general cargo that has been packaged in some way with the use of bags, boxes or drums. This cargo tends to have numerous origins, destinations and clients. Before containerization, economies of scale were difficult to achieve with break-bulk cargo as the loading and unloading process was very labor and time consuming.

bridge

A structure including supports erected over a depression or an obstruction, such as water, highway or railway, and having a track or passageway for carrying traffic or other moving loads, and having an opening measured along the center of the roadway of more than 20 feet between undercopings of abutments or spring lines of arches, or extreme ends of openings for multiple boxes; it may also include multiple pipes, where the clear distance between openings is less than half of the smaller contiguous opening.

British thermal unit (BTU)

The amount of energy required to raise the temperature of 1 pound of water 1 degree Fahrenheit (F) at or near 39.2 degrees F and 1 atmosphere of pressure.

bulk cargo

Refers to freight, both dry or liquid, that is not packaged such as minerals (oil, coal, iron ore) and grains. It often requires the use of specialized ships such as oil tankers as well as specialized transhipment and storage facilities. Conventionally, this cargo has a single origin, destination and client. It is also prone to economies of scale.

bulk carriers

All vessels designed to carry bulk cargo such as grain, fertilizers, ore and oil.

bulk terminal

A purpose-designed berth or mooring for handling liquid or dry commodities, in unpackaged bulk form, such as oil, grain, ore and coal. Bulk terminals typically are installed with specialized cargo handling equipment such as pipelines, conveyors, pneumatic evacuators, cranes with clamshell grabs, and rail lines to accommodate cargo handling operations with ships or barges. Commodity-specific storage facilities such as grain silos, petroleum storage tanks and coal stock yards are also located at these terminals.

bus (motorbus)

Any of several types of self-propelled vehicles, generally rubber-tired, intended for use on city streets, highways and busways, including but not limited to minibuses, 40- and 30-foot buses, articulated buses, double-deck buses and electrically powered trolley buses, used by public entities to provide designated public transportation service and by private entities to provide transportation service including, but not limited to, specified public transportation services. Self-propelled, rubber-tired vehicles designed to look like antique or vintage trolleys are considered buses.

bus, trolley

An electric, rubber-tired transit vehicle, manually steered, propelled by a motor drawing current through overhead wires from a central power source not on board the vehicle. Also known as "trolley coach" or "trackless trolley".

bus lane

A street or highway lane intended primarily for buses, either all day or during specified periods, but sometimes also used by carpools meeting requirements set out in traffic laws.

bus stop

A place where passengers can board or disembark from a bus, usually identified by a sign.

cable car

An electric railway operating in mixed street traffic with unpowered, individually controlled transit vehicles propelled by moving cables located below the street surface and powered by engines or motors at a central location not on board the vehicle.

cabotage

Transport between two terminals (a terminal of loading and a terminal of unloading) located in the same country irrespective of the country in which the mode providing the service is registered. Cabotage is often subject to restrictions and regulations. Under such circumstances, each nation reserves for its national carriers the right to move domestic freight or passenger traffic.

canal

An artificial open waterway constructed to transport water, to irrigate or drain land, to connect two or more bodies of water, or to serve as a waterway for watercraft.

capesize

Refers to a rather ill-defined standard which has the common characteristic of being incapable of using the Panama or Suez Canals, not necessarily because of their tonnage, but because of their size. These ships serve deepwater terminals handling raw materials, such as iron ore and coal. As a result, "Capesize" vessels transit via Cape Horn (South America) or the Cape of Good Hope (South Africa). Their size ranges between 80,000 and 175,000 dwt.

carbon dioxide (CO₂)

A colorless, odorless, non-poisonous gas that is a normal part of the ambient air. Carbon dioxide is a product of fossil fuel combustion.

carbon monoxide (CO)

A colorless, odorless, highly toxic gas that is a normal by-product of incomplete fossil fuel combustion. Carbon monoxide, one of the major air pollutants, can be harmful in small amounts if breathed over a certain period of time.

carpool

An arrangement where two or more people share the use and cost of privately owned automobiles in traveling to and from pre-arranged destinations together.

carrier

A company moving passengers or freight.

catchment area

Area or region whose economic, political, cultural, social, etc. influence is felt over a larger area; it is the radius of action of a given point. In transportation, it consists in the area under influence of a focal point towards which centripetal fluxes converge; an interception zone of several carriers. Also labeled as "area of influence" or "hinterland".

centrality

Focus on the terminal as a point of origin and destination of traffic. Thus, centrality is linked with the generation and attraction of movements, which are related to the nature and the level of economic activities within the vicinity of the concerned terminal. The function of centrality also involves a significant amount of intermodal activities.

charter

Originally meant a flight where a shipper contracted hire of an aircraft from an air carrier, but has usually come to mean any non-scheduled commercial service.

city logistics

The process for totally optimizing the logistics and transport activities by private companies in urban areas while considering the traffic environment, congestion and energy consumption within the framework of a market economy.

class I railroad

An American railroad with an annual gross operating revenue in excess of $250 million based on 1991 dollars.

Clean Air Act (CAA)

Federal legislation that sets national air quality standards.

coach service

Transport service established for the carriage of passengers at special reduced passenger fares that are predicated on both the operation of specifically designed aircraft space and a reduction in the quality of service regularly and ordinarily provided.

coal

A black or brownish-black solid, combustible substance formed by the partial decomposition of vegetable matter without access to air. The rank of coal, which includes anthracite, bituminous coal, subbituminous coal and lignite, is based on fixed carbon, volatile matter, and heating value. Coal rank indicates the progressive alteration, or coalification, from lignite to anthracite. Lignite contains approximately 9 to 17 million

British thermal unit (BTU) per ton. The heat contents of subbituminous and bituminous coal range from 16 to 24 million BTU per ton, and from 19 to 30 million BTU per ton, respectively. Anthracite contains approximately 22 to 28 million BTU per ton.

cold chain

A temperature-controlled supply chain linked to the material, equipment and procedures used to maintain specific shipments within the appropriate temperature range. Often relates to the distribution of food and pharmaceutical products.

combi

A type of aircraft whose main deck is divided into two sections, one of which is fitted with seats and one which is used for cargo.

commercial geography

Investigates the spatial characteristics of trade and transactions in terms of their cause, nature, origin and destination. It leans on the analysis of contracts and transactions.

commodity

Resources that can be consumed. They can be accumulated for a period of time (some are perishable while others can be virtually stored for centuries), exchanged as part of transactions or purchased on specific markets (such as a futures market). Some commodities are fixed, implying that they cannot be transferred, except for the title. This includes land, mining, logging and fishing rights. In this context, the value of a fixed commodity is derived from the utility and the potential rate of extraction. Bulk commodities are commodities that can be transferred, which includes for instance grains, metals, livestock, oil, cotton, coffee, sugar and cocoa. Their value is derived from utility, supply and demand (market price).

commodity chain (supply chain)

A functionally integrated network of production, trade and service activities that covers all the stages in a supply chain, from the transformation of raw materials, through intermediate manufacturing stages, to the delivery of a finished good to a market. The chain is conceptualized as a series of nodes, linked by various types of transactions, such as sales and intra-firm transfers. Each successive node within a commodity chain involves the acquisition or organization of inputs for the purpose of added value.

common carrier

A transportation company engaged in the business of handling persons or freight for compensation and for all customers impartially.

comparative advantages

The relative efficiencies with which countries (or any economic unit) can produce a product or service.

compressed natural gas (CNG)

Natural gas which is comprised primarily of methane, compressed to a pressure at or above 2,400 pounds per square inch and stored in special high-pressure containers. It is used as a fuel for natural gas-powered vehicles, mainly buses.

commuter

A person who travels regularly between home and work or school.

commuter bus service

Fixed route bus service, characterized by service predominantly in one direction during peak periods, limited stops, use of multi-ride tickets and routes of extended length, usually between the central business district and outlying suburbs. Commuter bus service may also include other service, characterized by a limited route structure, limited stops and a coordinated relationship to another mode of transportation.

commuter rail

Railroad local and regional passenger train operations between a central city, its suburbs and/or another central city. It may be either locomotive-hauled or self-propelled, and is characterized by multi-trip tickets, specific station-to-station fares, railroad employment practices and usually only one or two stations in the central business district. Also known as "suburban rail".

conference (liner)

An association of ship owners operating the same trade routes who operate under collective conditions such as tariff rates and shared capacity. They provide international liner services for the carriage of cargo on a particular route or routes within specified geographical limits and which has an agreement or arrangement within the framework of which they operate under uniform or common freight rates and any other agreed conditions with respect to the provision of liner services.

congestion

Occurs when transport demand exceeds transport supply in a specific section of the transport system. Under such circumstances, each vehicle impairs the mobility of others. Urban congestion mainly concerns two domains of circulation, private and public, often sharing the same infrastructures.

connecting carrier

A carrier that has a direct physical connection with another or forming a connecting link between two or more carriers.

consignee

A person or company to whom commodities are shipped. Officially, the legal owner of the cargo.

consolidated shipment

A method of shipping whereby an agent (freight forwarder or consolidator) combines individual consignments from various shippers into one shipment made to a destination agent, for the benefit of preferential rates. (Also called "groupage".) The consolidation is then de-consolidated by the destination agent into its original component consignments and made available to consignees. Consolidation provides shippers access to better rates than would be otherwise attainable.

constant dollars

Figures where the effect of change in the purchasing power of the dollar has been removed. Usually the data are expressed in terms of dollars of a selected year or the average of a set of years.

container

A large standard size metal box into which cargo is packed for shipment aboard specially configured oceangoing containerships and designed to be moved with common handling equipment enabling high-speed intermodal transfers in economically large units between ships, railcars, truck chassis and barges using a minimum of labor. The container, therefore, serves as the transfer unit rather than the cargo contained therein.

container on flatcar (COFC)

The movement of a container on a railroad flat car. This movement is made without the container being mounted on a chassis.

containerization

Refers to the increasing and generalized use of the container as a means of freight transport. As a standard and versatile means, the container has greatly contributed to intermodal transportation of merchandise; its widespread use, therefore, is responsible for profound mutations in the transport sector. Through reduction of handling time, labor costs and packing costs, container transportation allows considerable increases in speed of rotation along a circuit and thus entails a better optimization of time and money.

containership

A cargo vessel designed and constructed to transport, within specifically designed cells, portable tanks and freight containers which are lifted on and off with their contents intact. There are two types of containerships: full and partial. Full containerships are equipped with permanent container cells with little or no space for other types of cargo. Partial containerships are considered multipurpose container vessels, where one or more but not all compartments are fitted with permanent container cells, and the remaining compartments are used for other types of cargo. This category also includes container/car carriers, container/rail car carriers and container/roll-on/roll-off vessels.

conventional car

A single platform flat car designed to carry a trailer or container. Containers can only be single stacked on a conventional car. Conventional cars are equipped with one or two stanchions, depending on length, for shipment of one or two trailers.

corporate average fuel economy (CAFE) standards

CAFE standards were originally established by Congress for new automobiles, and later for light trucks, in Title V of the Motor Vehicle Information and Cost Savings Act (15 U.S.C. 1901, *et seq.*) with subsequent amendments. Under CAFE, automobile manufacturers are required by law to produce vehicle fleets with a composite sales-weighted fuel economy which cannot be lower than the CAFE standards in a given year; for every vehicle which does not meet the standard, a fine of $5.00 is paid for every one-tenth of a mpg below the standard.

corridor

A linear orientation of transport routes and flows connecting important locations that act as origins, destinations or points of transhipment. Corridors are multi-scalar entities depending on what types of flows are being investigated. Thus, they can be composed of streets, highways, transit routes, rail lines, maritime lines or air paths.

costs (transport)

Monetary measure of what the transport provider must pay to produce transportation services and comes as fixed (infrastructure) and variable (operating). They depend on a variety of conditions related to geography, infrastructure, administrative barriers, energy and on how passengers and freight are carried. Three major components, related to transactions, shipments and the friction of distance, impact on transport costs.

costs–insurance–freight (CIF)

The price of a good is a uniform delivered price for all customers everywhere, with no spatially variable shipping price, which implies that the average shipping price is built into the price of a good. The CIF cost structure can be expanded to include several rate zones.

cross-docking

A form of inventory management where goods are received at one door of the distribution center/sorting facility and shipped out through the other door in a very short time without putting them in storage. It consequently contributes in the reduction of operating costs with an increase in the throughput and with a reduction of inventory levels.

crude oil petroleum

A naturally occurring, oily, flammable liquid composed principally of hydrocarbons. Crude oil is occasionally found in springs or pools but usually is drilled from wells beneath the Earth's surface.

current dollars

The dollar value of a good or service in terms of prices current at the time the good or service is sold. This contrasts with the value of the good or service measured in constant dollars.

deadhead

Miles and hours that a vehicle travels when out of revenue service. This includes leaving and returning to the garage, changing routes, etc., and when there is no reasonable expectation of carrying revenue passengers. However, it does not include charter service, school bus service, operator training, maintenance training, etc. For non-scheduled, non-fixed-route service (demand responsive), deadhead mileage also includes the travel between the dispatching point and passenger pickup or drop-off.

deadweight tons

The lifting capacity of a ship expressed in long tons (2,240 lb), including cargo, commodities and crew. Reflects the weight difference between a fully loaded and an unloaded ship.

demand responsive

Non-fixed-route service utilizing vans or buses with passengers boarding and alighting at prearranged times at any location within the system's service area. Also called "Dial-a-Ride".

demand (transport)

The expression of the transport needs, even if those needs are satisfied, fully, partially or not at all. Similar to transport supply, it is expressed in terms of number of people, volume or tons per unit of time and space.

deregulation

Consists in a shift to a competitive economic climate by reorienting and/or suppressing regulatory mechanisms. Deregulation, however, does not necessarily refer to complete absence of free market regulation measures but rather to the promotion of competition-inducing ones (which can seek elimination of monopolies, for example). Particularly observed in the transport and telecommunications sectors.

distribution center (freight)

Facility or a group of facilities that perform consolidation, warehousing, packaging, decomposition and other functions linked with handling freight. Their main purpose is to provide value-added services to freight and are a fundamental component of freight distribution. DCs are often in proximity to major transport routes or terminals. They can also perform light manufacturing activities such as assembly and labeling.

dock

A feature built to handle ships. Can also refer to an enclosed port area used for maritime operations.

double stack
The movement of containers on articulated rail cars which enables one container to be stacked on another for better ride quality and car utilization.

downtime
A period during which a vehicle or a whole system is inoperative because of repairs or maintenance.

drayage
The movement of a container or trailer to or from the railroad intermodal terminal to or from the customer's facility for loading or unloading.

dry bulk cargo
Cargo which may be loose, granular, free-flowing or solid, such as grain, coal and ore, and is shipped in bulk rather than in package form. Dry bulk cargo is usually handled by specialized mechanical handling equipment at specially designed dry bulk terminals.

dwell time
The time a vehicle (bus, truck, train or ship) is allowed to load or unload passengers or freight at a terminal.

dynamic routing
In demand–response transportation systems, the process of constantly modifying vehicle routes to accommodate service requests received after the vehicle began operations, as distinguished from predetermined routes assigned to a vehicle.

economic evaluation (also called appraisal or analysis)
Refers to various methods for determining the value of a policy, project or program to help individuals, businesses and communities make decisions that involve tradeoffs. Economic evaluation is an important part of transportation decision making.

economies of agglomeration
Refer to the benefits of having activities locate (cluster) next to one another, such as the use of common infrastructures and services.

economies of scale
Cost reductions or productivity efficiencies achieved through size increase. The outcome is a decrease in the unit cost of production associated with increasing output. For example, freight rates usually decline as the volume of cargo tonnage shipped increases. Simplistically, the more passengers share the same taxi (up to the maximum size), the less their individual fares will be.

economies of scope
Cost savings resulting from increasing the number of different goods or services produced.

electronic data interchange (EDI)
Communication mode for inter- and intra-firm data exchange in the freight forwarding and logistics business.

energy
The capacity for doing work as measured by the capability of doing work (potential energy) or the conversion of this capability to motion (kinetic energy). Energy has several forms, some of which are easily convertible and can be changed to another form useful for work. Most of the world's convertible energy comes from fossil fuels

that are burned to produce heat that is then used as a transfer medium to mechanical or other means in order to accomplish tasks. Electrical energy is usually measured in kilowatt hours, while heat energy is usually measured in British thermal units.

energy intensity

In reference to transportation, the ratio of energy inputs to a process to the useful outputs form that process; for example, gallons of fuel per passenger-mile or BTU per ton-mile.

environmental impact assessment

A process for carrying out an appraisal of the full potential effects of a development project on the physical environment.

environmental management system

A set of procedures and techniques enabling an organization to reduce environmental impacts and increase its operating efficiency.

ethanol

An alternative fuel; a liquid alcohol fuel with vapor heavier than air; produced from agricultural products such as corn, grain and sugar cane.

European Union (EU)

Formerly the European Community (EC), became the European Union with the signing of the Maastricht Treaty in November 1993. A regional trade block composed of 27 European states. Its core institutions are known as the "institutional triangle" composed of the European Parliament (Strasbourg), the Commission (Brussels) and the EU Council (Brussels). Also there is the European Bank which manages the common currency.

exclusive right-of-way

A highway or other facility that can only be used by buses or other transit vehicles.

externality (external cost)

Economic cost not normally taken into account in markets or in decisions by market players.

fare

The price paid by the user of a transport service at the moment of use.

fare elasticity

The extent to which ridership responds to fare increases or decreases.

fare structure

The system set up to determine how much is to be paid by various passengers using a transit vehicle at any given time.

feeder

Short sea shipping service which connects at least two ports in order for the freight (generally containers) to be consolidated or redistributed to or from a deep-sea service in one of these ports. By extension, this concept may be used for inland transport services and air transportation.

ferryboat

A boat providing fixed-route service across a body of water, which can be short or long distance.

fixed cost

Costs that do not vary with the quantity shipped in the short run, i.e. costs that must be paid up-front to begin producing transportation services.

fixed route

Service provided on a repetitive, fixed-schedule basis along a specific route with vehicles stopping to pick up and deliver passengers or freight to specific locations; each fixed-route trip serves the same origins and destinations, unlike demand responsive. The terms apply to many modes of transportation, including public transit, air services and maritime services.

flag state

Country of registry of a sea-going vessel. A sea-going vessel is subject to the maritime regulations in respect of manning scales, safety standards and consular representation abroad of its country of registration.

flat car

A freight car having a floor without any housing or body above. Frequently used to carry containers and/or trailers or oversized/odd-shaped commodities. The three types of flat cars used in intermodal are conventional, spine and stack cars.

fleet

The vehicles in a transport system. Usually, "fleet" refers to highway and rail vehicles as well as to ships.

foreland

A maritime space with which a port performs commercial relationships. It includes overseas customers with which the port undertakes commercial exchanges.

forwarding agent/freight forwarder

Intermediary who arranges for the carriage of goods and/or associated services on behalf of a shipper.

fourth-party logistics provider (4PL)

Integrates the resources of producers, retailers and third-party logistics providers in view to build a system-wide improvement in supply chain management. They are non-asset based meaning that they mainly provide organizational expertise.

freight on board (FOB)

The price of a good is the combination of the factory costs and the shipping costs from the factory to the consumer. The consumer pays for the freight transport costs. Consequently, the price of a commodity will vary according to transportation costs.

free trade zone

A port or an area designated by the government of a country for duty-free entry of any non-prohibited goods. Merchandise may be stored, displayed, used for manufacturing, etc., within the zone and re-exported without duties.

freight consignee and handlers

Freight consignees are independent of shippers or producers. They are commissioned by the latter to accomplish all transport operations including storage, transport, management, sometimes re-expedition, etc. from origin to final destination. The notion of freight handler is broader. It comprises any actor involved in transport of freight from origin to destination including transport terminals and subcontractual services, for instance.

freight distribution center

See distribution center.

freight forwarder

An individual or company that accepts less-than-truckload (LTL) or less-than-carload (LCL) shipments from shippers and combines them into carload or truckload lots. Carriers collecting small shipments to be cumulatively consolidated and transported relying upon a single or several modes of transportation to a given destination. Functions performed by a freight forwarder may include receiving small shipments (e.g. less than container load) from consignors, consolidating them into larger lots, contracts with carriers for transport between ports of embarkation and debarkation, conducting documentation transactions and arranging delivery of shipments to the consignees.

freight village

A concentration (or a cluster) of freight-related activities within a specific area, commonly built for such a purpose, master planned and managed. These activities include distribution centers, warehouses and storage areas, transport terminals, offices and other facilities supporting those activities, such as public utilities, parking space and even hotels and restaurants. Although a freight village can be serviced by a single mode, intermodal facilities can offer direct access to global and regional markets.

fringe parking

An area for parking usually located outside the central business district (CBD) and most often used by suburban residents who work or shop downtown. Commonly corresponds to an access point of a transit system, such as a rail or subway station.

fuel cell

A device that produces electrical energy directly from the controlled electrochemical oxidation of the fuel, commonly hydrogen. It does not contain an intermediate heat cycle, as do most other electrical generation techniques.

gasohol

A blend of motor gasoline (leaded or unleaded) and alcohol (generally ethanol but sometimes methanol) limited to 10 percent by volume of alcohol. Gasohol is included in finished leaded and unleaded motor gasoline.

gasoline

A complex mixture of relatively volatile hydrocarbons, with or without small quantities of additives, obtained by blending appropriate refinery streams to form a fuel suitable for use in spark ignition engines. Motor gasoline includes both leaded or unleaded grades of finished motor gasoline, blending components and gasohol.

gateway

A location offering accessibility to a large system of circulation of freight, passengers and/or information. Gateways reap advantage of a favorable physical location such as highway junctions, confluence of rivers and seaboards, and have been the object of a significant accumulation of transport infrastructures such as terminals and their links. A gateway generally commands the entrance to and the exit from its catchment area. In other words, it is a pivotal point for the entrance and the exit of merchandise in a region, a country or a continent. Gateways tend to be locations where intermodal transfers are performed.

general cargo

General cargo consists of those products or commodities such as timber, structural steel, rolled newsprint, concrete forms, agricultural equipment that are not conducive to packaging or unitization. Break-bulk cargo (e.g. packaged products such as lubricants and cereal) are often regarded as a subdivision of general cargo.

geographic information system (GIS)
A special-purpose system composed of hardware and software in which a common spatial coordinate system is the primary means of reference. GIS contain subsystems for: data input; data storage, retrieval and representation; data management, transformation and analysis; and data reporting and product generation.

GIS-T
Acronym for Transportation-oriented Geographic Information System.

graph theory
A branch of mathematics concerned about how networks can be encoded and their properties measured.

great circle distance
The shortest path between two points on a sphere. The circumference inferred out of these two points divides the Earth in two equal parts, thus the great circle. The great circle distance is useful to establish the shortest path to use when traveling at the intercontinental air and maritime level. The great circle route follows the sphericity of the globe; any shortest route is the one following the curve of the planet, along the parallels.

gross domestic product (GDP)
A measure of the total value of goods and services produced by a domestic economy during a given period, usually one year. Obtained by adding the value contributed by each sector of the economy in the form of profits, compensation to employees and depreciation (consumption of capital). Only domestic production is included, not income arising from investments and possessions owned abroad, hence the use of the word domestic.

gross national product (GNP)
The total market value of goods and services produced during a given period by labor and capital supplied by residents of a country, regardless of where the labor and capital are located. GNP differs from GDP primarily by including the capital income that residents earn from investments abroad and excluding the capital income that nonresidents earn from domestic investment.

Handy and Handymax
Traditionally the workhorses of the dry bulk market, the Handy and more recent Handymax types remain popular ships with less than 50,000 dwt. This category is also used to define small-sized oil tankers.

headway
Time interval between vehicles moving in the same direction on a particular route.

heavy rail
An electric railway with the capacity for a "heavy volume" of traffic and characterized by exclusive rights-of-way, multi-car trains, high speed and rapid acceleration, sophisticated signaling and high platform loading.

high-occupancy-vehicle lane (HOV)
A highway or road lane reserved to vehicles that have a specific level of occupancy, with at least one passenger. Often used to alleviate congestion and favor carpooling.

hinterland

Land space over which a transport terminal, such as a port, sells its services and interacts with its clients. It accounts for the regional market share that a terminal has relative to a set of other terminals servicing this region. It regroups all the customers directly bounded to the terminal. The terminal, depending on its nature, serves as a place of convergence for the traffic coming by roads, railways or by sea/fluvial feeders.

hub

Central point for the collection, sorting, transhipment and distribution of goods and passengers for a particular area. This concept comes from a term used in air transport for passengers as well as freight. It describes collection and distribution through a single point such as the "hub-and-spoke" concept. Hubs tend to be transmodal (transfers within the same mode) locations.

inflation

Increase in the amount of currency in relation to the availability of assets, commodities, goods and services. Commonly the outcome of an indirect confiscation of wealth (theft) through an over-issuance of currency ("money printing") by central banks and governments. Although it directly influences prices, inflation is outside the supply–demand relationship and decreases the purchasing power, if wages are not increased accordingly. Almost all central banks have inflationary policies which enable governments to run deficits for decades by slowly devaluing the debt they contracted in the past.

infrastructure

(1) In transport systems, all the fixed components, such as rights-of-way, tracks, signal equipment, terminals, parking lots, bus stops, maintenance facilities, etc. (2) In transportation planning, all the relevant elements of the environment in which a transportation system operates.

integrated carriers

Carriers that have both air and ground fleets; or other combinations, such as sea, rail and truck. Since they usually handle thousands of small parcels an hour, they are less expensive and offer more diverse services than regular carriers.

intermediacy

Focus on the terminal as an intermediate point in the flows of passengers or freight. This term is applied to the frequent occurrence of places gaining advantage because they are between other places. The ability to exploit transhipment has been an important feature of many terminals.

intermodal terminal

A terminal which can accommodate several modes of transportation. They increasingly tend to be specializing at handling specific types of passengers or freight traffic, while they may share the same infrastructures.

intermodal transport

The movement of goods in one and the same loading unit or road vehicle, which uses successively two or more modes of transport without handling the goods themselves in changing modes. Enables cargo to be consolidated into economically large units (containers, bulk grain railcars, etc.) optimizing the use of specialized intermodal handling equipment to effect high-speed cargo transfer between ships, barges, railcars and truck chassis using a minimum of labor to increase logistic flexibility, reduce consignment delivery times and minimize operating costs.

intermodalism

In its broadest interpretation, intermodalism refers to a holistic view of transportation in which individual modes work together or within their own niches to provide the user with the best choices of service, and in which the consequences on all modes of policies for a single mode are considered. This view has been called balanced, integrated or comprehensive transportation in the past. In a narrower and more practical interpretation it refers to the carriage of freight across more than one mode of transport under one bill of lading or waybill. The shipper is provided with one freight rate from door-to-door regardless of the number of modal transfers during the journey.

International Air Transportation Association (IATA)

Established in 1945, a trade association serving airlines, passengers, shippers, travel agents and governments. The association promotes safety, standardization in forms (baggage checks, tickets, weight bills), and aids in establishing international airfares. The IATA headquarters are in Montreal, Canada.

International Civil Aviation Organization (ICAO)

A specialized agency of the United Nations whose objective is to develop the principles and techniques of international air navigation and to foster planning and development of international civil air transport. Headquartered in Montreal, Canada.

International Maritime Organization (IMO)

Established as a specialized agency of the United Nations in 1948. The IMO facilitates cooperation on technical matters affecting merchant shipping and traffic, including improved maritime safety and prevention of marine pollution. The IMO headquarters are in London, England.

International Organization for Standardization (ISO)

Worldwide federation of national standards bodies from some 100 countries, one from each country. ISO is a non-governmental organization established in 1947. The mission of ISO is to promote the development of standardization and related activities in the world with a view to facilitating the international exchange of goods and services, and to developing cooperation in the spheres of intellectual, scientific, technological and economic activity. ISO's work results in international agreements which are published as International Standards.

jet stream

A migrating stream of high-speed winds present at high altitudes.

jitney

Privately-owned, small or medium-sized vehicle usually operated on a fixed route but not on a fixed schedule.

just-in-time

The principle of production and inventory management in which goods arrive when needed for production or consumption. Warehousing tends to be minimal or non-existent, but in all cases is much more efficient and more limited in duration.

knot, nautical
The unit of speed equivalent to one nautical mile: 6,080.20 feet per hour or 1.85 kilometers per hour.

lading
Refers to the freight shipped; the contents of a shipment.

landbridge
An intermodal connection between two ocean carriers separated by a land mass, linked together in a seamless transaction by a land carrier.

landed cost
The dollar per barrel price of crude oil at the port of discharge. Included are the charges associated with the purchase, transporting and insuring of a cargo from the purchase point to the port of discharge. Not included are charges incurred at the discharge port (e.g. import tariffs or fees, wharfage charges and demurrage charges).

layover time
Time built into a schedule between arrival at the end of a route and the departure for the return trip, used for the recovery of delays and preparation for the return trip.

less than truckload (LTL)
A shipment that would not by itself fill the truck to capacity by weight or volume.

level of service
(1) A set of characteristics that indicate the quality and quantity of transportation service provided, including characteristics that are quantifiable and those that are difficult to quantify. (2) For highway systems, a qualitative rating of the effectiveness of a highway or highway facility in serving traffic, in terms of operating conditions. A rating of traffic flow ranging from A (excellent) through F (heavily congested), and compares actual or projected traffic volume with the maximum capacity of the intersection or road in question. (3) For paratransit, a variety of measures meant to denote the quality of service provided, generally in terms of total travel time or a specific component of total travel time. (4) For pedestrians, sets of area occupancy classifications to connect the design of pedestrian facilities with levels of service.

light-rail transit (LRT)
A fixed guideway transportation mode that typically operates on city streets and draws electric power from overhead wires; includes streetcars, trolley cars and tramways. Differs from heavy rail – which has a separated right of way, and includes commuter and inter-city rail – in that it has lighter passenger capacity per hour and more closely spaced stops.

lighter-aboard-ship (LASH)
A type of barge-carrying vessel equipped with an overhead crane capable of lifting barges of a common size and stowing them into cellular slots in athwartship position. LAS is an all-water technology analogous to containerization.

line haul costs
Costs that vary with distance shipped, i.e. costs of moving goods and people once they are loaded on the vehicles.

liner
Derived from the term "line traffic", which denotes operation along definite routes on the basis of definite, fixed schedules. A liner thus is a vessel that engages in this kind of transportation, which usually involves the haulage of general cargo as distinct from bulk cargo.

liquefied natural gas (LNG)

An alternative fuel; a natural gas cooled to below its boiling point of −260 degrees Fahrenheit so that it becomes a liquid; stored in a vacuum-type container at very low temperatures and under moderate pressure. LNG vapor is lighter than air.

load factor

The ratio of passengers or freight actually carried versus the total passenger or freight capacity of a vehicle or a route.

logistics

The process of designing and managing the supply chain in the wider sense. The chain can extend from the delivery of supplies for manufacturing, through the management of materials at the plant, delivery to warehouses and distribution centers, sorting, handling, packaging and final distribution to point of consumption. Derived from Greek *logistikos* (to reason logically), the word is polysemic. Nineteenth-century military referred to it as the art of combining all means of transport, revictualling and sheltering of troops. A more fitted meaning consists in the set of all operations required for goods (material or nonmaterial) to be made available on markets or to specific destinations. With increasing multimodal and containerized freight transport that complexify the coordination itineraries, logistics rely heavily on highly performing computerized information management. The term also applies to passenger transportation.

logistics center

Geographical grouping of independent companies and bodies which are dealing with freight transport (e.g. freight forwarders, shippers, transport operators, customs) and with accompanying services (e.g. storage, maintenance and repair), including at least a terminal. Also called "freight village".

logit model

A probabilistic model for representing a discrete choice behavior of individuals. On any choice occasion the individual is assumed to choose the mode of highest preference. Over repeated choice occasions preferences are assumed to have a probabilistic component. For the logit model this random component of preference is taken to have a double exponential distribution.

long ton

2,240 pounds.

Maglev (magnetic levitation)

Technology enabling trains to move at high speed above a guideway on a cushion generated by magnetic force.

manifest

A list of the goods being transported by a carrier.

maritime routes

Corridors of a few kilometers in width trying to avoid the discontinuities of land transport by linking ports, the main elements of the maritime/land interface. Maritime routes are a function of obligatory points of passage, which are strategic places, of physical constraints (coasts, winds, marine currents, depth, reefs, ice) and of political borders. As a result, maritime routes draw arcs on the Earth's water surface as intercontinental maritime transportation tries to follow the great circle distance.

maritime terminal

A designated area of a port, which includes but is not limited to wharves, warehouses, covered and/or open storage spaces, cold storage plants, grain elevators and/or bulk cargo loading and/or unloading structures, landings, and receiving stations, used for the transmission, care and convenience of cargo and/or passengers in the interchange of same between land and water carriers or between two water carriers.

market area

The surface over which a demand offered at a specific location is expressed. Commonly, a customer is assumed to go to a location where a product or service can be acquired or a part or a finished good has to be shipped from the place of production to the place of consumption.

materials management

Considers all the activities related in the manufacturing of commodities in all their stages of production along a supply chain. It includes production and marketing activities such as production planning, demand forecasting, purchasing and inventory management. It must insure that the requirements of supply chains are met by dealing with a wide array of parts for assembly and raw materials, including packaging (for transport and retailing) and, ultimately, recycling discarded commodities. All these activities are assumed to induce physical distribution demands.

MERCOSUR

A trade alliance between Argentina, Brazil, Paraguay and Uruguay, with Chile and Bolivia as associate members.

methanol

An alternative fuel; a liquid alcohol fuel with vapor heavier than air; primarily produced from natural gas.

microbridge

A cargo movement in which the water carrier provides a through service between an inland point and the port of load/discharge.

minibridge

A joint water, rail or truck container move on a single Bill of Lading for a through route from a foreign port to a US port destination through an intermediate US port or the reverse.

mobility

Refers to a movement of people or freight. It can have different levels linked to the speed, capacity and efficiency of movements.

modal share

The percentage of total passengers or freight moved by a particular type of transportation.

modal split (share)

(1) The proportion of total person trips that uses each of various specified modes of transportation. (2) The process of separating total person trips into the modes of travel used. (3) A term that describes how many people use alternative forms of transportation. It is frequently used to describe the percentage of people who use private automobiles, as opposed to the percentage who use public transportation.

mode, transport

The physical way a movement is performed.

model

An analytical tool (often mathematical) used by transportation planners to assist in making forecasts of land use, economic activity, travel activity and their effects on the quality of resources such as land, air and water.

monorail

An electric railway in which a rail car or train of cars is suspended from or straddles a guideway formed by a single beam or rail. Most monorails are either heavy rail or automated guideway systems.

motorway/highway

A road, specially designed and built for motor traffic, which does not serve properties bordering on it, and which: (1) is provided, except at special points or temporarily, with separate carriageways for the two directions of traffic, separated from each other, either by a dividing strip not intended for traffic, or exceptionally by other means; (2) does not cross at level with any road, railway or tramway track, or footpath; (3) is specially signposted as a motorway and is reserved for specific categories of road motor vehicles. Entry and exit lanes of motorways are included irrespective of the location of the signposts. Urban motorways are also included.

multimodal platform

A physical converging point where freight and/or passenger transhipment takes place between different modes of transportation, usually a transport terminal.

North American Free Trade Agreement (NAFTA)

Came into force on January 1, 1994. NAFTA binds Canada, the United States and Mexico over respect of a series of common economics rules. Beside the liberalization of exchange of goods and services, the NAFTA regulates investments, intellectual property, publics markets and the non-tariff barrier. The NAFTA is a result of a tradition of trade negotiations between Canada and the USA that became explicit with the 1989 Free Trade Agreement (FTA) and the 1991 Canada–US Trade Agreement (CUSTA).

net tonnage

The net or register tonnage of a vessel is the remainder after deducting from the gross tonnage of the vessel the tonnage of crew spaces, master's accommodation, navigation spaces, allowance for propelling power, etc. It is expressed in tons of 100 cubic feet.

network

Framework of routes within a system of locations, identified as nodes. A route is a single link between two nodes that are part of a larger network that can refer to tangible routes such as roads and rails, or less tangible routes such as air and sea corridors.

network analysis

The pattern of transportation systems, the location of routes or rails, the location of intersections, nodes and terminals can be considered as a network. However, on the analytic side, more attention is paid to the whole system rather than to single routes or terminals. Network analysis aims at identifying flows, shortest distances between two given points, or the less expensive road to take for transporting goods between those points. To facilitate the task, networks have been approximated by the use of the graph theory relying on topology.

nitrogen oxides

A product of combustion of fossil fuels whose production increases with the temperature of the process. It can become an air pollutant if concentrations are excessive.

ocean bill of lading

A receipt for the cargo and a contract for transportation between a shipper and the ocean carrier. It may also be used as an instrument of ownership which can be bought, sold or traded while the goods are in transit.

oceanic airspace

Airspace over the oceans of the world, considered international airspace, where oceanic separation and procedures per the ICAO are applied. Responsibility for the provisions of air traffic control service in this airspace is delegated to various countries, based generally upon geographic proximity and the availability of the required resources.

off-peak period

Non-rush periods of the day when travel activity is generally lower and less transit service is scheduled. Also called "base period".

offshore hub

A port terminal that dominantly serves transmodal operations, implying limited connections in relation to its total traffic with its hinterland. They are mainly used to feedering, relay and interlining between maritime shipping routes. The term offshore can be misleading as many ports performing this function are located at standard port locations.

operating cost

Costs that vary with the quantity shipped in the short run. (1) Fixed operating cost: refers to expenditures that are independent of the amount of use. For a car, it would involve costs such as insurance costs, fees for license and registration, depreciation and finance charges. (2) Variable operating cost: expenditures which are dependent on the amount of use. For a car, it would involve costs such as the cost of gasoline, oil, tires and other maintenance.

pallet

A raised platform, normally made of wood, facilitating the handling of goods. Pallets are of standard dimensions.

pandemic

An epidemic of infectious disease that spreads through human populations across a large area, even worldwide.

Panamax

A maritime standard corresponding to about 65,000 deadweight tons or 4,000 TEU. Refers to a ship with dimensions that allow it to pass through the Panama Canal: maximum length 295 m, maximum beam overall 32.25 m, maximum draught 13.50 m.

park and ride

An access mode to transit in which patrons drive private automobiles or ride bicycles to a transit station stop or carpool/vanpool waiting area and park their vehicle in the area provided for the purpose. They then ride the transit system or take a car- or vanpool to their destinations.

particulates

Carbon particles formed by partial oxidation and reduction of the hydrocarbon fuel. Also included are trace quantities of metal oxides and nitrides, originating from engine wear, component degradation and inorganic fuel additives. In the transportation sector, particulates are emitted mainly from diesel engines.

passenger-km (or passenger-mile)

The total number of miles (km) traveled by passengers on vehicles; determined by multiplying the number of unlinked passenger trips times the average length of the trips.

payload

Weight of commodity being hauled. Includes packaging, pallets, banding, etc., but does not include the truck, truck body, etc.

peak oil

A theory concerning oil production initially published by the geophysicist King Hubbert in 1956, that assumes due to the finite nature of oil reserves production will at some point reach maximum output. Once peak production has been reached, production declines and prices go up until oil resources are depleted or too costly to have a widespread use.

peak period (hour)

Represents a time period of high usage of a transport system. For transit, it refers to morning and afternoon time periods when ridership is at its highest.

peak/base ratio

The number of vehicles operated in passenger or freight service during the peak period divided by the number operated during the base period.

pendulum service

Involves a set of sequential port calls along a maritime range, commonly including a transoceanic service from ports in another range and structured as a continuous loop. They are almost exclusively used for container transportation with the purpose of servicing a market by balancing the number of port calls and the frequency of services.

physical distribution

The collective term for the range of activities involved in the movement of goods from points of production to final points of sale and consumption. It must insure that the mobility requirements of supply chains are entirely met. Physical distribution comprises all the functions of movement and handling of goods, particularly transportation services (trucking, freight rail, air freight, inland waterways, marine shipping and pipelines), transhipment and warehousing services (e.g. consignment, storage, inventory management), trade, wholesale and, in principle, retail. Conventionally, all these activities are assumed to be derived from materials management demands.

piggyback trailers

Trailers which are designed for quick loading on railcars.

pipeline

A continuous pipe conduit, complete with such equipment as valves, compressor stations, communications systems and meters for transporting natural and/or supplemental gas from one point to another, usually from a point in or beyond the producing field or processing plant to another pipeline or to points of utilization. Also refers to a company operating such facilities.

planning

Refers to a process that allows people's needs, preferences and values to be reflected in decisions. Planning occurs at many different levels, from day-to-day decisions made by individuals and families, to major decisions made by governments and businesses that have comprehensive, long-term impacts on society. Management can be considered a short-term form of planning, while planning can be considered a longer-term form of management.

platform/modular manufacturing
A strategy in which a multinational corporation retains its core competencies, namely its research and development, retailing, marketing and distribution, while subcontracting much of the manufacturing to the lowest bidders.

policy (transport)
The development of a set of constructs and propositions that are established to achieve particular objectives relating to social, economic and environmental development, and the functioning and performance of the transport system.

port
A harbor area where marine terminal facilities are located for transferring cargo between ships and land transportation.

port authority
An entity of state or local government that owns, operates or otherwise provides wharf, dock and other marine terminal investments at ports.

port holding
An entity, commonly private, that owns or leases port terminals in a variety of locations. It is also known as a port terminal operator.

port of entry
A port at which foreign goods are admitted into the receiving country. Also refers to an air terminal or land access point (customs) where foreign passengers and freight can enter a country.

primary transportation
Conveyance of large shipments of petroleum raw materials and refined products usually by pipeline, barge or ocean-going vessel. All crude oil transportation is primary, including the small amounts moved by truck. All refined product transportation by pipeline, barge, or ocean-going vessel is primary transportation.

product life cycle
Defined as the period that starts with the initial product design (research and development) and ends with the withdrawal of the product from the marketplace. A product life cycle is characterized by specific stages, including research, development, introduction, maturity, decline and obsolescence.

propane
An alternative fuel; a liquid petroleum gas (LPG) which is stored under moderate pressure and with vapor heavier than air; produced as a by-product of natural gas and oil production.

public transportation
Passenger transportation services, usually local in scope, that are available to any person who pays a prescribed fare. They operate on established schedules along designated routes or lines with specific stops and are designed to move relatively large numbers of people at one time.

radio frequency identification device (RFID)
A technology that uses small devices attached to objects that transmit data to a receiver. An alternative to bar coding used for identification and tracking purposes, notably for items shipped in units (boxes, containers, etc.), but can also be attached to an individual item. Main technical advantages include data storage capacity, read/write capability and no line-of-sight requirements during scanning.

rail, commuter

Railroad local and regional passenger train operations between a central city, its suburbs and/or another central city. It may be either locomotive-hauled or self-propelled, and is characterized by multi-trip tickets, specific station-to-station fares, railroad employment practices and usually only one or two stations in the central business district. Also known as "suburban rail".

rail, heavy

An electric railway with the capacity for a "heavy volume" of traffic and characterized by exclusive right of way, multi-car trains, high speed and rapid acceleration, sophisticated signaling and high platform loading. Also known as "rapid rail", "subway", "elevated (railway)" or "metropolitan railway (metro)".

rail, high-speed

A rail transportation system with exclusive right of way which serves densely traveled corridors at speeds of 124 miles per hour (200 km/h) and greater.

rail, light

An electric railway with a "light volume" traffic capacity compared to heavy rail. Light rail may use shared or exclusive right of way, high or low platform loading and multi-car trains or single cars. Also known as "streetcar", "trolley car" and "tramway".

Railroad

All forms of non-highway ground transportation that run on rails or electro-magnetic guideways, including: (1) commuter or other short-haul rail passenger service in a metropolitan or suburban area, and (2) high-speed ground transportation systems that connect metropolitan areas, without regard to whether they use new technologies not associated with traditional railroads. The term does not include rapid transit operations within an urban area that are not connected to the general railroad system of transportation.

rapid transit

Rail or motorbus transit service operating completely separate from all modes of transportation on an exclusive right of way.

rate

The price of transportation services paid by the consumer. They are the negotiated monetary cost of moving a passenger or a unit of freight between a specific origin and destination. Rates are often visible to the consumers since transport providers must provide this information to secure transactions.

reefer ship

General cargo ship with 80 percent or more insulated cargo space.

ridesharing

A form of transportation, other than public transit, in which more than one person shares the use of the vehicle, such as a van or car, to make a trip. Also known as "carpooling" or "vanpooling".

ridership

The number of rides taken by people using a public transportation system in a given time period.

Roll-on/roll-off (RORO) vessel

Ships which are especially designed to carry wheeled container trailers, or other wheeled cargo, and use the roll-on/roll-off method for loading and unloading.

rolling stock
The vehicles used in a transit system, including buses and rail cars.

semi-trailer
A non-powered vehicle for the carriage of goods, intended to be coupled to a motor vehicle in such a way that a substantial part of its weight and of its load is borne by the motor vehicle.

Shimbel index
Measures the minimum number of links necessary to connect one node with all other nodes in a defined graph.

shipper
The company sending goods.

short sea shipping
Commercial waterborne transportation that does not transit an ocean. It is an alternative form of commercial transportation that utilizes inland and coastal waterways to move commercial freight from major domestic ports to its destination.

shunting
Operation related to moving a rail vehicle or set of rail vehicles within a railway installation (station, depot, workshop, marshalling yard, etc.). It mainly concerns the assembly and disassembly of unit trains.

shuttle
A public or private vehicle that travels back and forth over a particular route, especially a short route or one that provides connections between transportation systems, employment centers, etc.

Silk Road
Historical trade route linking the Eastern Mediterranean basin to Central and East Asia. Named as such because of many prized commodities, namely silk, tea and jade, that were carried from China. Was operational between the first century BC and the sixteenth century.

single-occupant vehicle (SOV)
A vehicle with one occupant, the driver, who is sometimes referred to as a "drive alone".

site
The geographical characteristics of a specific location.

situation
The relationships a location has in regard to other locations.

source loading
This refers to the loading of a shipment, commonly in a container, at the location where the goods it carries are produced. The shipment remains untouched until it reaches its destination, thus conferring a level of integrity in the supply chain.

spatial interaction
A realized movement of people, freight or information between an origin and a destination. It is a transport demand/supply relationship expressed over a geographical space. Spatial interactions cover a wide variety of movements such as journeys to work, migrations, tourism, the usage of public facilities, the transmission of information or capital, the market areas of retailing activities, international trade and freight distribution.

spatial structure

The manner in which space is organized by the cumulative locations of infrastructure, economic activities and their relations.

Suezmax

The standard which represents the limitations of the Suez Canal. Before 1967, the Suez Canal could only accommodate tanker ships with a maximum of 80,000 dwt. The canal was closed between 1967 and 1975 because of the Israel–Arab conflict. Once it reopened in 1975, the Suezmax capacity went to 150,000 dwt. An enlargement to enable the canal to accommodate 200,000 dwt tankers is being considered.

supply chain

See commodity chain.

supply chain management (SCM)

The management of the whole commodity/supply chain, from suppliers, manufacturers, retailers and the final customers. To achieve higher productivity and better returns, SCM mainly tries to reduce inventory, increase transaction speeds and satisfy the needs of customers in terms of cost, quantity, quality and delivery as much as possible.

supply (transport)

The capacity of transportation infrastructures and modes, generally over a geographically defined transport system and for a specific period of time. Therefore, supply is expressed in terms of infrastructures (capacity), services (frequency) and networks. The number of passengers, volume (for liquids or containerized traffic) or mass (for freight) that can be transported per unit of time and space is commonly used to quantify transport supply.

tanker

An ocean-going ship specially designed to haul liquid bulk cargo in world trade, particularly oil.

tare weight

(1) The weight of a container and the material used for packing. (2) As applied to a car/trailer, the weight of the car/trailer exclusive of its contents.

tariff

A general term for any listing of rates or charges. The tariffs most frequently encountered in foreign trade are: tariffs of international transportation companies operating on sea, land and in the air; tariffs of international cable, radio and telephone companies; and the customs tariffs of the various countries that list goods that are duty free and those subject to import duty, giving the rate of duty in each case.

terminal

Any location where freight and passengers either originate, terminate or are handled in the transportation process. Terminals are central and intermediate locations in the movements of passengers and freight. They often require specific facilities to accommodate the traffic they handle.

terminal costs

Costs of loading and unloading. They do not vary with distance shipped.

third-party logistics provider (3PL)

An asset-based company that offers logistics and supply chain management services to its customers (manufacturers and retailers). It commonly owns distribution centers and transport modes.

threshold

The minimum and vital market size required to support a given type of economic activity. A mean number of passengers per trip can be identified to sustain profitability of a coach line, for example. A threshold thus rests on a level of demand and can play a determining role in organizing both freight and passenger transport structures on the basis of demographic dynamics, geographic relations to markets and intensity of economic activities.

ton

A unit of measurement of weight. Frequently used in freight transport statistics, a metric ton is equivalent to 1,000 kilograms or 2,205 pounds. A short ton is equivalent to 2,000 pounds or 0.908 metric tons (in the United States the term ton is commonly used but implies short ton). A long ton, a term not as frequently used, is equivalent to 2,240 pounds or 1.06 metric tons.

ton-km (or ton-mile)

Measure expressing the realized freight transport demand. Although both the passenger-km and ton-km are most commonly used to measure realized demand, the measure can equally apply for transport supply.

track gauge

The distance between the internal sides of rails on a railway line. It is generally 1.435 m. Other gauges are generally used in some European countries: for instance, 1.676 m in Spain and Portugal, 1.524 m in the Russian Federation.

trailer on flat car (TOFC)

A rail trailer or container mounted on a chassis that is transported on a rail car. Also known as piggyback.

tramp

An ocean-going vessel that does not operate along a definite route or on a fixed schedule, but rather calls at any port where cargo is available.

transaction costs

Costs required for gathering information, negotiating and enforcing contracts and transactions. Often referred to as the cost of doing business.

transactions

In the business domain, the word transaction is synonymous with exchange and refers to a commercial operation. Transactions generate varying costs, depending on the stakes, the competition, the context of the economic market, etc.

transhipment

The transfer of goods from one carrier to another and/or from one mode to another.

transit system

An organization (public or private) providing local or regional multi-occupancy vehicle passenger service. Organizations that provide service under contract to another agency are generally not counted as separate systems.

transloading

The transhipment of loads from truck to rail and vice versa. It is done to exploit the respective advantages of trucking and rail, namely to avoid long-distance trucking. Also refers to the moving of the contents of a container, say a 40-foot maritime container, into another container, such as a 53-foot domestic container, or a regular truckload.

transmodal transportation

The movements of passengers or freight within the same mode of transport. Although "pure" transmodal transportation rarely exists and an intermodal operation is often required (e.g. ship to dockside to ship), the purpose is to insure continuity within the network.

transport geography

Subdiscipline of geography concerned about movements of freight, people and information. It seeks to link spatial constraints and attributes with the origin, the destination, the extent, the nature and the purpose of movements.

transportability

The ease of movement of passengers, freight or information. It is related to transport costs as well as to the attributes of what is being transported (fragility, perishability, price). Political factors can also influence transportability such as laws, regulations, borders and tariffs. When transportability is high, activities are less constrained by distance.

trip assignment

In planning, a process by which trips, described by mode, purpose, origin, destination and time of day, are allocated among the paths or routes in a network by one of a number of models.

trip generation

In planning, the determination or prediction of the number of trips produced by and attracted to each zone.

twenty-foot equivalent unit (TEU)

A standard unit based on an ISO container 20-foot long (6.10 m), used as a statistical measure of traffic flows or capacities. One standard 40-foot ISO Series 1 container equals 2 TEUs.

unit load

Packages loaded on a pallet, in a crate or any other way that enables them to be handled as a unit.

unlinked passenger trips

The number of passengers who board public transportation vehicles. A passenger is counted each time he/she boards a vehicle even though he/she may be on the same journey from origin to destination.

upstream/downstream

Refers to the relative location of a given activity along a supply chain.

urban form

The spatial imprint of an urban transport system as well as the adjacent physical infrastructures and socio-economic activities. Jointly, they confer a level of spatial arrangement to cities.

variable cost

A cost that varies in relation to the level of operational activity.

very large crude carrier (VLCC)

A crude oil carrying ship of between 150,000 and 320,000 dwt. They offer good flexibility because many terminals can accommodate their draft. They are used in ports that have depth limitations, mainly around the Mediterranean, West Africa and the North Sea. They can be ballasted through the Suez Canal.

vessel
> Every description of watercraft, used or capable of being used as a means of transportation on the water.

warehouse
> A place for the reception, delivery, consolidation, distribution and storage of freight.

waterway
> River, canal, lake or other stretch of water that by natural or man-made features is suitable for navigation.

waybill
> A document covering a shipment and showing the forwarding and receiving station, the names of consignor and consignee, the car initials and number, the routing, the description and weight of the commodity, instructions for special services, the rate, total charges, advances and waybill reference for previous services and the amount prepaid.

weight
> Gross: the weight of the goods including packing, wrappers or containers, both internal and external. The total weight as shipped. Net: the weight of the goods themselves without the inclusion of any wrapper. Tare: the weight of the packaging or container. Weight/measurement ton: in many cases, a rate is shown per weight/measurement ton, carrier's option. This means that the rate will be assessed on either a weight ton or measurement ton basis, whichever will yield the carrier the greater revenue. Weight ton: metric measure equals 1,000 kilograms; in English measure a short ton is 2,000 pounds, a long ton is 2,240 pounds.

wharf
> A landing place where vessels may tie up for loading and unloading of cargo.

World Trade Organization (WTO)
> The WTO was established on January 1, 1995, as a result of the Uruguay Round negotiations (1986–94). The seat of the WTO is located in Geneva, Switzerland. The WTO has 148 member countries as of 2004. It performs various functions including administering WTO trade agreements, organizing forums for trade negotiations, handling trade disputes, monitoring national trade policies, providing technical assistance and training for developing countries, and it cooperates with other international organizations.

yard
> A system of auxiliary tracks used exclusively for the classification of passenger or freight cars according to commodity or destination; assembling of cars for train movement; storage of cars; or repair of equipment.

Index

access 6, 22, 318
accessibility 6, 14, 62, 63, 68–72, 318;
  attractiveness 72; connectivity 68, 69;
  emissiveness 72; geographic/potential 70–2
aframax 318
agglomeration economies see economies of
  agglomeration
air carrier 290, 318
Air Deregulation Act (1978) 141
air space 137, 318
air transport 14, 17, 52, 54, 66, 136–42, 199,
  263, 318; alliances 141; freedom rights
  139–41; liberalization process 141; low-cost
  carriers 141, 153
airport 165–6, 170; sites 175–6
alternative fuels 265–8, 319
Amtrak 153, 291, 319
Antwerp rule 169
appraisal/analysis see economic evaluation
arterial street 319
automated identity system (AIS) 305
automated transport systems 27
automobile: car sharing 300; cordon pricing
  301–2, 313; dependency on 225, 232, 244,
  248; development 51; fair pricing 301;
  ownership 91, 239, 241, 246–9; park and
  ride 300, 338; parking management 300;
  priority lanes for buses/high occupancy
  vehicles 250, 331; production 54; tolls 121,
  247, 301; traffic calming 300
average vehicle occupancy (AVO) 319
average vehicle ridership (AVR) 319

balance of payments 93, 319
balance of trade 319
barge 47, 128, 134, 147, 150, 162, 212, 263,
  319
barrel 100, 262, 267, 320
barriers to movement: absolute 10; relative 11
base fare 320
benefit–cost ratio 307
benzene and volatile components (BTX) 270
berth 172, 320

bicycles 50, 226, 229, 300
bill of lading 147, 320, 338
BLT (Build-Lease-Transfer) 124, 313
BOT (Build-Operate-Transfer) 124, 313
Bremen rule 169
bridge 129, 301, 313, 320
British thermal unit (Btu) 320
bulk cargo 130, 166, 202, 320, 327
bulk carriers 134, 172, 320
bulk terminal 321
Burgess concentric model 235–6
bus (motorbus) 242–3, 291, 300, 321

cable car 321
cabotage 100, 133, 140, 321
canal 9, 11, 13, 24, 47, 49–50, 65, 86–7, 133,
  291, 321
Capesize 321
capital costs 48, 102, 128, 129, 135, 245
carbon dioxide (CO₂) 270, 295, 322
carbon monoxide (CO) 270, 322
carpool 23, 248, 250, 322, 331, 341
carrier 322
catchment area 56, 322, 330
central business district (CBD) 60, 176, 227–8,
  230–2
Central Place Theory 58, 117, 178
charter 24, 322
Chicago: O'hare airport 175–6; Interlining 189;
  Chicago Area Transportation Study (CATS)
  253; Willow Springs 81–2
chlorofluorocarbons (CFCs) 270
circumnavigation 16
city logistics 258–9, 322
Civil Aeronautics Board 290
class I railroad 322
Clean Air Act (CAA) 322
climate 10
clusters 57, 171, 199, 208, 219
coach service 47, 322
cold chain 205, 323
combi 323
combinatory costs 153–4